MATHEMATICS FOR SHEET METAL FABRICATION

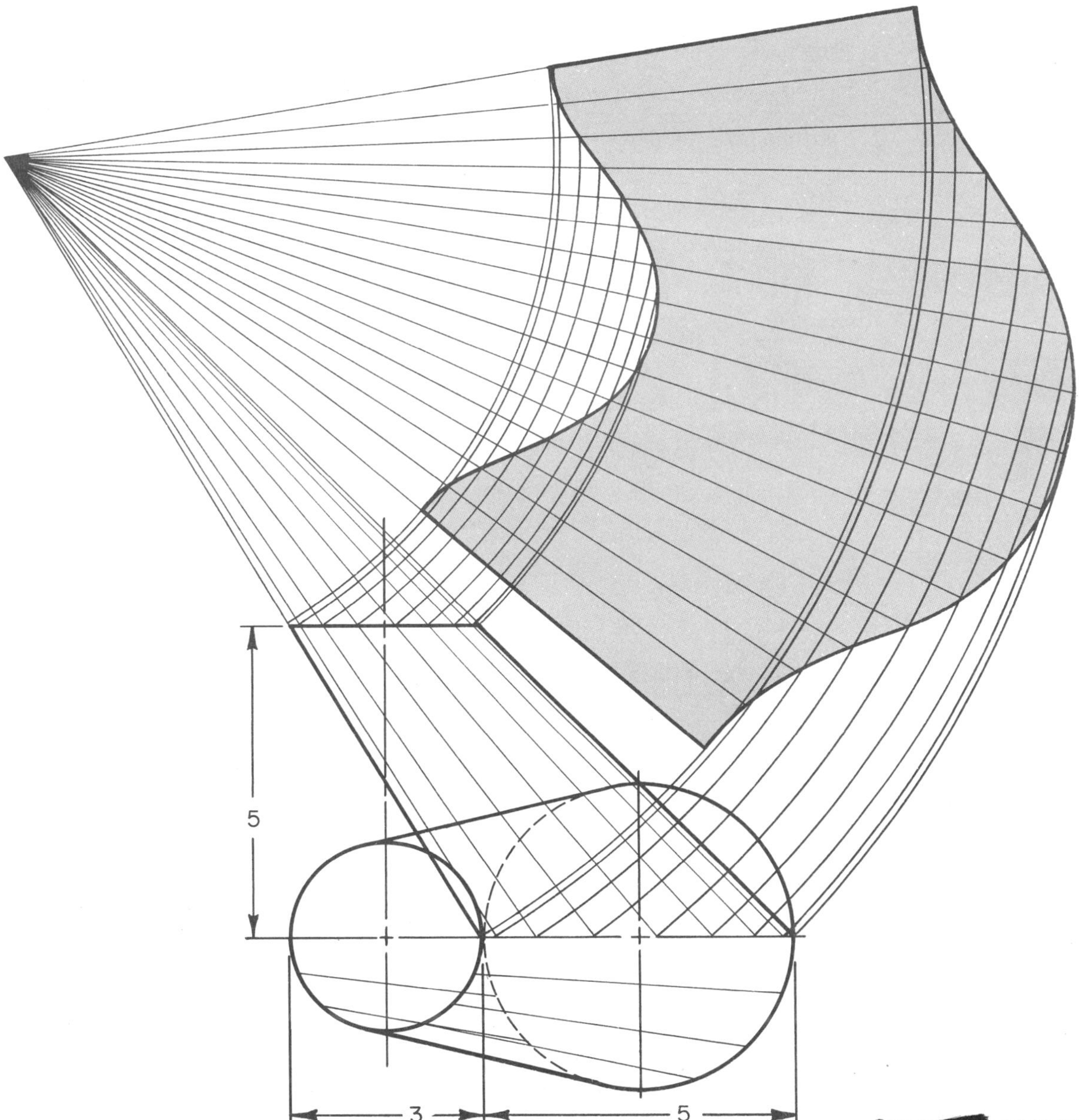

DELMAR PUBLISHERS INC.
2 COMPUTER DRIVE, WEST – BOX 15-015
ALBANY, NEW YORK 12212

DELMAR PUBLISHERS INC.

Library of Congress Catalog Card Number:
79-118846
ISBN: 0-8273-0295-9

Printed in the United States of America
Published simultaneously in Canada
by Nelson Canada,
A Division of International Thomson Limited

NOTICE TO THE READER

PREFACE

Mathematics for Sheet Metal Fabrication is a complete revision of *Sheet Metal Mathematics.* The new book is a basic text related directly to the present-day sheet metal industry. It is concerned with the fundamental principles and concepts of mathematics needed by a tradesman who is required to solve problems and make measurements and layouts. These mathematical principles and concepts are related to the general fields of arithmetic, algebra, plane geometry, and trigonometry.

The mathematical areas covered include: fractions and decimals; linear measure; equations; formulas; averages, percentages, costs, and wages; ratio and proportion; perimeters; area and volume measure; geometry; trigonometry; and graphs. They were selected after a careful analysis of the sheet metal industry by a group of competent tradesmen. As a result, students, apprentices and craftsmen, through the use of this text, can acquire a broad mathematical background needed for mastery of trade practices. A working knowledge of mathematics is becoming increasingly important to skilled workers who wish to keep pace with the rapid progress in technology.

Appreciation is expressed to Mr. Robert E. Busse, Instructor, Mr. Levy J. Siega, Instructor, and Mr. Harry V. Clendenning, Senior Instructor of the Related Subjects Department of the Industrial Division, Northern Alberta Institute of Technology, Edmonton, Alberta, Canada, for making a careful analysis of the content and arrangement of the text which resulted in a complete revision and the addition of pertinent new material furnished by them. We believe that the new organization is a more logical approach to the study of mathematics and that the new material enhances the value of the text.

Acknowledgment is made to the Lufkin Rule Company for the use of certain illustrations on the rule.

By combining the material from the original book, which was prepared by a group of men who had extensive trade experience and a working knowledge of the industry, with the contribution of our colleagues in Canada, it has been possible to provide a comprehensive text on sheet metal mathematics.

Albany, New York

Arthur F. Ahr
Managing Editor

TO THE STUDENT

The production of sheet metal parts and units depends largely on the skill of the craftsman in accurately performing hand processes, machine processes, and measurement and layout work.

To develop these skills, it is important for the beginner to be able to solve everyday trade problems quickly and accurately. Arithmetic and mathematical principles are described in this text in a series of units to aid you in developing these skills. Each new principle is applied in one or more illustrative problems. Examples of these problems are broken down into step-by-step descriptions for ease in interpreting and understanding. They contain the information needed to work the mathematical problems included in the assignment portion of the unit.

The knowledge gained in studying these principles and working the problems will be a definite asset to you when you are required to make a practical application of them on a specific job.

Study each unit carefully and work out each illustrative problem or example in the unit before attempting to solve the problems. Master the meaning of each new term, symbol and formula; review each illustration, and complete each unit before proceeding with the next.

In solving the arithmetic problems and geometric constructions in the different assignments, it is not necessary to rewrite or redraw each step as performed in the illustrative problems. However, your solution to each problem should include the following parts:

1. A sketch or drawing when required
2. A list of steps performed
3. A statement of the rules and formulas used
4. Computations
5. A check on all operations

Whenever possible, each step in the problem should be marked with the proper unit of measurement. By so doing, greater accuracy is assured. The final answer should also be labeled with the correct unit of measurement followed by the abbreviation (Ans.). Always check your computations by using a different method, going over your work, or by actual measurement.

The development of careful work habits is essential. Your work and solutions will be rated on: (1) interpretation of problem, (2) accuracy, (3) procedure, (4) arrangement, and (5) neatness. The illustrative problems contained in each unit should serve as a guide to the solution of problems.

The sections are identified at the top of left-hand pages and the units at the top of right-hand pages. Following the last section is an appendix which contains valuable information on: terms, abbreviations and symbols; formulas (arranged by sections); useful tables and the powers of roots and square roots. Instructions for using the appendix are given on the introductory page to this section.

CONTENTS

SECTION 1 FRACTIONS AND DECIMALS

SECTION 2 LINEAR MEASURE

SECTION 3 AVERAGES; PERCENTAGES; COSTS; AND WAGES

SECTION 4 GEOMETRY

SECTION 5 EQUATIONS

SECTION 6 RATIO AND PROPORTION

SECTION 7 PERIMETER

SECTION 8 AREA MEASURE

SECTION 9 VOLUME MEASURE

SECTION 10 FORMULAS: SPECIAL PROBLEMS

SECTION 1 FRACTIONS AND DECIMALS

UNIT 1 INTRODUCTION TO COMMON FRACTIONS

The sheet metal worker must learn to solve some of the problems of his trade by use of common fractions. In measuring material, for instance, he must often use not only the major divisions of a rule which are known as inches, but also the fractional divisions of an inch, such as 1/2 in., 1/4 in., and 1/8 in. These divisions are found on the rule.

The word *digit* refers to the numbers 1, 2, 3, 4, 5, 6, 7, 8, 9, and 0 (zero). In fractional work, as in whole numbers, the following symbols and processes are used:

SYMBOL	WORD	MEANING
+	plus	addition
−	minus	subtraction
x	multiply	multiplication
$\overline{)\quad}$ ÷ $\frac{?}{?}$?/?	divide Note that the fraction indicates the process of division.	division
=	equals	same quantity

EXPLANATION OF FRACTIONAL PARTS

If a line is divided into eight parts, one part is $\frac{\text{one part}}{\text{total parts}}$, or 1/8 of the total length of the line.

Two parts = 2/8 of the total length; three parts = 3/8; four parts = 4/8; five parts = 5/8; six parts = 6/8; seven parts = 7/8; eight parts = 8/8 or the total length.

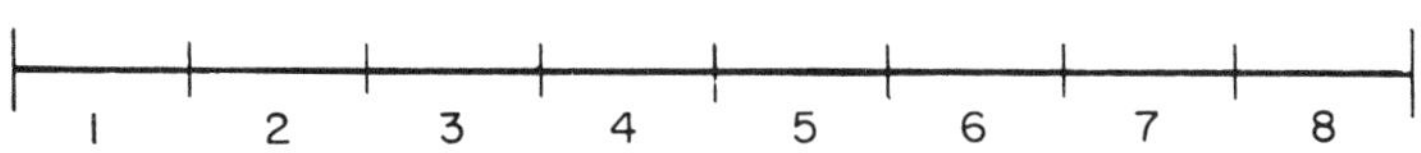

If a line is divided into 16 parts, then one part is 1/16 of the total length of the line, and 1/2 of one part is 1/32 of the total length of the line. Similarly, 10 parts are 10/16 of the total length of the line.

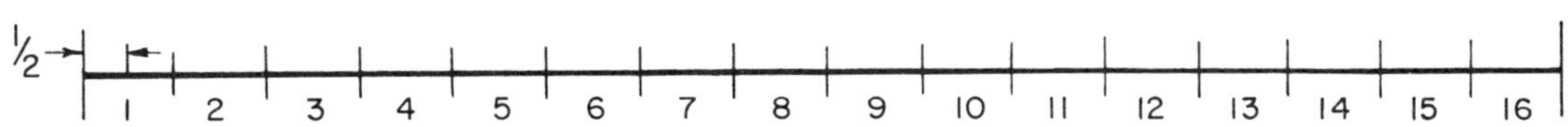

DEFINITIONS OF COMMON FRACTIONS

- A *fraction* is a part of a whole quantity, such as 3/4 and 7/32. Thus 3/4 shows that three of the four parts of a whole are considered; 7/32 shows that seven of the thirty-two parts are considered.
- The *denominator* is the number that shows the number of equal parts into which the whole has been divided. The denominator is written below the line.
- The *numerator* is the number which shows how many equal parts of the whole have been taken to make a fraction. The numerator is written above the line.
- The *terms* of a fraction are the denominator and the numerator.
- A *common fraction* is a fraction which has both terms expressed.
- A *proper fraction* is a fraction whose numerator is less than its denominator, as 3/8, 5/16, etc.
- An *improper fraction* is a fraction whose numerator is larger than its denominator, as 3/2, 5/4, 15/8, etc.
- A *unit fraction* is a fraction whose numerator is one.
- A *mixed number* is a number composed of a whole number and a fraction, as 3 3/4, 7 1/2, etc.
- A *complex fraction* is a fraction in which one or both of its terms are fractions or mixed numbers. Examples:

$$\frac{3/4}{6} \qquad \frac{15/3}{32} \qquad \frac{4\,1/4}{7\,5/8}$$

REDUCTION OF COMMON FRACTIONS AND MIXED NUMBERS

Reduction of Fractions

RULE: The value of a fraction is not changed by multiplying the numerator and denominator by the same quantity. The value of a fraction is not changed by dividing the numerator and denominator of a fraction by the same number

$$1/2 = \frac{1 \times 4}{2 \times 4}, \text{ or } \frac{4}{8} \text{ Ans.} \qquad 8/12 = \frac{8 \div 4}{12 \div 4}, \text{ or } \frac{2}{3} \text{ Ans.}$$

When both the numerator and the denominator cannot be divided by the same number, the fraction is in its lowest terms. An improper fraction can be reduced to a mixed number by dividing the numerator by the denominator. $95/8 = 95 \div 8$, or 11 7/8 Ans.

Reduction of Mixed Numbers

RULE: A mixed number can be changed to an improper fraction by multiplying the whole number by the denominator and adding the numerator. For example, when changing 2 3/4 to an improper fraction, multiply 2 by 4, add 3, and place the result over 4.

$$2\frac{3}{4} = \frac{2 \times 4 + 3}{4}, \text{ or } \frac{11}{4} \text{ Ans.}$$

ASSIGNMENT

A. Fractional Parts

1. Name fractional parts of the total length of the line below. One part = ? Two parts = ? Three parts = ?

2. Name the fractional parts of the total length of the line below as indicated. Four parts = ? Six parts = ? Nine parts = ?

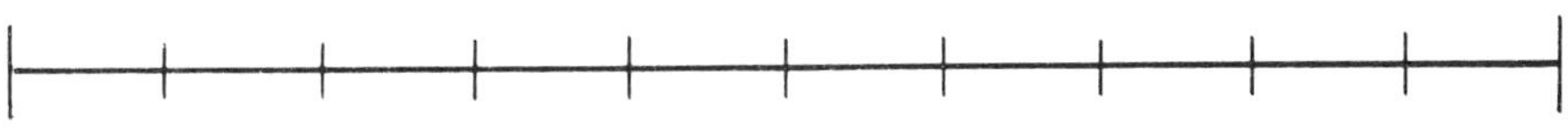

B. Definitions of Common Fractions

Write each fraction or mixed number in a vertical column on another sheet of paper. Write the *correct* name of the fraction or mixed number opposite each fraction or mixed number.

Unit fraction	$\frac{3/4}{16}$
Complex fraction	7 3/4
Proper fraction	1/8
Mixed number	3/8

C. Reduction of Common Fractions and Mixed Numbers

1. Reduce to halves: 2/4; 16/8; 4/8; 32/8
2. Reduce to eighths: 1/2; 1/4; 3/4; 7/2; 4/16; 10/16
3. Reduce to sixty-fourths: 3/8; 15/32; 7/16; 24/128
4. Reduce to lowest terms: 4/8; 12/32; 5/10; 10/16; 28/32; 10/64
5. Change to mixed numbers: 5/2; 25/2; 7/4; 21/4; 9/8; 79/8; 21/16; 105/16; 43/32; 321/32; 81/64; 6/5; 21/10; 81/50
6. Change to improper fractions: 1 1/2; 2 1/4; 8 3/4; 1 5/8; 4 3/16; 2 9/32; 4 63/64; 2 3/5; 5 4/5; 4 3/10; 2 17/50
7. Arrange in order from the smallest size to the largest size, twist drills having the following fractional sizes: 3/8"; 1/4"; 7/16"; 23/64"; 17/32"; 1/2"; 11/32"; 27/64"
8. How many 32nds, 8ths, 64ths, 16ths are there in 1 5/8"? **Hint:** Change 1 5/8 to an improper fraction and then to fractions having denominators of 32, 8, 64, and 16.

UNIT 2 FRACTIONAL DIVISIONS OF THE STEEL RULE

The steel rule is used to measure and to lay out lengths in inches and fractional parts of inches. The steel rule is also used as a straightedge.

The fractional divisions of an inch are found by dividing the inch into equal parts. The common divisions are halves, quarters, eighths, sixteenths, thirty-seconds, and sixty-fourths. For example, there are 8 one-eighth-inch divisions in one inch. Inches are designated with a double prime ("). The length, three inches, is written as 3". The abbreviation for inch, or inches is *in.*

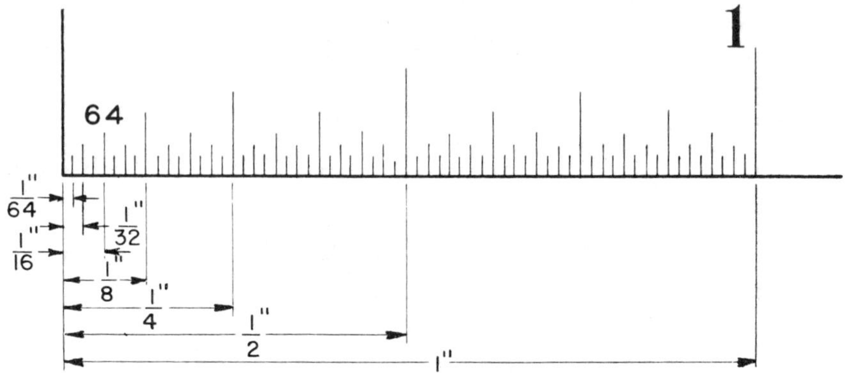

READING FOUR TYPES OF DIVISIONS

Determine the values of the divisions before reading a scale

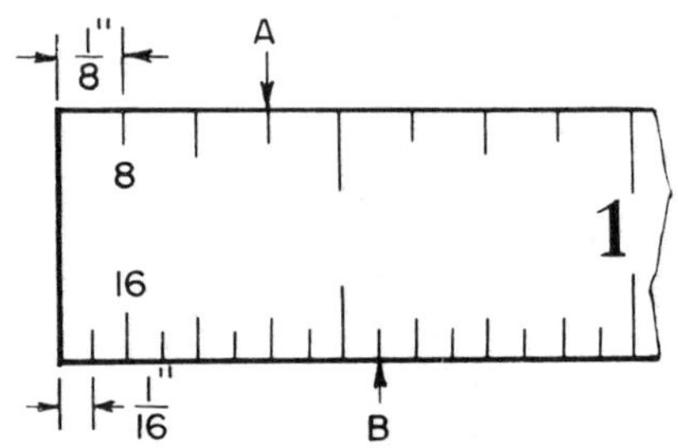

8ths Scale

$A = \frac{3''}{8}$ Count 3 one-eighth-inch divisions.

16ths Scale

$B = \frac{9''}{16}$ Count 9 one-sixteenth-inch divisions or count one-sixteenth-inch division beyond the 1/2-inch (8/16") division.

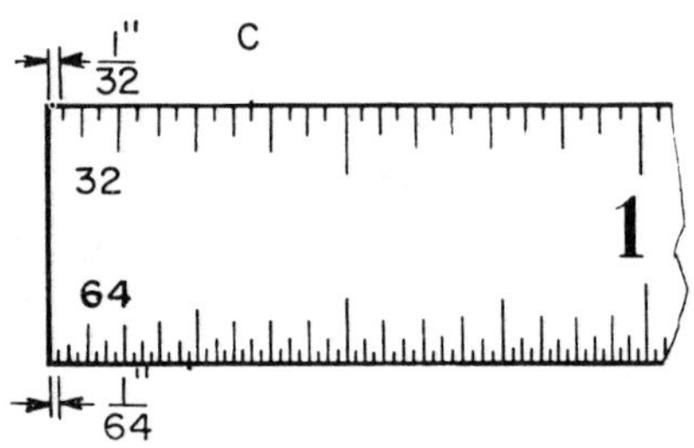

32nds Scale

$C = \frac{11''}{32}$ Count 11 thirty-second-inch divisions or subtract 1 thirty-second-inch division from the 3/8-inch (12/32") division.

64ths Scale

$D = \frac{15''}{64}$ Count 15 sixty-fourth-inch divisions or count one-sixty-fourth-inch division back from the 1/4-inch (16/64") division.

MEASURING WITH A RULE

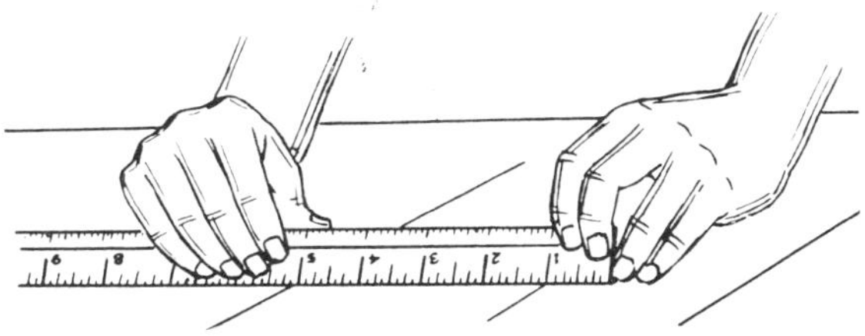

Note: The measuring of some jobs, such as the one shown, requires the use of two rules.

1. Tip the rule if necessary so that the correct scale is in contact with the work to be measured. (It is more accurate to start at the 1″ division.)
2. Read the graduations on the rule. (Subtract 1″ from the reading if the measurement is started at the 1″ division.)

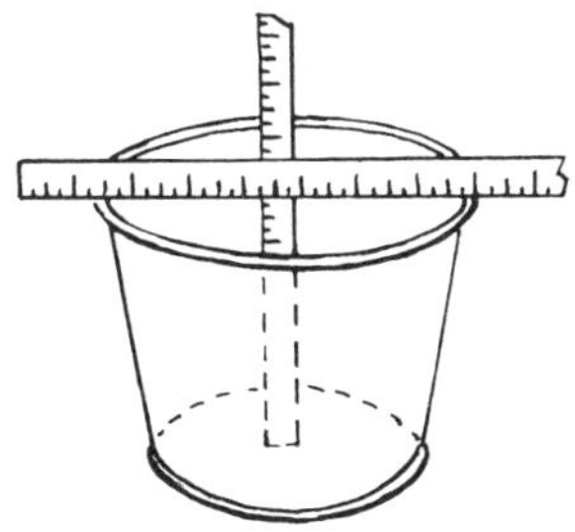

LAYING OUT WITH A RULE

1. Tip the rule if necessary to obtain accurate measurements.
2. Be sure the pencil is tipped in drawing the line so that the line drawn will be close to the rule.

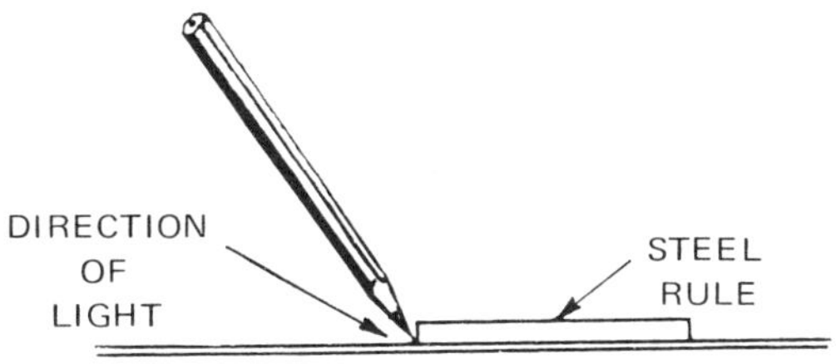

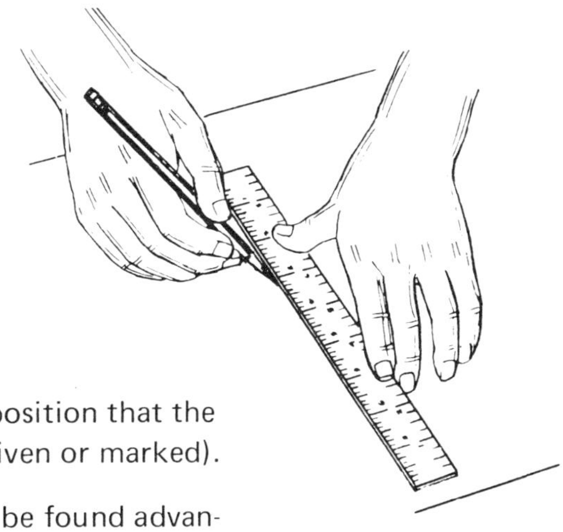

3. Make certain that the rule is placed in such a position that the line to be drawn will pass through the points (given or marked).

   Note: When long rules are being used, it will be found advantageous to put the pencil on one point and swing the rule until the other end is centered at the other point.

ASSIGNMENT

A. Reading the Fractional Divisions

Read the distance from the start of the rule to letters A, B, C, and D in each problem.

1. 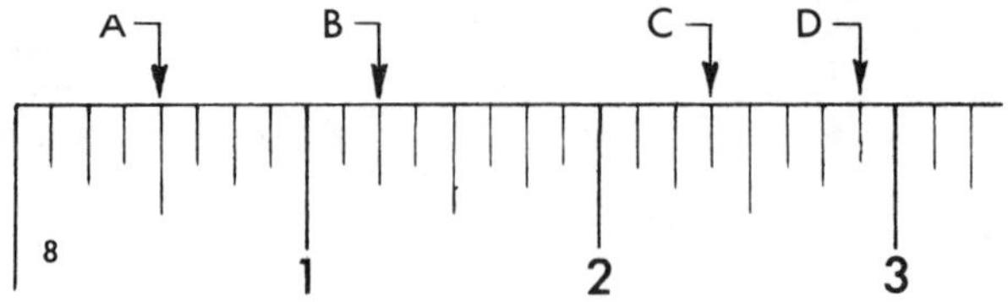

2. 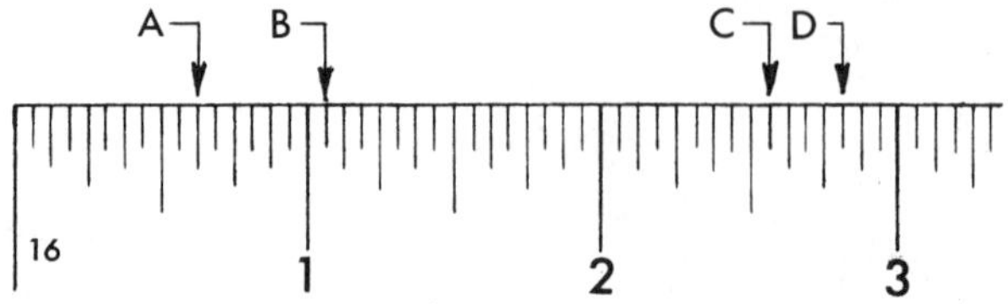

3. 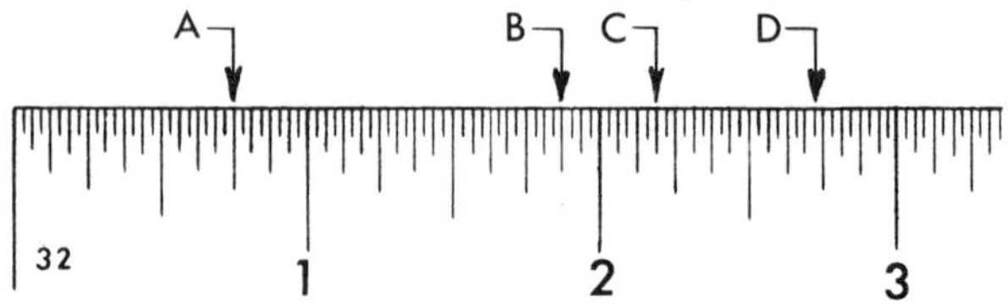

B. Measuring Lines

1. Measure this line to the nearest eighth of an inch.
2. Measure this line to the nearest sixteenth of an inch.
3. Measure this line to the nearest sixteenth of an inch.
4. Measure this line to the nearest thirty-second of an inch.
5. Measure this line to the nearest thirty-second of an inch.
6. Measure this line to the nearest sixty-fourth of an inch.
7. Measure this line to the nearest sixty-fourth of an inch.

C. Drawing Lines

Draw lines of the following lengths:

1. 2″	3. 4 5/16″	5. 3 13/32″	7. 1 43/64″
2. 3 1/8″	4. 1 13/16″	6. 2 29/32″	8. 2 57/64″

UNIT 3 ADDITION OF COMMON FRACTIONS

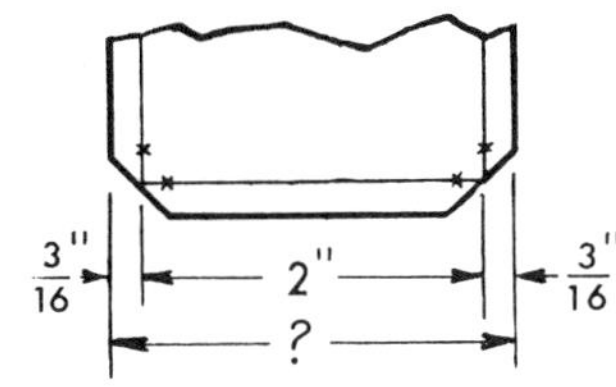

In order to determine the length of materials — sheet metal, wire, band iron, etc. — needed for a job, it is often necessary to add fractional dimensions. Addition, the symbol for which is (+), is the process of finding the sum of two or more numbers.

Fractions cannot be added unless they have a *common denominator,* such as 5/8, 7/8, and 12/8. This is a denominator into which all the denominators will divide. It will save work to use the lowest (least) common denominator which may be found by trial. The least common denominator of 1/6 and 1/4 is 12.

RULE 1: To add fractions, reduce them to a least common denominator. Then add the numerators, write their sum over the common denominator, and reduce the result to its lowest terms.

Example: Add 3/4, 7/8, and 11/16

1. Reduce the fractions to the least common denominator (16).

$$\frac{3 \times 4}{4 \times 4} = \frac{12}{16} \qquad \frac{7 \times 2}{8 \times 2} = \frac{14}{16} \qquad \frac{11 \times 1}{16 \times 1} = \frac{11}{16}$$

2. Add the fractions by adding the numerators and writing this result over the common denominator.

   Note: When it is possible, reduce the result to a mixed number in its lowest terms.

   Note: If the fractions of the above problem are in inches, the answer can be checked by measuring the overall length x on the drawing.

$$\frac{12}{16} + \frac{14}{16} + \frac{11}{16} = \frac{37}{16} = 2\frac{5}{16} \text{ Ans.}$$

   Check: To check the addition, add the numbers in the reverse direction. To check the accuracy of the fractions 12/16 and 14/16, reduce them to their lowest terms.

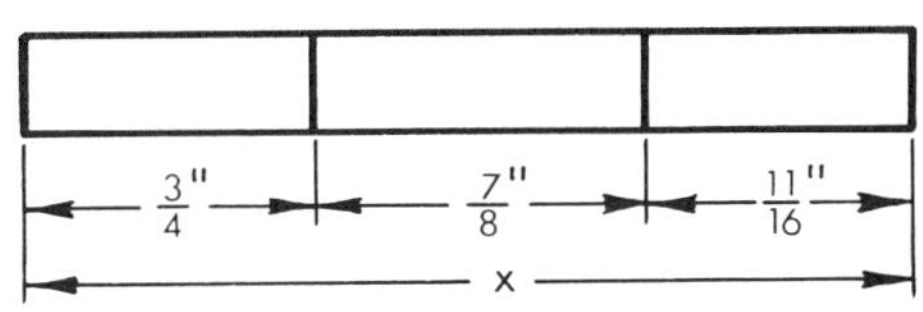

RULE 2: To add mixed numbers, add the whole numbers and the fractions separately, and then combine the results.

ASSIGNMENT

Solve and check the following problems, reducing the results to the lowest terms.

1. 1/4 + 3/4	3. 9/16 + 13/16	5. 29/32 + 27/32
2. 3/8 + 1/8	4. 11/16 + 7/16	6. 63/64 + 37/64

7. 21/50 + 39/50
8. 79/100 + 53/100
9. 1/4 + 1/2
10. 1/4 + 3/8
11. 2/3 + 1/2
12. 5/16 + 5/8
13. 19/32 + 13/16
14. 57/64 + 29/32
15. 17/50 + 3/10
16. 7/20 + 3/16
17. 1/2 + 4 3/4
18. 5/8 + 3 9/16
19. 7 5/8 + 3/4
20. 2 15/16 + 1 1/2
21. 3 31/32 + 2 9/16 + 4 3/8
22. 4 27/64 + 2 9/32 + 7 7/16
23. 9 21/50 + 5 37/100 + 6 17/25
24. 19 69/100 + 11 3/10 + 5 15/50

25. Find the distance x in the stretchout.

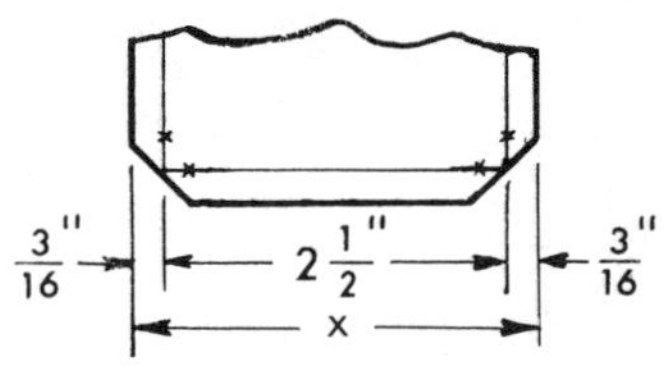

26. Compute the overall length of the stretchout.

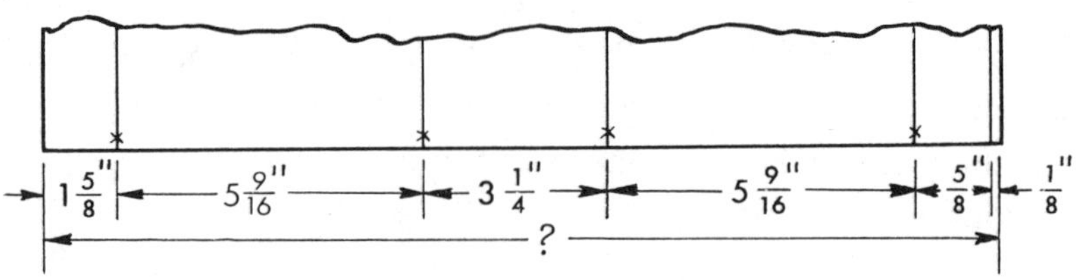

27. Find the distance x.

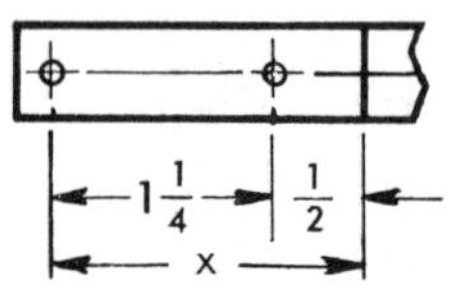

28. Compute the overall length of the pipe below.

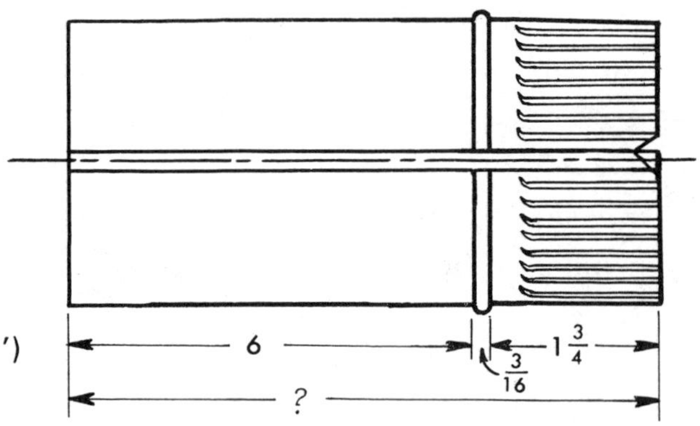

Note: In many cases, the inch marks (") are omitted from drawings.

UNIT 4 SUBTRACTION OF COMMON FRACTIONS

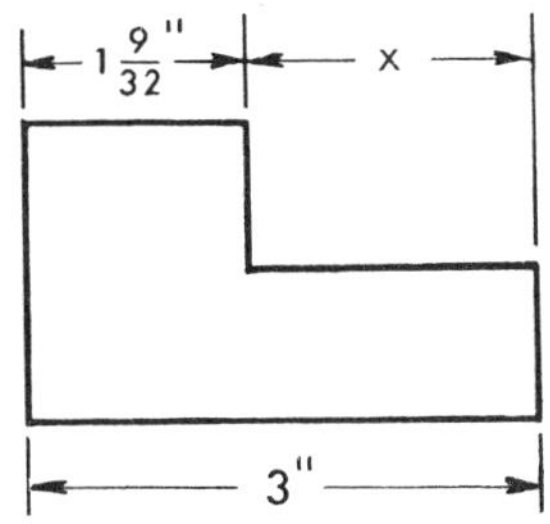

If a dimension which is not on the drawing of the job is needed, it may be necessary to subtract fractional dimensions to obtain that dimension. Subtraction is the process of taking one number from another.

RULE 1: To subtract fractions, reduce the fractions to the least common denominator and subtract the numerators only. Write the difference over the common denominator and reduce the resulting fraction to its lowest terms.

Example: Subtract 1 1/8 from 2 9/16

Reduce the fractions to the least common denominators. Subtract the whole numbers. Subtract the fractions by subtracting the numerators and placing this result over the common denominator.

$$\begin{array}{rcl} 2\,\frac{9 \times 1}{16 \times 1} & = & 2\,\frac{9}{16} \\ -1\,\frac{1 \times 2}{8 \times 2} & = & -1\,\frac{2}{16} \\ \hline & \text{Ans.} & 1\,\frac{7}{16} \end{array}$$

Note: If the numbers of the above problem are in inches, the answer can be checked by measuring the distance x.

Check: The answer can be checked by adding 1 7/16 and 1 1/8. This equals 2 9/16.

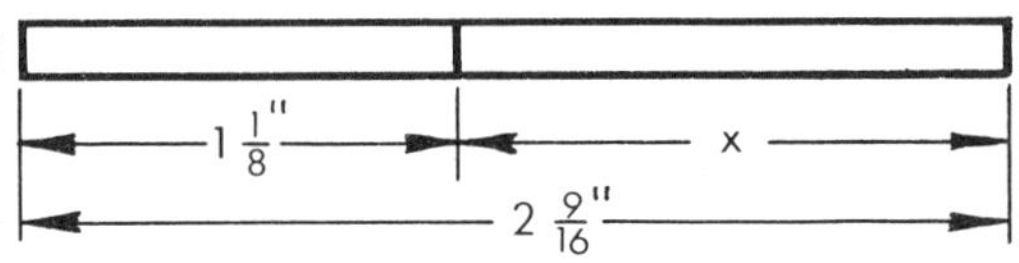

RULE 2: To subtract mixed numbers when the fraction of the mixed number to be subtracted is larger, reduce the fractions to the least common denominator, borrow one unit to make up a fraction larger than the one being subtracted, and then subtract in the usual way.

Example: Subtract 4 3/8 from 12 9/64

Reduce the fractions to the least common denominators. Borrow 64/64 (or 1) from 12 and add this fraction to 9/64. Subtract the whole numbers and the fractions separately.

$$\begin{array}{rcccl} 12\,\frac{9}{64} & = & 11\,\frac{64}{64} + \frac{9}{64} & = & 11\,\frac{73}{64} \\ -4\,\frac{3}{8} & = & -4\,\frac{24}{64} & = & -4\,\frac{24}{64} \\ \hline & & & \text{Ans.} & 7\,\frac{49}{64} \end{array}$$

ASSIGNMENT

A. Subtracting Common Fractions

Solve and check the following problems, reducing the results to the lowest terms.

1. 3/4 - 1/4	7. 47/50 - 3/50	13. 5/8 - 9/64	19. 7 9/16 - 29/32
2. 7/8 - 3/8	8. 98/100 - 21/100	14. 2/3 - 1/2	20. 5 1/6 - 2 1/2
3. 15/16 - 3/16	9. 1/2 - 1/4	15. 9/10 - 24/50	21. 3 21/64 - 2 29/32
4. 29/32 - 13/32	10. 1/2 - 1/3	16. 51/100 - 11/50	22. 5 - 3 17/32
5. 63/64 - 29/64	11. 3/4 - 5/8	17. 2 1/4 - 1/2	23. 4 49/50 - 3 3/10
6. 9/10 - 3/10	12. 9/16 - 5/32	18. 3 5/8 - 1 3/4	24. 8 51/100 - 3 41/50

25. Compute the missing dimension on the drawing below.

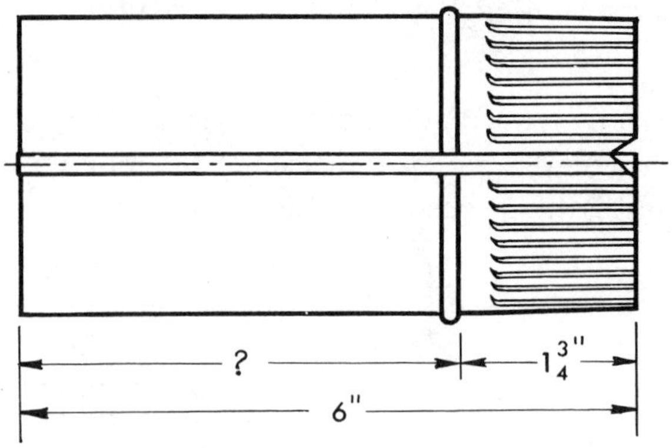

26. Compute the missing dimension on the drawing below.

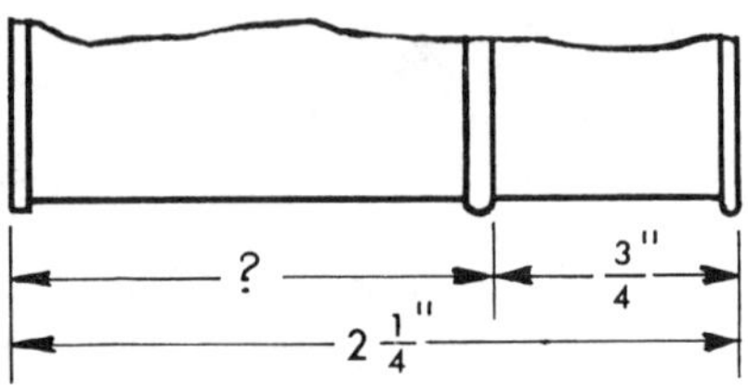

27. Compute the missing dimension on the drawing below.

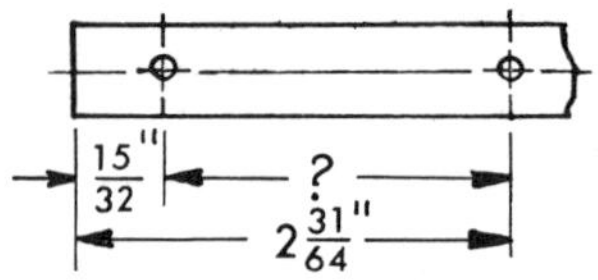

B. Adding and Subtracting Common Fractions

Solve and check each of the following problems, reducing the results to the lowest terms.

1. 1/2 + 1/3 − 2/5
2. 5/12 + 23/16 − 1/8
3. 1 1/8 + 7 5/16 − 31/64 + 2 1/4
4. 3 3/50 + 7 9/10 + 54/100 − 4 2/5
5. 52 1/4 + 5/8 + 13/16 − 31/32
6. 10 1/2 − 3/4 − 1 3/5
7. How much material will remain after the following pieces have been cut from a piece of band iron 24 in. long: 3 1/4 in., 5 5/8 in., 4 1/2 in., 5 11/16 in.? (Allow 1/16 in. for each cut.)
8. Compute the dimensions x, y, and z. Check the answers by measurement.

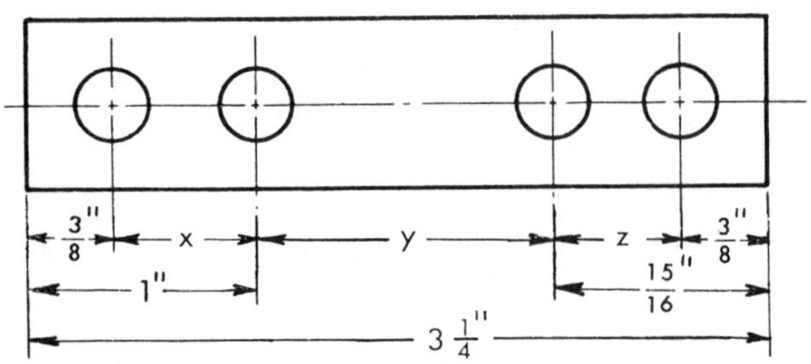

9. Compute x.

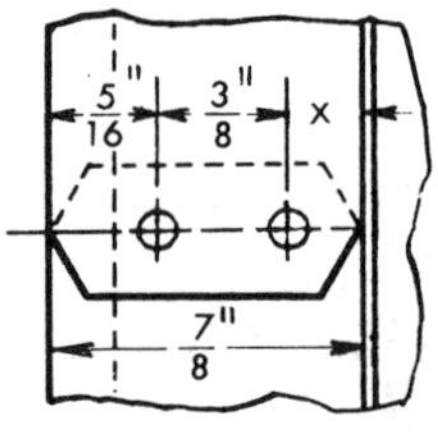

10. Compute the dimension x.

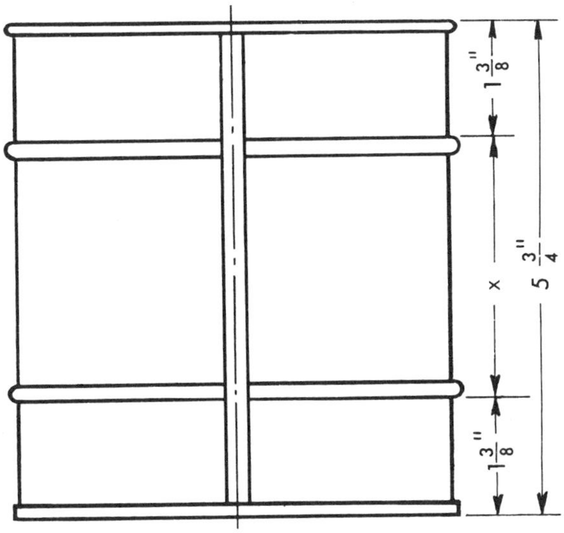

11. Compute dimensions x and y.

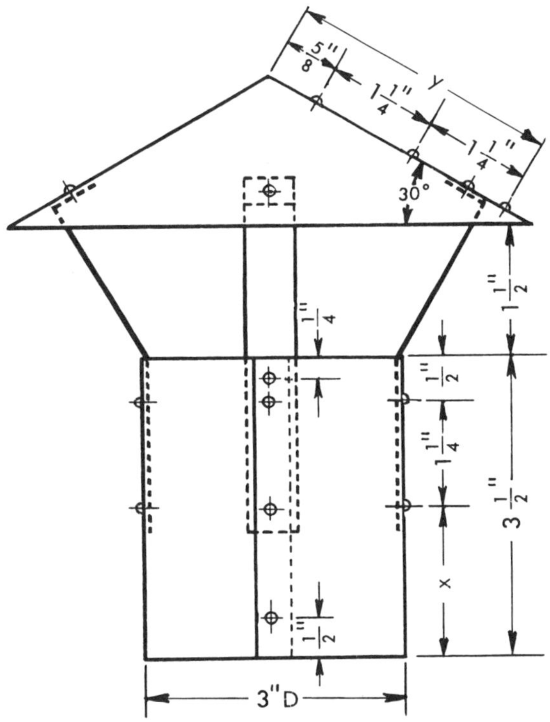

UNIT 5 MULTIPLICATION OF COMMON FRACTIONS

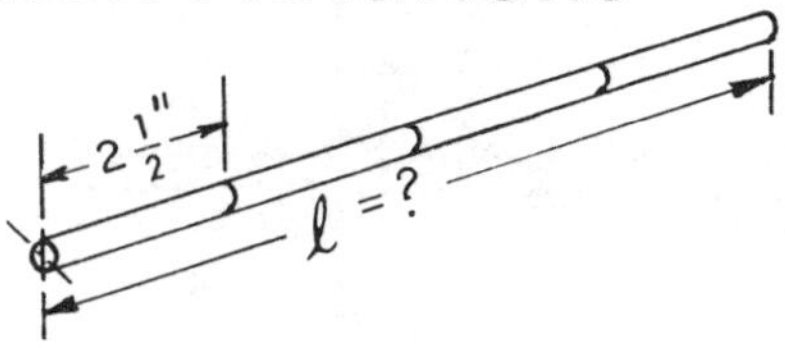

Multiplication of fractions is necessary to find the total length of material, such as sheet metal, wire, etc., needed for a job.

Multiplication, the symbol for which is x, is the process of increasing a number by its own value a certain number of times and is a short method of adding. The word *of* means multiplication.

RULE 1: To find the product of two or more fractions, multiply the numerators and multiply the denominators. Write the product of the numerators over the product of the denominators. Reduce the resulting fraction to its lowest terms.

Example: Multiply 3/4 by 8/9

1. Multiply the numerators together and the denominators together. Write results in fractional form. $\frac{3 \times 8}{4 \times 9} = \frac{24}{36}$

2. Reduce the answer to its lowest terms. $\frac{24 \div 12}{36 \div 12} = \frac{2}{3}$ Ans.

Check: Repeat the operations, or use the cancellation method.

CANCELLATION

Cancellation consists of dividing the factors that are common to the numerators and the denominators of the given fractions. *Factors* are parts of a number which, when multiplied together, will give that number. For example, 3 and 2 are factors of 6.

1. Divide by 3 to obtain 1 in the numerator and 3 in the denominator. $\frac{\cancel{3}^{1}}{4} \times \frac{8}{\cancel{9}_{3}} =$

2. Divide by 4 to obtain 2 in the numerator and 1 in the denominator. $\frac{\cancel{3}^{1}}{\cancel{4}_{1}} \times \frac{\cancel{8}^{2}}{\cancel{9}_{3}} =$

3. Multiply the numerators and the denominators. Write the results in fractional form. $\frac{1 \times 2}{1 \times 3} = \frac{2}{3}$ Ans.

RULE 2: Change mixed numbers to improper fractions before multiplying.

Example: $2\frac{1}{2} \times 3\frac{1}{3} = \frac{5}{\cancel{2}_{1}} \times \frac{\cancel{10}^{5}}{3}$, or $\frac{25}{3}$, or $8\frac{1}{3}$ Ans.

ASSIGNMENT

Solve and check each of the following problems, reducing the results to the lowest terms.

1. 1/2 x 2/3
2. 3/8 x 4/5
3. 9/16 x 8/1
4. 15/32 x 1/3
5. 10 x 9/64
6. 8 x 41/100
7. 1/2 of 1/4
8. 3/4 of 7/8
9. 5/8 of 13/64
10. 2/3 of 9/16
11. 16/25 x 15/32
12. 27/50 x 2/3
13. 2 5/8 x 8
14. 4 x 1 15/32
15. 1 1/4 x 2 5/8
16. 3 9/16 x 2/3
17. 2 21/50 x 25/64
18. 2 9/32 x 1 1/2
19. Compute the distance x on the drawing below.

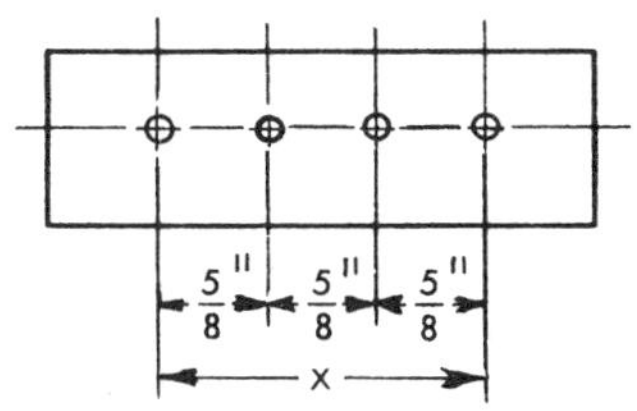

20. Compute the overall length of the pattern shown below.

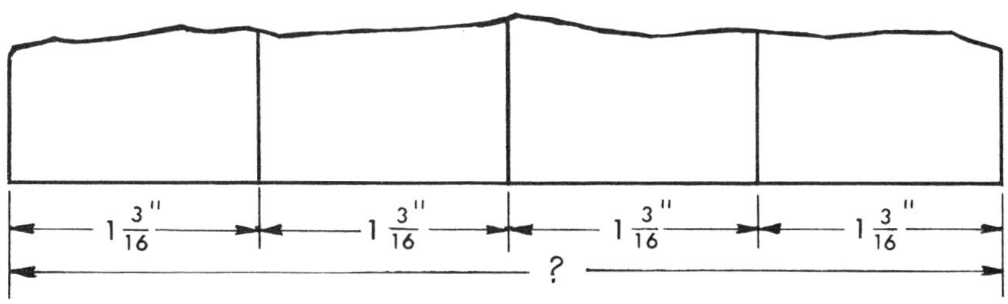

21. Find the total length of 8 pieces of wire, each 7 3/16 in. long.
22. Find the total length of 10 pieces of wire, each 3 9/32 in. long.
23. Compute the total length of band iron required if 6 pieces are needed, each of which is 4 1/4 in. long. (Allow 1/16 in. waste for each cut.)

    Hint: Add 1/16 in. to the length of one piece before multiplying by number of pieces.
24. What is the total length of angle iron required if 10 pieces are needed, each of which should be 5 3/8 in. long? (Allow 1/16 in. for each cut.)
25. Compute the total length of band iron required if 20 pieces are needed, each of which should be 6 9/16 in. long. (Allow 3/32 in. for each cut.)

UNIT 6 DIVISION OF COMMON FRACTIONS

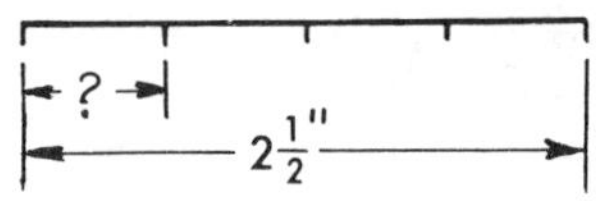

Division of fractions is used when it is necessary to find the number of pieces which can be obtained from a length of wire, band iron, or sheet metal, or when it is necessary to divide a line into equal parts.

Division is the process of determining how many times one number is contained in another. It is indicated by the division sign ÷, such as 6 ÷ 2, by a bracket and an extended line, such as $2\overline{)6\quad}$, or by a line between the two numbers, such as 6/2 or $\frac{6}{2}$. The divisor is the number used to divide, the dividend is the number to be divided, and the quotient is the answer. In the fraction 6/2, 2 is the divisor (denominator), 6 is the dividend (numerator), and 3 is the quotient.

Hence, it will be noted that a fraction expresses the process of division. For example, 4/4 indicates that a unit is divided into four equal parts. The fraction 3/4 indicates that a unit has been divided into four parts and that one of the four parts has been taken away. When fractions are divided, the problem is to find out *how many fractional parts are contained in a unit.* 3 ÷ 1/2 is another way of expressing, "How many halves are there in three?" To find the answer, multiply the whole number by the numerator and the denominator of the fraction. This kind of multiplication is called *inversion* because the fraction is turned "upside down."

$$3 \div 1/2 = \frac{3}{1} \times \frac{2}{1} = \frac{6}{1} = 6$$

There are six halves in three.

Likewise, when a fraction is divided by another fraction the dividend (numerator) of one fraction is multiplied by the divisor (denominator) of the other fraction. 7/8 ÷ 3/4 is another way of saying "How many three-quarters are there in seven-eighths?"

$$7/8 \div 3/4 = \frac{7}{8} \times \frac{4}{3} = \frac{28}{24} = \frac{7}{6} \text{ or } 1\ 1/6$$

When a whole number or a fraction is divided by a fraction, the answer is always *larger* than the quantity being divided because we use the processes of inversion and multiplication.

RULE 1: To divide fractions, invert the divisor and then multiply the fractions.

Example: Divide 5/7 by 3/4

1. Write as a division problem. $\frac{5}{7} \div \frac{3}{4}$
2. Replace the division sign with a multiplication sign and invert the divisor. $\frac{5}{7} \times \frac{4}{3}$
3. Multiply as before and reduce the resulting fraction to its lowest terms. $\frac{5}{7} \times \frac{4}{3} = \frac{20}{21}$ Ans.

Check: Multiply the divisor (3/4) by the quotient (20/21) to obtain the dividend (5/7), or repeat the operations.

RULE 2: Mixed numbers should be changed to improper fractions before dividing.

Example: How many pieces of wire 3 1/2 in. long can be obtained from 20 in. of stock?

1. Write as a division problem. $20 \div 3\ 1/2$
2. Change the mixed number to an improper fraction. $3\frac{1}{2} = \frac{3 \times 2 + 1}{2}$, or $\frac{7}{2}$
3. Invert the divisor and multiply. Ans. is 5 pieces; the 5/7 is waste. $20 \times \frac{2}{7} = \frac{40}{7}$, or $5\frac{5}{7}$

ASSIGNMENT

Solve the problems below and check each answer.

1. 9/16 ÷ 3/4
2. 3/8 ÷ 9/32
3. 21/32 ÷ 3/5
4. 1/2 ÷ 1/3
5. 3/4 ÷ 3/8
6. 7/10 ÷ 49/50
7. 17/20 ÷ 34/100
8. 3/8 ÷ 8
9. 9/16 ÷ 4
10. 1/2 ÷ 16
11. 7/32 ÷ 14
12. 15/64 ÷ 5
13. 21/64 ÷ 7
14. 11/20 ÷ 10
15. 27/50 ÷ 20
16. 4 ÷ 1/2
17. 20 ÷ 5/8
18. 10 ÷ 5/2
19. 7 ÷ 7/8
20. 9 ÷ 29/50
21. 11 ÷ 22/100
22. 15 ÷ 45/64
23. 1/2 ÷ 1 1/4
24. 1/4 ÷ 3 1/8
25. 1/16 ÷ 1 1/64
26. 2 1/2 ÷ 3 1/4
27. 2 1/10 ÷ 1 13/50
28. How many pieces of wire 2 1/2 in. long can be obtained from 15 in. of stock?
29. How many pieces of wire 5 3/4 in. long can be obtained from 46 in. of stock?
30. How many pieces of band iron 3 7/16 in. long can be obtained from 30 in. of stock? (Allow 1/16 in. for saw cut.) **Hint:** Add the saw-cut allowance to the length of one piece before dividing.
31. Find the distance center to center of each hole on the drawing below.

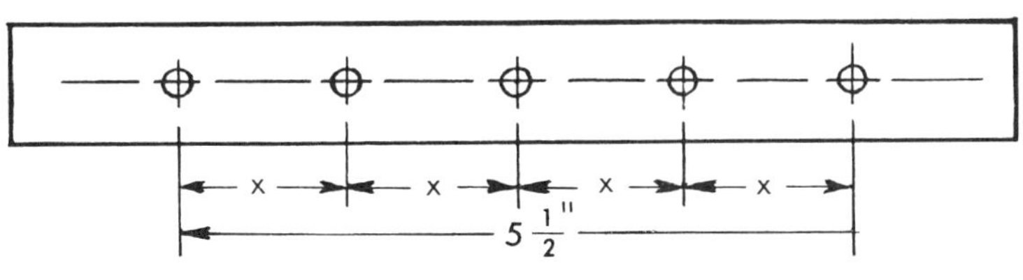

32. How many pieces can be cut from a strip of steel 325 in. long if each stamping requires 2 1/2 in. of steel?
33. How many pieces of angle iron 8 7/16 in. long can be obtained from a piece that is 76 1/2 in. long? Allow 1/16 in. for the saw cut.

UNIT 7 SUMMARY OF COMMON FRACTIONS

There are various ways of expressing problems involving the four fundamental operations: addition, subtraction, multiplication, and division. These are indicated with their proper symbols below:

Addition (+)

1. Plus
2. Sum of
3. Added to
4. Increased by
5. Together we have
6. More than

Subtraction (–)

1. Minus
2. Difference between
3. Subtracted from
4. Diminished by
5. Greater than
6. Less than

Multiplication (x)

1. Times
2. Product of
3. Multiplied by
4. Of

Division (÷ , $\frac{x}{x}$, x/x , $\overline{)\quad}$, /)

1. Divided by
2. Quotient of
3. Fractional part of
4. One number over another

When the four fundamental operations are performed, it is important to know the unit of measurement or the name of the answer. Whenever possible, the unit of measurement should be written after each number as the operations are being performed. When the unit of measurement is being determined, the following rules will be helpful.

1. When inches or other units of measurements are *added,* the answer will be in the same unit.

   in. + in. = in.
   5 in. + 5 in. = 10 in.

2. When inches or other units of measurements are *subtracted,* the answer will be in the same unit.

   in. – in. = in.
   6 in. – 2 in. = 4 in.

3. When inches or other units of measurements are *multiplied* by a number with no dimension, the answer will be in the same unit.

   in. x no. = in.
   2 in. x 3 = 6 in.

4. When inches or other units of measurements are *divided* by a number with no dimension, the answer will be in the same unit.

   $$\frac{\text{in.}}{\text{no.}} = \text{in.} \qquad \frac{6 \text{ in.}}{3} = 2 \text{ in.}$$

5. When inches or other units of measurements are *divided* by inches or the same unit, the answer will be a number only.

   $$\frac{\text{in.}}{\text{in.}} = \text{no.} \qquad \frac{6 \text{ in.}}{3 \text{ in.}} = 2$$

ASSIGNMENT

1. If 6 1/2 in. are increased by 2 3/4 in. and then diminished by 4 7/8 in., what is the final answer?
2. If 2 1/2 in. are multiplied by 4 and then divided by 8, what is the final answer?
3. If 4 in. are divided by 8 and the quotient increased by 1 1/16 in., what is the final answer?
4. If 8 in. are divided by 4 and the quotient decreased by 3/32 in., what is the final answer?
5. Find the distance x on the drawing if the middle hole is to be centered between the outer holes. Check by measurement.

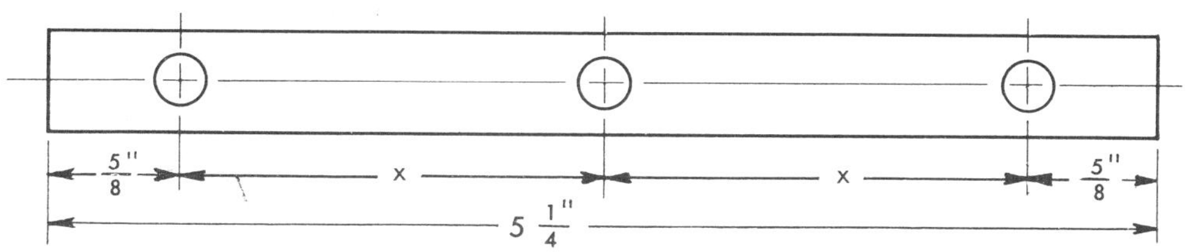

6. How many pieces of band iron 8 7/16 in. long can be obtained from two lengths of 120 in. each? (Allow 1/16 in. for saw cut.)
7. How many 16ths are there in 1 5/8 in. minus 3/4 in.? How many 64ths?
8. Find 1/2 of 2 3/4 in.; 1/4 of 2 3/4 in.; 1/8 of 2 3/4 in.
9. Divide 3 1/2 in. into 2 parts; into 4 parts; into 8 parts; into 16 parts.
10. Compute the rivet spacing indicated below.
11. Compute the distances x and y.

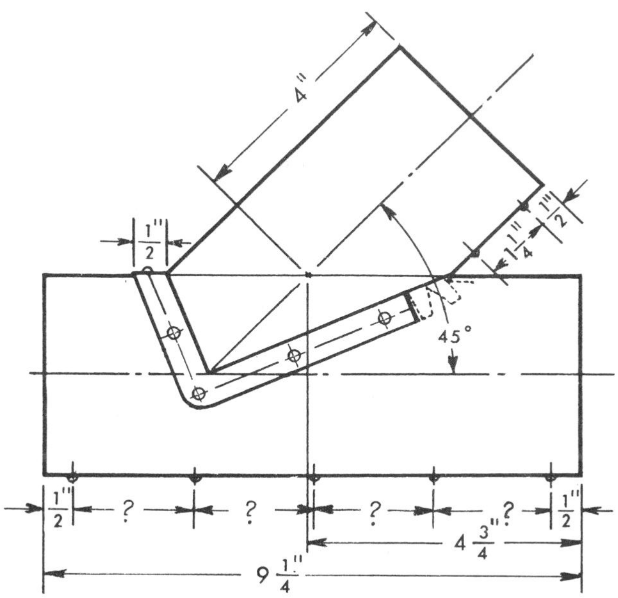

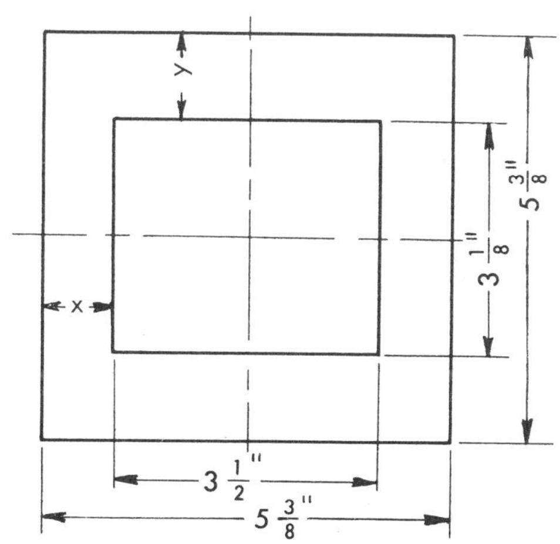

UNIT 8 INTRODUCTION TO DECIMAL FRACTIONS

Although parts of inches may be expressed in fractional form, it is sometimes more convenient to express them in decimal form, such as: .1 in. (one-tenth of an inch), .01 in. (one-hundredth of an inch), .001 in. (one-thousandth of an inch).

There is an increasing tendency, especially in industrial sheet metal work, to use measuring devices graduated in decimal fractions, such as the decimal steel rule and the micrometer. Decimal fractions are more convenient to use in most computations.

DECIMAL PARTS

If a line is divided into 10 parts, one part is one part divided by the total parts, or 1/10 of the total length of the line.

Two parts are equal to 2/10 of the total length.
Three parts are equal to 3/10 of the total length.
Four parts are equal to 4/10 of the total length.
Five parts are equal to 5/10 of the total length.
Six parts are equal to 6/10 of the total length.
Seven parts are equal to 7/10 of the total length.
Eight parts are equal to 8/10 of the total length.
Nine parts are equal to 9/10 of the total length.
Ten parts are equal to 10/10 or equal to the total length.

The drawing below shows the relation between the various fractions.

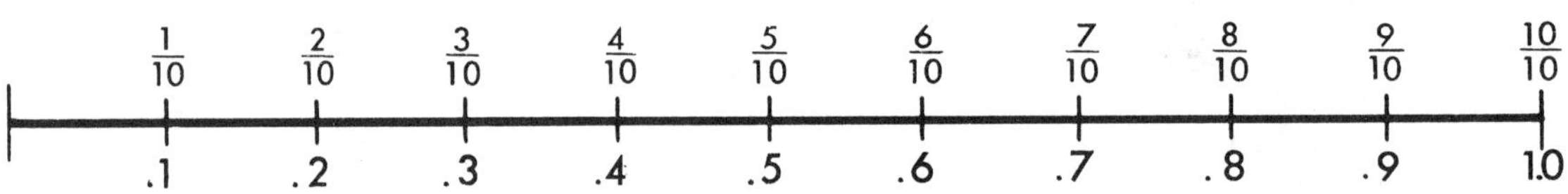

If a line is divided into 10 parts, and the first part is again divided into 10 parts, then one small part is 1/10 of 1/10, or 1/100 of the line.

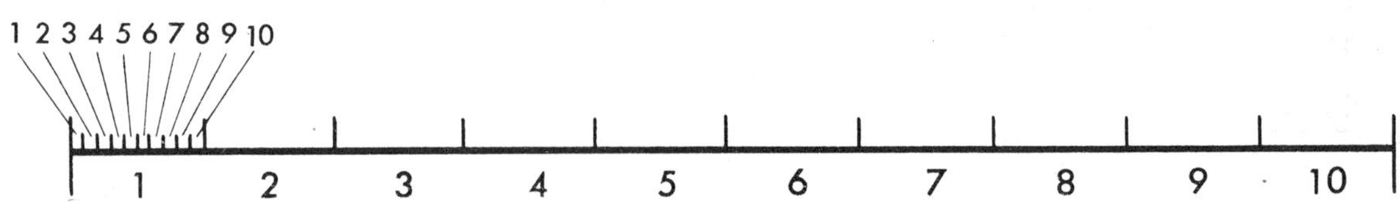

Similarly, two small parts are 2/100 of the total length of the line, and 9 small parts are 9/100 of the total length of the line.

DEFINITIONS OF DECIMAL FRACTIONS

A decimal fraction is a fraction in which the denominator is 10, 100, 1000, 10000, etc., usually not expressed, but signified by a point placed at the left of the numerator.

Examples:

$\frac{7}{10}$	$= .7$	Read as seven *tenths.*
$\frac{7}{100}$	$= .07$	Read as seven *hundredths.*
$\frac{7}{1000}$	$= .007$	Read as seven *thousandths.*
$\frac{7}{10000}$	$= .0007$	Read as seven *ten-thousandths.*
$\frac{7}{100000}$	$= .00007$	Read as seven *hundred-thousandths.*
$\frac{250}{100}$	$= 2.50$ or 2.5	Read as two and fifty *hundredths* or two and five *tenths.*

Note: Another way of writing a decimal fraction such as .46 is 0.46 — the zero before the decimal point emphasizes the fact that the decimal fraction is less than 1.

Value of decimal places:

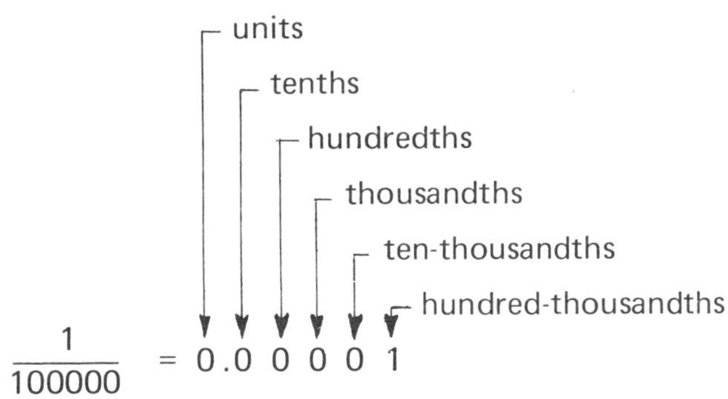

CONVERSION OF COMMON FRACTIONS TO DECIMAL FRACTIONS

1. Denominators less than 100 – Change 3/25 to hundredths.
   a. Divide the denominator 25 into 100. $\frac{100}{25} = 4$
   b. Multiply the numerator and the denominator by the result (4). $\frac{3 \times 4}{25 \times 4} = \frac{12}{100}$, Ans. or .12
2. Denominators greater than 100 – Change 8/200 to hundredths.

   Divide the denominator and the numerator by the number needed to give 100 in the denominator. $\frac{8 \div 2}{200 \div 2} = \frac{4}{100}$, Ans. or .04

ASSIGNMENT

A. Decimal Parts

If the total length of each line is equal to 1, what is the decimal value of the distance represented by each letter on each of the two drawings below?

1.

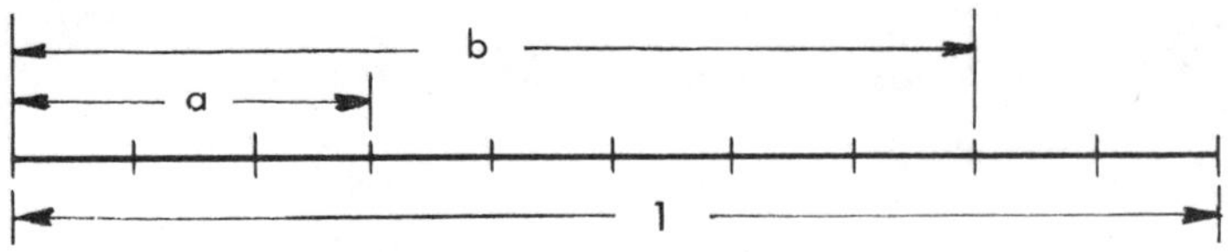

2.

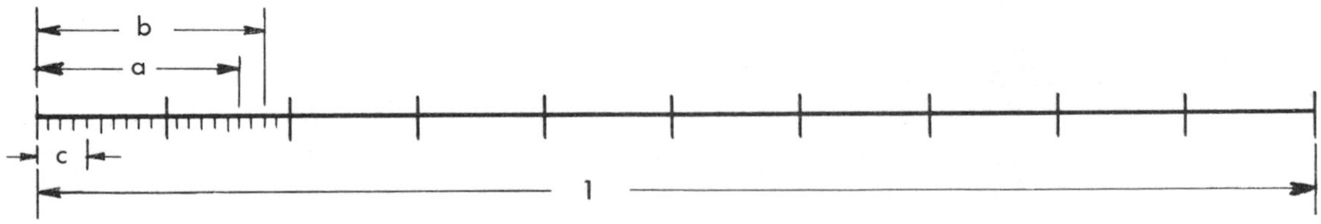

B. Definitions of Decimal Fractions

1. Write the following numbers in the form of words:

a. 651
b. 89.0
c. .3
d. .50
e. .21
f. 0.372
g. 5.258
h. 4.114
i. 102.06

2. Write the following words in the form of numbers:

a. Ninety
b. Forty-nine
c. Nine hundred and nine
d. Six-tenths
e. Sixteen-hundredths
f. Two thousand one hundred fifty-two ten-thousandths
g. Forty-two thousand seven hundred twenty-one hundred-thousandths
h. Four hundred and thirty-one thousandths

C. Reduction to Decimal Fractions

1. Change 7/25 to hundredths
2. Change 31/50 to hundredths
3. Change 53/50 to thousandths
4. Change 82/200 to hundredths
5. Change 82/200 to thousandths
6. Change 80/800 to tenths

UNIT 9 THE STEEL RULE: DECIMAL DIVISIONS

There are rules which are divided according to the decimal scale. The drawing below shows a rule divided into 50ths and 100ths of an inch. Since such scales reduce the possibility of error arising from the necessity of changing common fractions to decimals, they are used by some industrial firms.

$\frac{1}{10}''$	=	.1″	=	one-tenth of an inch
$\frac{1}{50}''$	=	.02″	=	one-fiftieth of an inch
$\frac{1}{100}''$	=	.01″	=	one-hundredth of an inch
$\frac{1}{1000}''$	=	.001″	=	one-thousandths of an inch

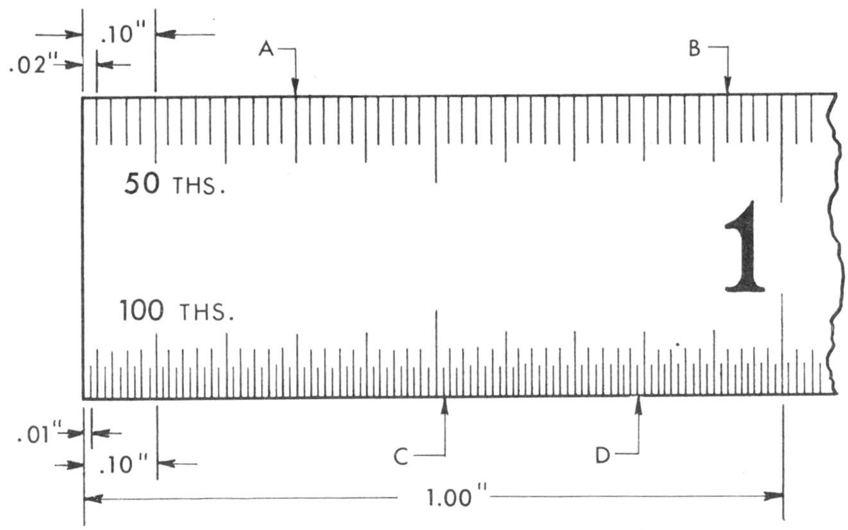

To read the distances from the end of the rule (above) to points A, B, C, and D, proceed as follows:

A	=	.3″		Count three one-tenth-inch divisions.
B	=	.9″ + .02″	= .92″	Count nine one-tenth-inch divisions and add the value of the smallest division.
C	=	.5″ + .01″	= .51″	Count five one-tenth-inch divisions and add the value of the smallest division.
D	=	.7″ + .09″	= .79″	Count seven one-tenth-inch divisions and add the value of nine of the smallest divisions, or count one small division back from .8″.

ASSIGNMENT

A. Reading the Decimal Divisions

Read the distances from the start of the rule to the letters on the rule.

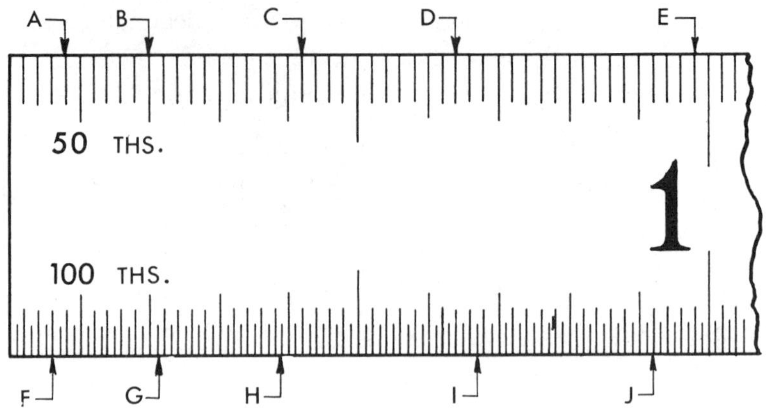

B. Measuring Lines

1. Measure the line below to the nearest 10th of an inch.
2. Measure the line below to the nearest 10th of an inch.
3. Measure the line below to the nearest 50th of an inch.
4. Measure the line below to the nearest 50th of an inch.
5. Measure the line below to the nearest 50th of an inch.
6. Measure the line below to the nearest 50th of an inch.
7. Measure the line below to the nearest 100th of an inch.
8. Measure the line below to the nearest 100th of an inch.

C. Drawing Lines

Draw lines of the following lengths:

1. 1.4 in.
2. 2.9 in.
3. 1.44 in.
4. 3.82 in.
5. 2.48 in.
6. 2.92 in.
7. 1.99 in.
8. 3.13 in.

UNIT 10 ADDITION AND SUBTRACTION OF DECIMAL FRACTIONS

When the dimensions of a job are expressed in decimal form, it is often necessary to add or subtract decimal fractions to find the total length of the job, or the distance between certain points on a job.

RULE: To add or subtract decimals, place the numbers so that the decimal points are under each other, and then add or subtract as with whole numbers.

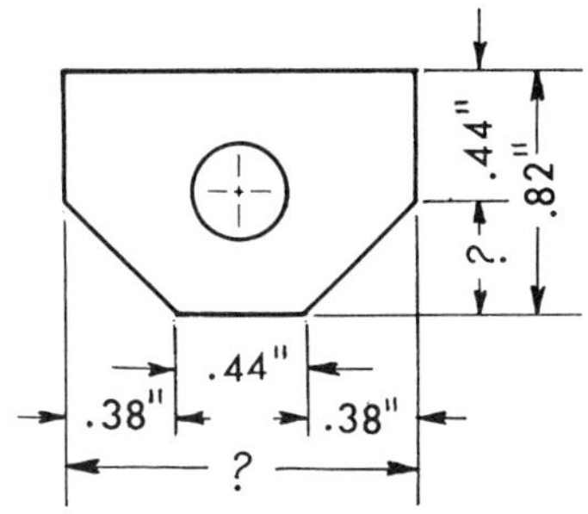

ADDITION

Example: Add 3.25, 864.0725, 647, and 0.875

1. Write the given numbers, arranging them so that the decimal points are under each other.

```
     3.25
   864.0725
   647.
 +   0.875
```

2. To avoid errors, add zeros to the numbers with the fewer places so that all the numbers will have an equal number of places after the decimal point

```
     3.2500
   864.0725
   647.0000
 +   0.8750
```

3. Proceed as in addition of whole numbers, placing the decimal point in the sum under the other decimal points.

```
           3.2500
         864.0725
         647.0000
       +   0.8750
(Sum)   1515.1975  Ans.
```

Check: Add the numbers in reverse order.

SUBTRACTION

Example: From 42.63 take 18.275

1. Write numbers so that the decimal points are under each other and subtrahend is under the minuend.

```
(Minuend)       42.63
(Subtrahend)  - 18.275
```

2. If one of the numbers has fewer figures after the decimal point than the other, add zeros to it.

```
(Minuend)       42.630
(Subtrahend)  - 18.275
```

3. Subtract as with whole numbers, placing the decimal point in the remainder under the other decimal points.

```
   42.630
 - 18.275
   24.355  Ans.
```

Check: Minuend equals sum of subtrahend and remainder.

ASSIGNMENT

A. Addition

Solve and check the following problems:

1. 283.5 + 69.2
2. The sum of 384.73 and 63.14
3. 1.738 + .0087 + 7.58
4. 932.07 + 93.207 + 3.1416 + .93207
5. .067 + 37.762 + 762.37 + 7.9
6. .5957 + .4058 + 2.9530 + 3.3146 + .49
7. Find the total thickness of the following two sheets. .0179 in. and .1046 in.
8. Find the total thickness of a piece of sheet metal .0784 in. thick and a piece of band iron .25 in. thick.
9. Find the total thickness of two pieces of sheet metal, each .0239 in. thick, and one piece of band iron .30 in. thick.
10. Compute the overall length of the gage on the drawing below.

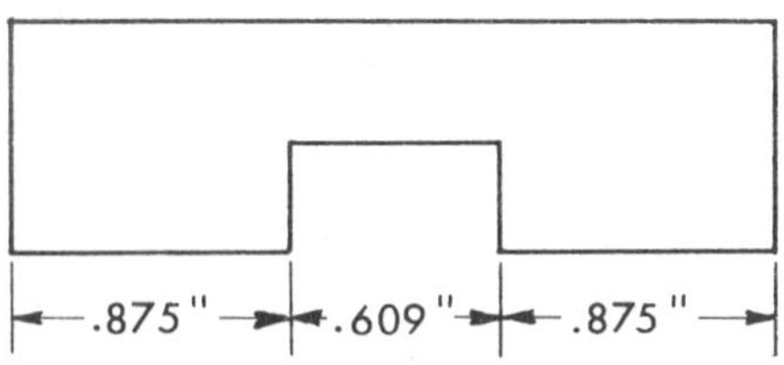

11. Compute the height of the stiffener shown below.

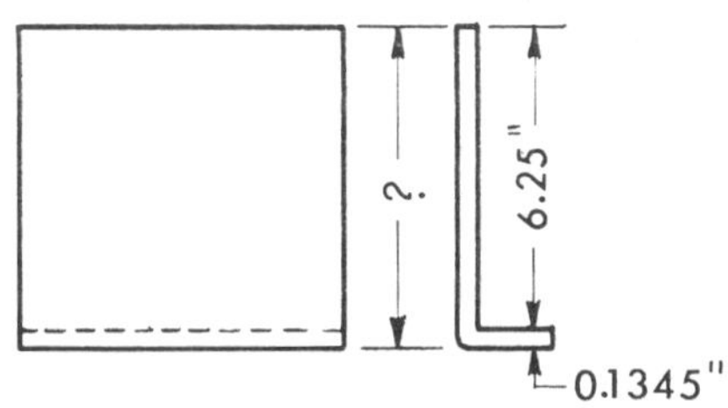

B. Subtraction

Solve and check the following problems:

1. 25.1 - 12
2. 67.4 - 32.1
3. 129.9 - 58.1
4. 35.1 - 29.6
5. 198.23 - 19.12
6. 12.328 - 8.769
7. Subtract 21.811 from 98
8. From 123.1 take 79.52
9. Take 0.6347 from .8765
10. Take 2.574 from 5.25
11. What is 1.0 diminished by .051?
12. What is 5.0 diminished by .0199?
13. What is the difference between 47.5621 and 23.89?
14. Find the difference in thickness between two sheets of metal .0179 in. and .0149 in.

15. Compute the distances x and y of the strip on the drawing. Check your answer by measuring with a rule (100th scale).

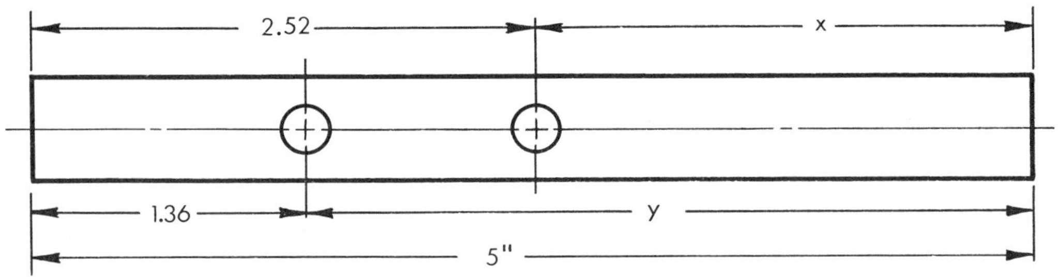

16. Compute the distance x.

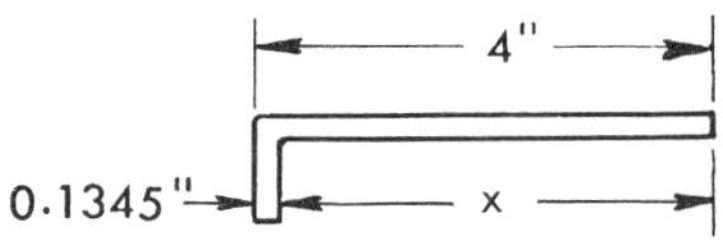

17. Compute the length x.

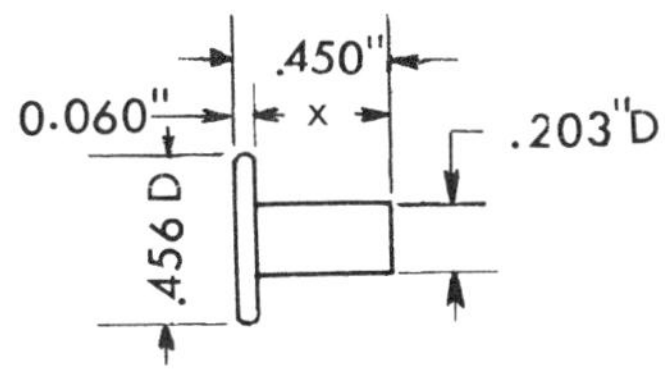

C. Addition and Subtraction of Decimals

Solve and check the following problems.

1. 2.490 + 3.9821 - 5.62
2. 4.598 + 11.295 - 8.9218
3. Compute the distance between the centers of the circles in the channel.
4. 110.412 + 231.22 - 298.4172
5. 258.32 + 4.213 - 9.20
6. Compute the distances x and y.

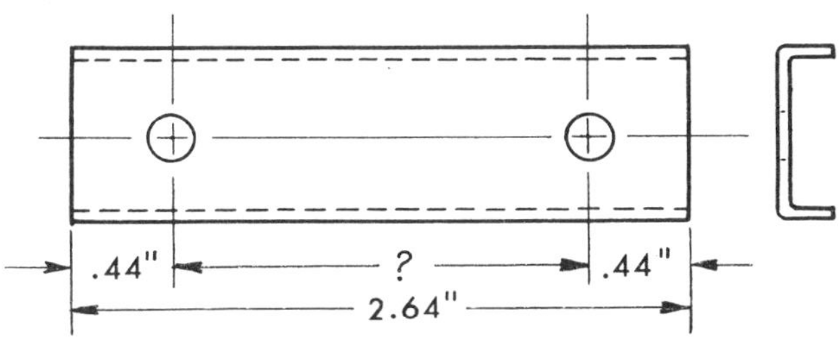

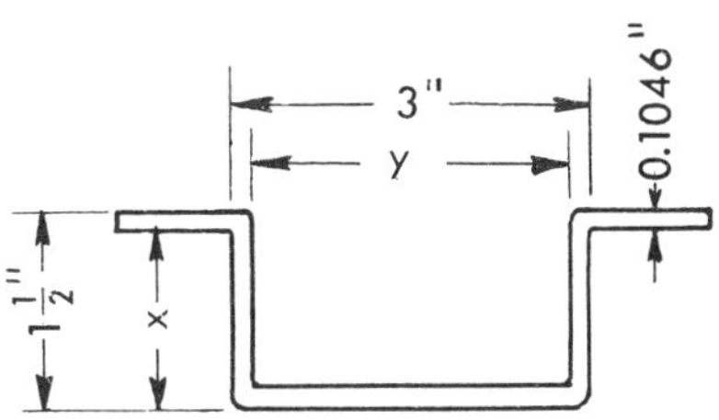

7. Compute the distances x and y on the drawing below.

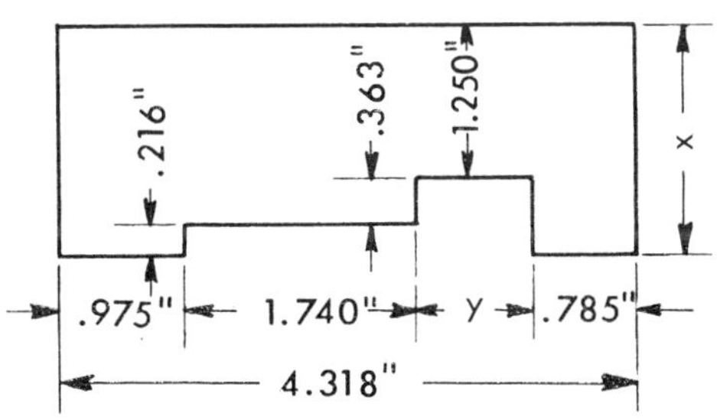

UNIT 11 MULTIPLICATION OF DECIMAL FRACTIONS

The multiplication of decimals is used for finding the total length of material needed for a job; for finding the height to which a pile of metal sheets will reach, etc.

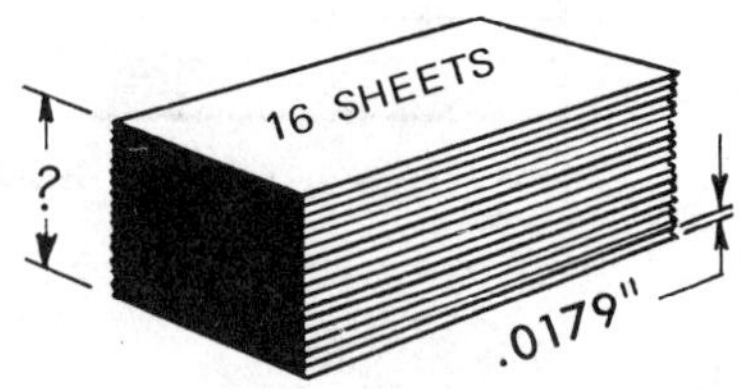

RULE: To multiply decimals, multiply as in whole numbers, and point off as many decimal places in the product as there are decimal places in both the multiplicand and the multiplier.

Example 1: Multiply 43.286 by 6.04

1. Write the number with the most digits first.

```
 43.286
 x 6.04
```

2. Multiply as with whole numbers.

```
(Multiplicand)       43.286
(Multiplier)         x 6.04
                     173144
                   2597160
(Product)          26144744
```

3. Beginning at the right, point off in the product the sum of the decimal places in the multiplicand and the multiplier.

```
(3 places)                 43.286
(2 places)                 x 6.04
                           173144
                         2597160
(5 places)          Ans. 261.44744
```

Check: Either interchange the multiplier and multiplicand and rework the problem, or repeat the operations.

Example 2: Multiply .1875 by .02

1. Multiply.
2. Prefix two zeros to make the necessary six places.

```
(4 places)            .1875
(6 places)             x .02
               Ans. .003750
```

Note: When multiplying numbers, such as 25.020 by 4.690, drop the zeros at the extreme right before proceeding.

```
 25.020 =  25.02
x 4.690 = x 4.69
```

ASSIGNMENT

Solve and check the following problems.

1. 2 x 1.5
2. 1.5 x 3
3. 4.5 x 7.5
4. 12.3 x 7.1
5. 1.1 x 1.1
6. .2 x .15

7. .15 x .3
8. .45 times .75
9. 1.23 times .71
10. 1.1 multiplied by .11
11. 2.87 x 3.1416
12. 1.001 x 5.30
13. What is the product of .149 and .125?
14. What is .1875 multiplied by 2.8625?
15. .6875 times .501
16. .3740 times .875
17. What is 2.0313 times .6250?
18. What is the total thickness of 3 sheets of metal, each .0097 in. thick?
19. Find the total height of a stack of 250 sheets of metal, each .0149 in. thick.
20. Find the total height of a stack of 851 sheets of metal, each .0097 in. thick.
21. By multiplication, find the distance between the centers of the outer holes in the bracket shown below, and check the result by addition.

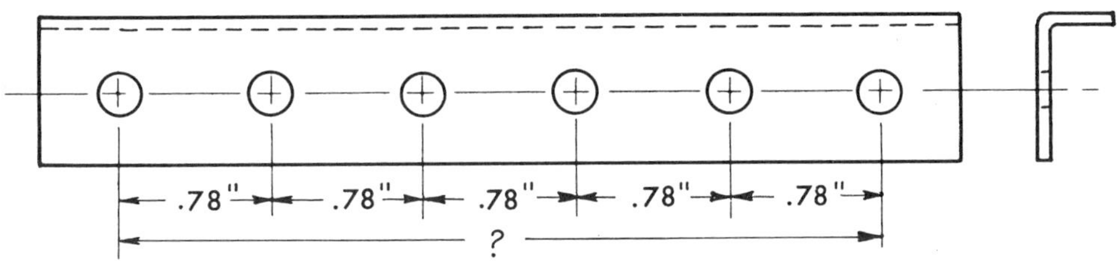

22. If 9 strips, each 3.125 in. wide, must be cut from one piece of black iron, what must be the total width of the black iron?
23. Find the total height of a stack of 85 sheets of tin plate, each .0152 in. thick.

UNIT 12 DIVISION OF DECIMAL FRACTIONS

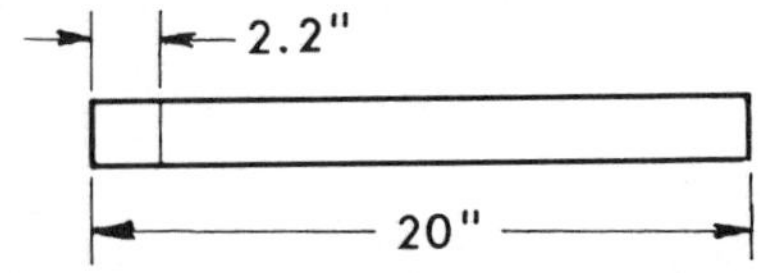

The division of decimals is used to find the number of pieces of wire or band iron which can be obtained from a piece of stock or to find the number of pieces that can be cut from a sheet.

RULE: To divide decimals, divide as in whole numbers, and point off in the quotient (answer) the number of decimal places equal to the difference between the number of decimal places in the dividend and those in the divisor.

Example: Divide 0.78 by 0.964 to 4 decimal places.

1. Place the dividend inside and the divisor outside the symbol as shown.

   (Divisor) .9 6 4) .7 8 (Dividend)

2. Make a whole number out of the divisor by moving the decimal point and use a caret (^) to indicate the new decimal place.

   .9 6 4^) .7 8

3. Move the decimal point in the dividend the same number of places to the right indicating the position with a caret. If necessary, add zeros to the dividend.

   .9 6 4^) .7 8 0^

4. Divide as in whole numbers, adding as many zeros after the caret in the dividend as are necessary to give the required number of places in the quotient. Place the decimal point in the quotient directly above the caret in the dividend.

```
(Quotient)      .8 0 9 1  Ans.
.9 6 4^) .7 8 0^0 0 0 0
           7 7 1 2
               8 8 0 0
               8 6 7 6
                 1 2 4 0
                   9 6 4
           .0 0 0 0 2 7 6  (Remainder)
```

Check: Multiply the quotient by the divisor and add the remainder, or repeat the operations.

Note: If there are extra zeros in the divisor, drop these before proceeding with the problem. If .20900 is the divisor, use .209.

ASSIGNMENT

Solve and check the following problems.

1. 10 ÷ 1
2. 10 ÷ 10
3. 10 ÷ 100
4. 10 ÷ 1000
5. 10 ÷ 10000
6. 100 ÷ 1
7. 100 ÷ 10
8. 100 ÷ 100
9. 100 ÷ 1000
10. 100 ÷ 10000
11. 1000 ÷ 10000
12. 10000 ÷ 100
13. 10.8 ÷ .2
14. 15.62 ÷ .2
15. 4.4 ÷ .4
16. 2.8/.02
17. .24/2
18. 2.4/3

19. 2.5 ÷ 2.5
20. 25/2.5
21. 2.5/25
22. 24/1.2
23. .20/.4
24. 1.96/.2
25. $\frac{1.6}{.32}$
26. 1.20/.15
27. Divide 39 by .03
28. Divide .122 by .002
29. .0021/.007
30. $\frac{1932}{2.1}$
31. .464/.016
32. Find the quotient of 267 divided by .3.
33. How many sheets will there be in a stack of sheets 3 in. high? The thickness of each sheet is .0598 in.
34. How many sheets will there be in a stack of sheets 20 in. high if the thickness of each sheet is .0097 in.?
35. A steel sheet 2.5 in. wide is to be sheared into four strips of equal width. How wide will each strip be?
36. A piece of band iron 24.375 in. long is to be divided into equal spaces of 1.625 in. each. How many spaces will there be?
37. Find the width of one space on the drawing below. Check by measurement.

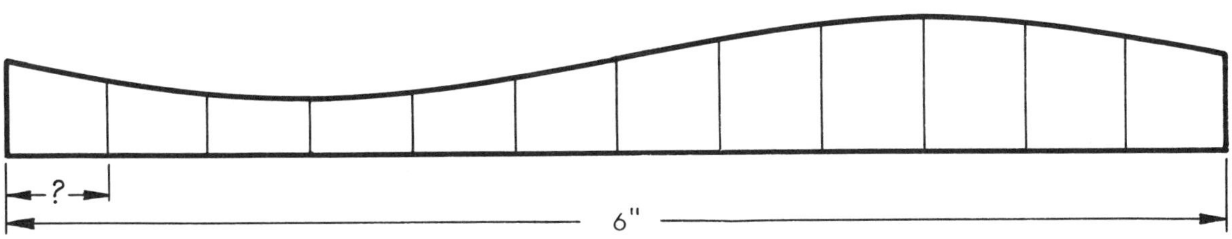

UNIT 13 ROUNDING OFF DECIMALS

To save time in working a problem, no more decimal places should be held than are indicated by the accuracy of the measurements. In sheet metal work, accuracy to a degree of more than one thousandth of an inch (.001 in. – three decimal places) is seldom required. In order to hold the decimal places to a required number, the numbers must be rounded off. *Rounding off* means expressing a number with fewer digits. Example: .15934 in. rounded off to 3 places is .159 in.

ROUNDING OFF DECIMALS

RULE: Decide how many decimal places are needed in the answer. If the number following the last decimal place required is 5 or larger, add 1. If this number is less than 5, do not add 1. Drop all numbers following the last decimal place required.

Example: Round off .672853 to three decimal places.

1. Three decimal places are required.
2. The number following the last decimal place required is 8; therefore 1 is added to the 2, and .672 becomes .673.
3. Drop .000853, and the final result is .673. Ans.

NUMBER OF PLACES IN COMPUTATIONS

Hold one more decimal place during the computations than the number of decimal places required in the answer. For instance, if the measurements are correct to three places, hold four places throughout the computations and three places in the answer. This is a good general rule to follow; the exceptions to this rule will be given later.

Example: Multiply 12.125 by 2.510 and then by 6.875. Three decimal places are required in the answer.

```
   12.125
   x 2.51
  -------
    12125
   60625
  24250
  -------
  30.43375  = 30.4338
```

```
       30.4338
       x 6.875
    ----------
       1521690
      2130366
     2434704
    1826028
    ----------
    209.2323750 = 209.232 Ans.
```

ASSIGNMENT

A. Rounding Off Decimals

1. Round off the following numbers to three decimal places:

a. .7844	c. .6836	e. .0036	g. .29438
b. 1.90042	d. 1.73445	f. .5499	h. 1.46859

2. Round off each decimal fraction to four places in the decimal equivalent table on page 37.

B. Number of Places in Computations

1. Add the following numbers and express the sum correctly to three decimal places:

   4.604 1.6385 1.8291 14.96 103.154

2. A thickness gage has 8 tempered steel leaves of the following thicknesses:

   .002 .004 .006 .008 .010 .015625 .03125 .0625

   What is their total thickness (three decimal places)?

3. Subtract 3.7869 from 18.432 and express the difference correctly to three decimal places.
4. Multiply 14.037 by 8.72 and express the product correctly to two decimal places.
5. Multiply 10.120 by 2.864 and then multiply the result by 2.46. Hold three places in the computations and two places in the answer.
6. Multiply 12.125 by 8.875 and then multiply the result by 2.693. Hold four decimal places throughout the computations and three decimal places in the answer.
7. Divide 104 by 14.6 and express the result correctly to two decimal places.
8. Divide 12.21 by 2.05 and then divide the result by 3.42. Hold three places in the computations and two places in the answer.
9. Divide 16.278 by 2.425 and then divide the result by 3.411. Hold four places in the computations and three places in the answer.
10. 193.74 ÷ 2.59 (1 place)
11. 2850 ÷ .098 (2 places)
12. .2850 ÷ .0029 (1 place)
13. 1.462 ÷ 3.40 (2 places)
14. 156.200 ÷ .043 (2 places)
15. 1.006 ÷ 1.0162 (4 places)

UNIT 14 CHANGING FRACTIONS AND DECIMALS

It is frequently necessary to change dimensions expressed in the form of common fractions to dimensions expressed in decimal form, or vice versa. Changing common fractions to decimal form is done by division, and changing decimals to common fractions is done by multiplication. These methods can be used to change other common fractions and decimals such as those involved in dollars, and weights.

CHANGING A COMMON FRACTION TO A DECIMAL FRACTION

RULE: Divide the numerator of the common fraction by the denominator. Add zeros after the decimal point, if necessary.

Example: Change 15/16 in. to a decimal fraction.

$16\overline{)15}$

1. Write the common fraction as a division problem.

2. Since 16 does not go into 15, place a decimal point after the 5, add zeros and divide. In this case, the division was continued until there was no remainder.

$$\begin{array}{r} 0.9375 \text{ in. Ans.} \\ 16\overline{)15.0000} \\ \underline{144} \\ 60 \\ \underline{48} \\ 120 \\ \underline{112} \\ 80 \\ \underline{80} \end{array}$$

CHANGING A DECIMAL FRACTION TO A COMMON FRACTION

RULE: Multiply the given decimal by the fractional divisions required. Form a common fraction with this result (rounded off) as a numerator and the divisions required as a denominator.

Example: Change 0.385 in. to a common fraction (closest 64ths).

1. Multiply the given number by the closest divisions needed; in this case by 64. The result is 24.64.

$$\begin{array}{r} .385 \\ \underline{\times 64} \\ 1540 \\ \underline{2310} \\ 24.640 \end{array}$$

2. Round off 24.64 to a whole number. 24.64, or 25

3. Put this result, 25, over 64. $\frac{25}{64}$ in. Ans.

ASSIGNMENT

A. Changing Common Fractions to Decimal Fractions

Change the following common fractions to decimal fractions of four places:

1. 1/2 2. 1/4 3. 3/4 4. 1/8 5. 3/8

6. 5/8	9. 5/16	12. 9/32	15. 19/64	18. 4/5
7. 7/8	10. 15/16	13. 31/32	16. 2/3	19. 1/25
8. 1/16	11. 1/32	14. 1/64	17. 5/6	20. 3/48

B. Changing Decimal Fractions to Common Fractions

Change the following decimal fractions to the common fractions indicated and reduce each fraction to its lowest terms.

16ths	32nds	64ths
1. .1875	6. .03125	11. .04688
2. .6875	7. .28125	12. .20313
3. .4375	8. .71875	13. .95313
4. .8127	9. .906	14. .5156
5. .1880	10. .7815	15. .4844

C. Trade Problems

1. Change the diameters of the following tinners' rivets to the fractional form (nearest 64th of an inch):

a. 10 oz. – .095″	b. 1 1/2 lb. – .130″	c. 4 lb. – .176″
14 oz. – .109″	2 lb. – .144″	5 lb. – .186″
1 lb. – .112″	3 lb. – .160″	6 lb. – .203″

2. You are required to make a templet of the shape shown, and you have a rule graduated in 64ths. Change all dimensions to the fractional form, accurate to the nearest 64th of an inch.

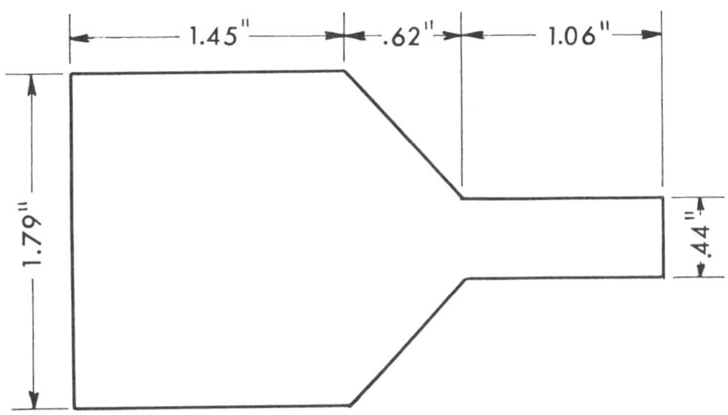

UNIT 15 OTHER USES OF FRACTIONS AND DECIMALS

There are various special methods in arithmetic which involve fractions and decimals. These methods are useful in solving certain trade problems, and, therefore, should be understood by the mechanic.

FINDING A WHOLE NUMBER IF A PART IS GIVEN

Example 1: If 3/4 of a number is 360, what is the whole number?

1. Divide the given number by the numerator of the fraction.

   $\frac{360}{3} = 120$ **Note:** 120 is 1/4 of the whole number.

2. Multiply the quotient by the denominator of the fraction. $120 \times 4 = 480$ Ans.

   Note: In order to find 4/4 (the whole), it is necessary to first find 1/4 and then multiply it by 4.

Example 2: If .8 of a number is 40, what is the whole number?

1. If .8 = 40
2. Then, .1 = 40/8, or 5
3. And 1.0, or the whole = 10 x 5, or 50 Ans.

   Note: 40 divided by .8 = 50 Ans.

DIVIDING NUMBERS

Example: Divide 48 into one part and three parts, all equal.

1. Find the total parts.
2. Find the fractional value of each group.
3. Multiply the whole by each fraction and add the answers as a check.

1 part + 3 parts	=	4 parts
First group	=	1/4 of whole
Second group	=	3/4 of whole
1/4 x 48	=	12 Ans.
3/4 x 48	=	36 Ans.
		48 Check

COMPLEX FRACTIONS

Example: $\frac{1/2 + 1/4}{1/2 \times 1/4} = ?$

1. Do the work indicated in the numerator. $\frac{1}{2} + \frac{1}{4} = \frac{2}{4} + \frac{1}{4}$, or $\frac{3}{4}$

2. Do the work indicated in the denominator. $\frac{1}{2} \times \frac{1}{4} = \frac{1}{8}$

3. Divide the numerator by the denominator. $\frac{3}{4} \div \frac{1}{8} = \frac{3}{\cancel{4}_{1}} \times \frac{\cancel{8}^{2}}{1}$, or 6 Ans.

ASSIGNMENT

A. Finding a Whole Number if a Part is Given.

1. If 3/4 of a job costs $9,000, what will the entire job cost?
2. If 2/3 of a job takes 6 hr., how long will the entire job take?
3. If 7/12 of a coil of wire costs $42.00, what will the coil cost?
4. If .6 of a coil of wire weighs 2 lb., what will the coil weigh?

B. Dividing Numbers

1. Divide an 18-in. piece of band iron into two lengths, one piece 1/3 and the other 2/3 of the total length.
2. Babbitt metal is 4 parts copper, 89 parts tin, and 7 parts antimony. How many pounds of each are in 20 lb. of babbitt?
3. On a certain job, one man worked 7 hours, and another worked 15 hours. If the job paid $91.30 and each man was paid the same hourly rate, how much did each man receive?
4. An alloy contains by weight 2/3 copper and 1/3 zinc. How many pounds of each metal will be required for a 75-lb. sheet?
5. A victory solder is composed of metals in the following proportions: 13 parts tin, 2 parts silver, 85 parts lead. Using 100 lb. of solder as a basis, find the saving in tin when the victory solder is compared with a solder made of 50% tin and 50% lead.
6. The ratio of tin to lead in certain solder is 1 to 2 and the total weight of solder wanted is 90 lb. How much lead is required? How much tin?
7. If the ratio of the rate of a journeyman to the rate of an apprentice is 2 to 1, and if both together receive $6.30 an hour, what is the rate of the apprentice; of the journeyman?

C. Complex Fractions

1. $\dfrac{3/4 + 1/4}{3/4 - 1/4} = ?$

2. $\dfrac{1/4 - 1/8}{1/4 + 1/8} = ?$

3. $\dfrac{5/8 - 1/4}{1/8 + 3/4} = ?$

4. $\dfrac{3/4 \times 1/3}{3/8 + 1/2} = ?$

5. $\dfrac{5/8 \times 4}{2.5 \times 5} = ?$

6. $\dfrac{10.50}{1.5 \times 2 + 2.25} = ?$

7. Three holes are to be drilled in a piece of band iron 6 in. long, the outer holes being 3/8 in. from each end. What will be the spacing of the holes? **Note:** Write the problem as a complex fraction and solve this fraction.
8. If 2 1/2 lengths, each 12′ long of 1/8″ x 1/2″ band iron and 4 1/2 lengths each 15′ long of 1/8″ x 1/2″ band iron cost $6.90, what is the cost of the band iron per lineal foot?

UNIT 16 THE DECIMAL EQUIVALENT TABLE

The speed of computation may be increased and the possibility of error reduced if a decimal equivalent table is used to convert common fractions to decimal fractions and decimal fractions to common fractions.

Decimal equivalent tables similar to the one shown on the next page are available on charts, on the backs of scales, and in handbooks.

SELECTION OF VALUES IN TABLES

Example 1: Select the decimal equivalent of 1/64 in. In the left-hand column of the table on the next page, we find 1/64, and opposite it, .015625. Since 1/64 is in inches, then .015625 must be in inches also.

Note: Divide 1 by 64 to prove that this value is correct.

Example 2: Select the fractional equivalent of .109375 in. Look in the right-hand column until .109375 is found; read the fractional value to the left. The value is 7/64 in.

Note: Multiply .109375 by 64 to prove that this value is correct.

SELECTION OF DECIMAL VALUES NOT IN TABLE

Example: Select the fractional equivalent of .813 in.

13/16	.8125	← .813
53/64	.828125	

1. Examine the table to find the two decimal equivalents between which .813 falls.

2. Find the difference between the greater number and .813, and find the difference between .813 and the smaller number.

$$\begin{array}{r} .828125 \\ -\ \underline{.813000} \\ .015125 \end{array} \qquad \begin{array}{r} .8130 \\ -\ \underline{.8125} \\ .0005 \end{array}$$

3. Since .0005 is smaller than .015125, look up the fractional equivalent of .8125, which is 13/16. Therefore, the nearest fractional equivalent to .813 in. is 13/16 in.

ASSIGNMENT

Find the decimal equivalent of the common fraction or mixed number; when decimals are given, find the nearest common fraction or mixed number (closest 64th of an inch).

A. Values in Table

1. 1/2
2. 1/4
3. 3/4
4. 5/8
5. 7/8
6. 1/16
7. 5/16
8. 11/16
9. 7/32
10. 63/64
11. .125
12. .375
13. .8125
14. .3125
15. .234375
16. .84375
17. 29/32
18. .953125
19. .7185
20. 23/64

B. Values Not in Table

1. .639
2. .830
3. .418
4. .21
5. .52
6. .755
7. .98
8. .70
9. .907
10. .331
11. .481
12. .0125
13. .033
14. 3.156
15. 2.436
16. 2.995
17. 1.09675
18. 2 61/64
19. 1.125
20. 5 11/16

C. Trade Problems

1. Find the decimal equivalent of the following drills:

a. 1/64″
b. 1/32″
c. 3/64″
d. 5/32″
e. 5/64″
f. 7/64″

g. 

2. Find the nearest fractional equivalent in 64ths of an inch of the following drills:

a. .0635″ b. .1065″ c. .1200″ d. .2130″

3. Find the nearest fractional equivalent in 64ths of an inch of the following solid punches:

a. .355″
b. .305″
c. .280″
d. .250″
e. .225″
f. .213″
g. .188″
h. .168″
i. .145″
j. .128″
k. .114″
l. .093″

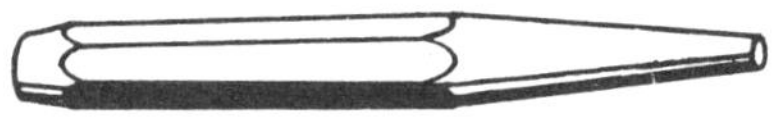

DECIMAL EQUIVALENT TABLE

Fraction	Decimal	Fraction	Decimal	Fraction	Decimal	Fraction	Decimal
1/64	.015625	17/64	.265625	33/64	.515625	49/64	.765625
1/32	.03125	9/32	.28125	17/32	.53125	25/32	.78125
3/64	.046875	19/64	.296875	35/64	.546875	51/64	.796875
1/16	.0625	5/16	.3125	9/16	.5625	13/16	.8125
5/64	.078125	21/64	.328125	37/64	.578125	53/64	.828125
3/32	.09375	11/32	.34375	19/32	.59375	27/32	.84375
7/64	.109375	23/64	.359375	39/64	.609375	55/64	.859375
1/8	.125	3/8	.375	5/8	.625	7/8	.875
9/64	.140625	25/64	.390625	41/64	.640625	57/64	.890625
5/32	.15625	13/32	.40625	21/32	.65625	29/32	.90625
11/64	.171875	27/64	.421875	43/64	.671875	59/64	.921875
3/16	.1875	7/16	.4375	11/16	.6875	15/16	.9375
13/64	.203125	29/64	.453125	45/64	.703125	61/64	.953125
7/32	.21875	15/32	.46875	23/32	.71875	31/32	.96875
15/64	.234375	31/64	.48438	47/64	.734375	63/64	.984375
1/4	.250	1/2	.500	3/4	.750	1	1.000

UNIT 17 DIVIDING LINES

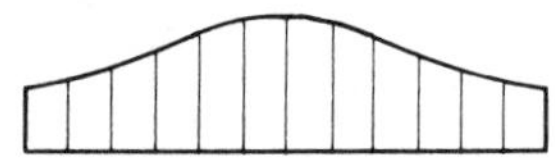

The sheet metal worker is often called upon to divide lines into equal parts. This occurs when he must divide a stretchout line into equal parts and locate rivets in a given space.

Dividing lines can be done by construction or computation. When computation is used, express the length of line as a decimal; then divide by the number of spaces required. Finally, change the decimal to a fractional equivalent which is on a scale. Computation is used when an odd number of spaces is required. Both methods are accurate enough for most layout work when performed with care. More accurate methods are given on pages 93 and 94.

EQUAL PARTS BY BISECTION

Example: Bisect (divide into two equal parts) a line 3 in. long.

Step 1

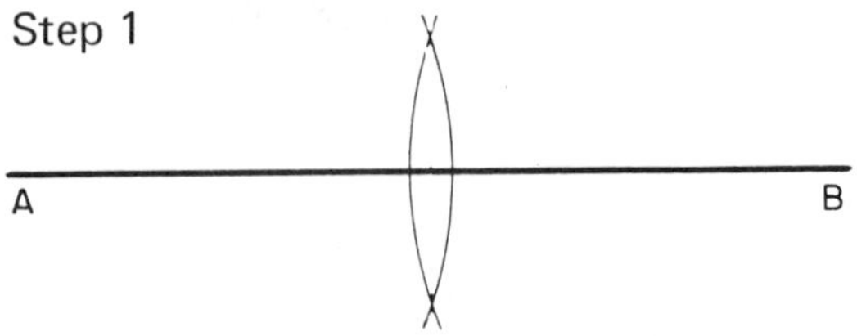

Draw a 3-in. line, and draw intersecting arcs from points A and B.

Step 2

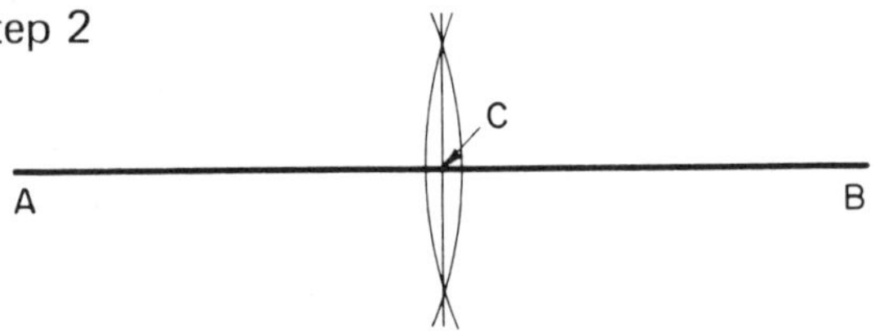

Draw a line through intersections of arcs. AC = CB. Check by computation and measurement.

Note: If this process is repeated, the line may be divided into additional parts, such as 4, 8, 16, 32, etc.

EQUAL PARTS BY BISECTION AND TRIAL

Example: Divide a line 3 in. long into six equal parts.

Step 1

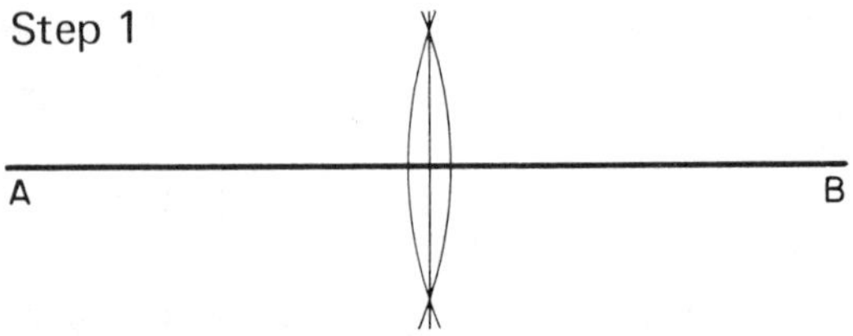

Draw a 3-in. line and divide it into two equal parts by bisection.

Step 2

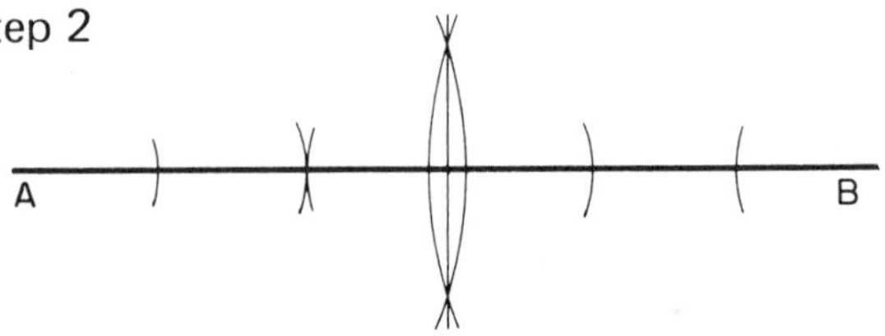

Divide each half into three parts by changing the compass opening until the parts are equal. Check as above.

ASSIGNMENT

Express answers in fractional form (closest 64th of an inch).

1. Divide a line 6 in. long into 4 equal parts by construction, and check the answer by computation and measurement.

2. Divide a line 5 1/4 in. long into 8 equal parts by construction, and check the answer by computation and measurement.
3. Divide a line 4 in. long into 6 equal parts by construction, and check the answer by computation and measurement.
4. Divide a line 5 1/2 in. long into 6 equal parts by construction, and check the answer by computation and measurement.
5. Divide a line 7 in. long into 12 equal parts by construction, and check the answer by computation and measurement.
6. Divide a line 6 1/2 in. long into 12 equal parts by construction, and check the answer by computation and measurement.
7. Find the distance between adjacent elements on the stretchout. (The elements are equally spaced.)

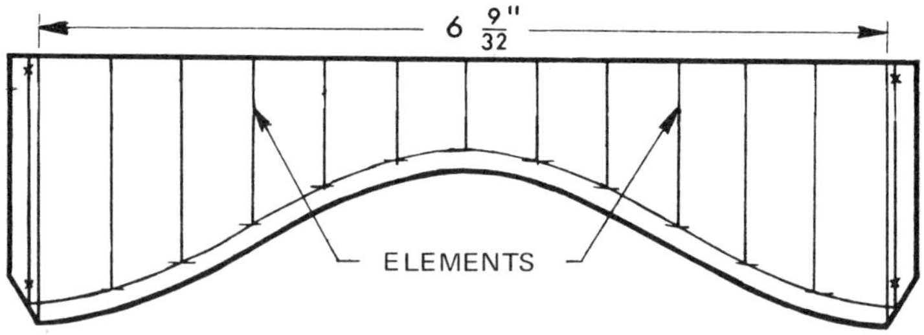

8. There are to be 6 equal pieces cut from a plate 12 9/16 in. wide. How wide will each piece be?
9. There are to be 9 equal pieces cut from a plate 28 1/2 in. wide. How wide will each piece be?
10. If 5 rivets are to be spaced equally in a length of 14 in., what will be the rivet spacing?

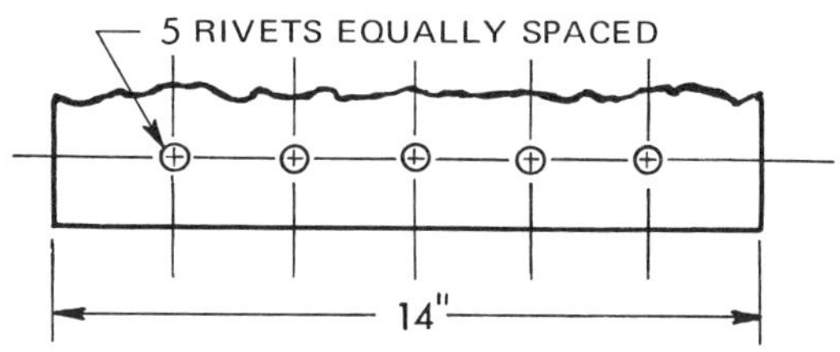

11. If 7 rivets are to be spaced equally in a length of 25 in., what will be the rivet spacing?

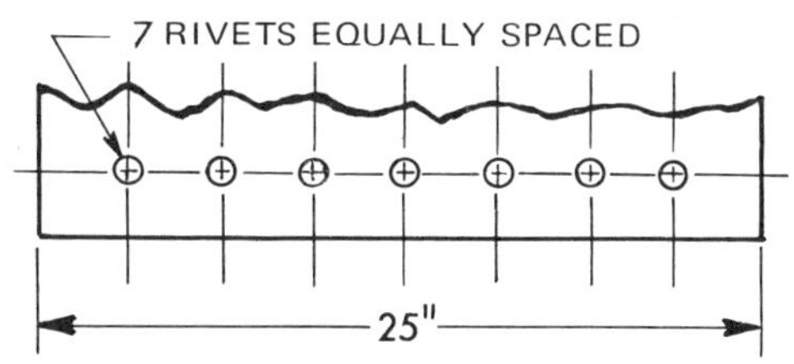

12. Compute the rivet spacing on the drawing opposite.

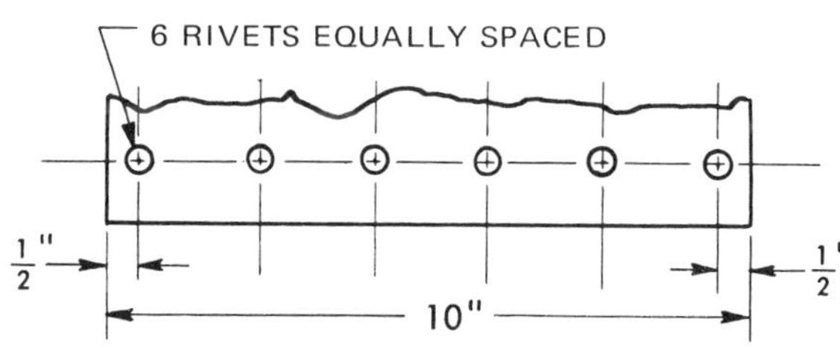

UNIT 18 USE OF TABLES – TRADE DATA

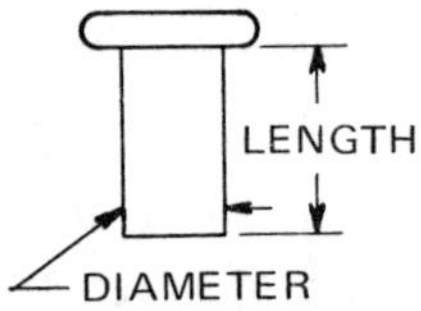

Tables of gages of sheet metal and of rivet sizes are useful to the sheet metal worker in computing the length and diameter of rivets needed for the job. In the use of tables, the following procedure should be observed:

1. Determine which table to use by reading the titles and noting the contents of tables which pertain to the problem.
2. Study the headings of the columns in the tables.
3. Select the data needed.

SHEET METAL TABLES

The sheet metal worker uses tables like those shown in tables 17-1 and 17-2 to find the thickness of metal if the gage is given and to find the gage of metal if the thickness is given.

TABLE 17-1 GAGES AND THICKNESSES OF SHEET METAL: BLACK IRON

U.S. Gage	Thickness Inches	U.S. Gage	Thickness Inches	U.S. Gage	Thickness Inches
38	.0060	26	.0179	14	.0747
37	.0064	25	.0209	13	.0897
36	.0067	24	.0239	12	.1046
35	.0075	23	.0269	11	.1196
34	.0082	22	.0299	10	.1345
33	.0090	21	.0329	9	.1495
32	.0097	20	.0359	8	.1644
31	.0105	19	.0418	7	.1793
30	.0120	18	.0478	6	.1943
29	.0135	17	.0538	5	.2092
28	.0149	16	.0598	4	.2242
27	.0164	15	.0673	3	.2391

TABLE 17-2 GAGES AND THICKNESS OF TIN PLATE

Base Wt. Lb.	Trade Terms	Approx. U.S. Ga.	Approx. Thick. In.	Base Wt. Lb.	Trade Terms	Approx. U.S. Ga.	Approx. Thick. In.
100	ICL	30.5	.0113	155	2X	27	.0175
107	IC	30	.0120	175	3X	26	.0197
128	IXL	29	.0144	195	4X	25	.0220
135	IX	28	.0152	215	5X	24	.0246

SELECTION OF RIVETS

A rivet must be long enough so that a properly shaped head can be formed. In order that this may be done, the rivet must project beyond the material being riveted a distance equal to 1 1/2 times the diameter of the rivet. Therefore, the length of rivet is equal to the sum of the thicknesses of the material (the grip) to be riveted and 1 1/2 times the diameter of the rivet.

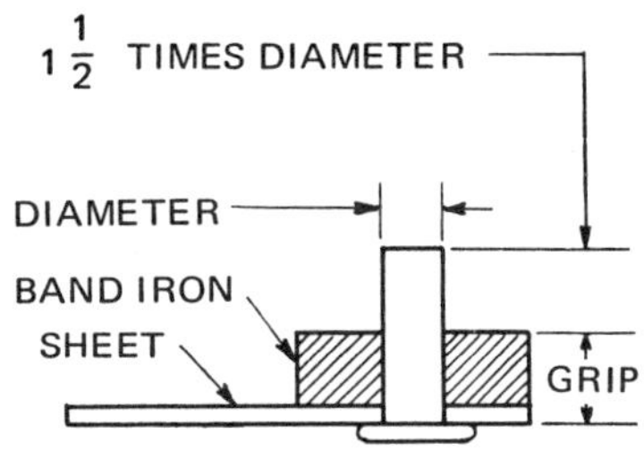

Flathead Rivets

Example: Compute the length (closest 32nd) of a flathead rivet, 3/16 in. in diameter, needed to rivet one thickness of 24-gage metal and one thickness of 1/4-in. band iron.

Add: .0239 in. (24-gage metal thickness from table)
.2500 in. (1/4-in. band iron)
.2813 in. (1 1/2 x 3/16 in.)
.5552 in. or 9/16 in. Ans.

Tinners' Rivets

Table 17-3 was computed for tinners' rivets and two thicknesses of sheet metal.

TABLE 17–3 SELECTION OF RIVETS ACCORDING TO METAL GAGE

Gage of Metal	Rivet Size	Gage of Metal	Rivet Size
30-gage	10 oz.	24-gage	2 lb.
28-gage	14 oz.	22-gage	2 1/2 lb.
26-gage	1 lb.	20-gage	3 lb.

Example: Select a tinners' rivet from the table for two thicknesses of 20-gage sheet metal.
Answer: 3-lb. rivet

TABLE 17–4 DIMENSIONS OF COATED TINNERS' RIVETS

Size	Diameter Inches	Length Inches	Size	Diameter Inches	Length Inches
8 oz.	.089	5/32	3 1/2 lb.	.165	21/64
10 oz.	.095	11/64	4 lb.	.176	11/32
12 oz.	.105	3/16	5 lb.	.186	3/8
14 oz.	.109	3/16	6 lb.	.203	25/64
1 lb.	.112	13/64	7 lb.	.220	13/32
1 1/4 lb.	.120	7/32	8 lb.	.224	7/16
1 1/2 lb.	.130	15/64	9 lb.	.230	29/64
1 3/4 lb.	.134	1/4	10 lb.	.238	15/32
2 lb.	.144	17/64	12 lb.	.259	1/2
2 1/2 lb.	.148	9/32	14 lb.	.284	33/64
3 lb.	.160	5/16	16 lb.	.300	17/32

ASSIGNMENT

A. Sheet Metal Tables

1. Find the thickness in inches of each of the following black iron gages: 32; 26; 22; 16; 10.

2. Find the thickness in inches of each sheet of the following weights of tin plate: 100 lb.; 107 lb.; 1X; 2X.

3. Find the gages of the following thicknesses of black iron: .0478 in.; .1046 in.; .0120 in.; .0747 in.

4. Find the base weights of the following thicknesses of tin plate: .0144 in.; .0197 in.; .0246 in.

B. Selection of Rivets

1. Find the lengths of the flathead rivets to the closest 32nd of an inch in the following problems:

   a. One thickness of 26-gage sheet metal and one piece of 12-gage band iron; rivet diameter = 1/8 in.

   b. One thickness of 26-gage sheet metal and one piece of 1/8-in. band iron; rivet diameter = 5/32 in.

   c. One thickness of 22-gage sheet metal and one piece of 3/16-in. band iron; rivet diameter = 3/16 in.

   d. Two thicknesses of 28-gage sheet metal and one piece of 1/8-in. band iron; Rivet diameter = 5/32 in.

   e. Two thicknesses of 26-gage sheet metal and two thicknesses of 1/8-in. band iron; rivet diameter = 1/4 in.

   f. Three thicknesses of 26-gage sheet metal and one piece of 3/16-in. band iron; rivet diameter = 3/16 in.

   g.

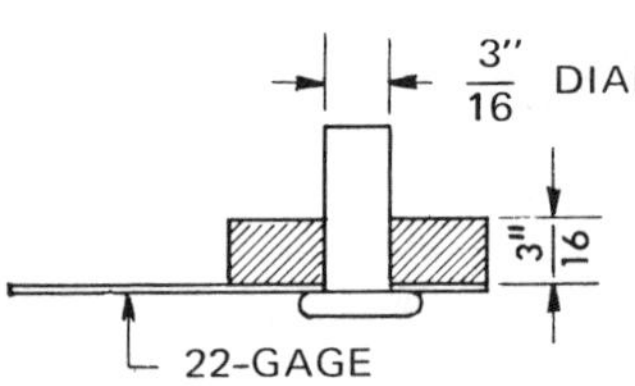

   h.

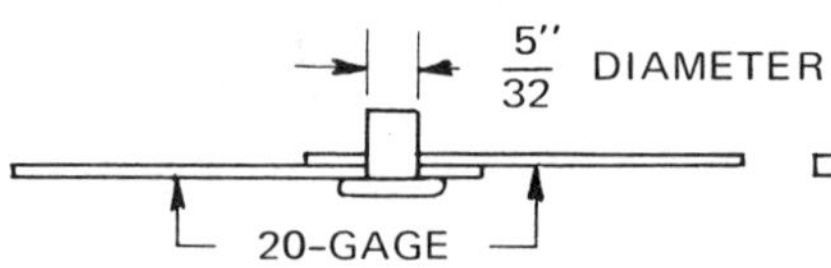

   i.

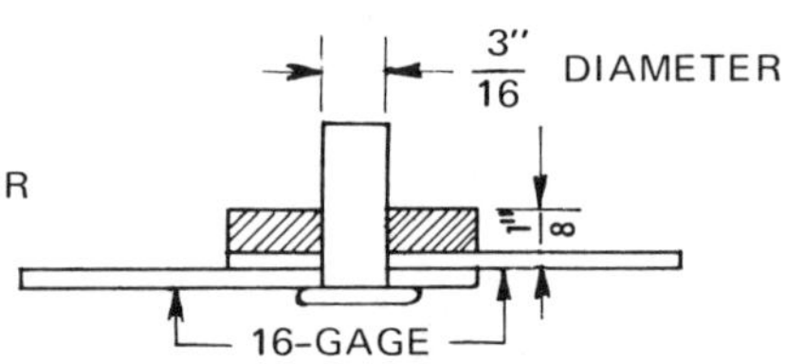

2. Find the size, diameter, and length of tinners' rivets for two thicknesses of the following pieces of metal:

   a. 28-gage   b. 24-gage   c. 20-gage

UNIT 19 THE MICROMETER

The smallest measurement which can be made with the use of the steel rule is one 64th of an inch in common fractions and one 100th of an inch in decimal fractions. To measure more closely than these divisions (to thousandths and to ten-thousandths of an inch), a micrometer is used. If a dimension in common fractions is to be measured with the micrometer, it must be converted to its decimal equivalent before the measurement can be made.

To use a micrometer caliper correctly, see *Measurement and Layout,* (Delmar Publishers, Inc.).

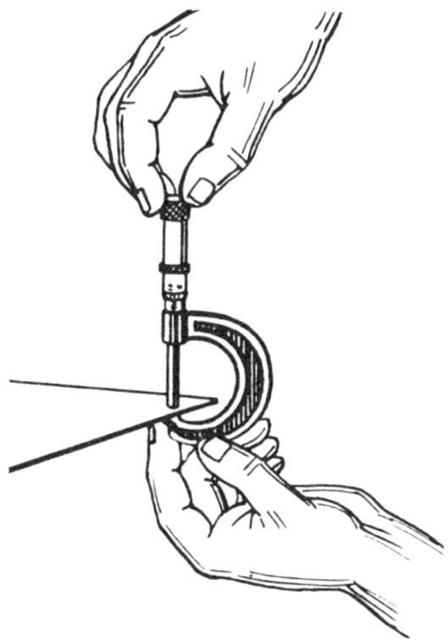

DESCRIPTION OF THE MICROMETER CALIPER

The principal parts of the micrometer caliper are the fixed parts — the frame, the barrel, and the anvil; and the movable parts — the thimble and the spindle. The thimble rotates the spindle which moves in the threaded portion inside the barrel.

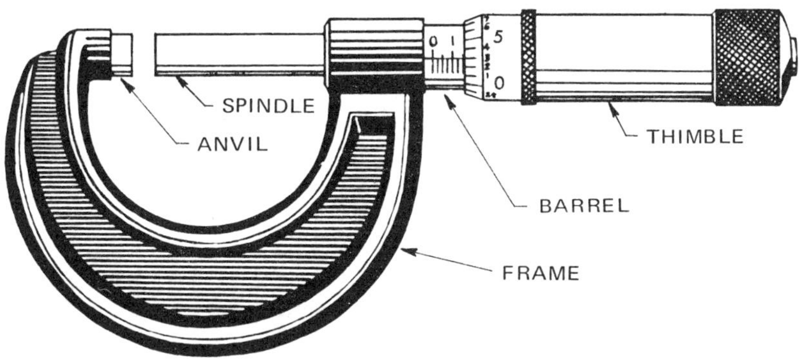

Turning the thimble provides an opening between the anvil and the end of the spindle where the work is measured. The size of the work is indicated by the graduations on the barrel and the thimble.

READING THE MICROMETER SCALES

The lines on the barrel marked 1, 2, 3, 4, etc., indicate measurements of 10ths or .100 in., .200 in., 300 in., .400 in., etc., respectively.

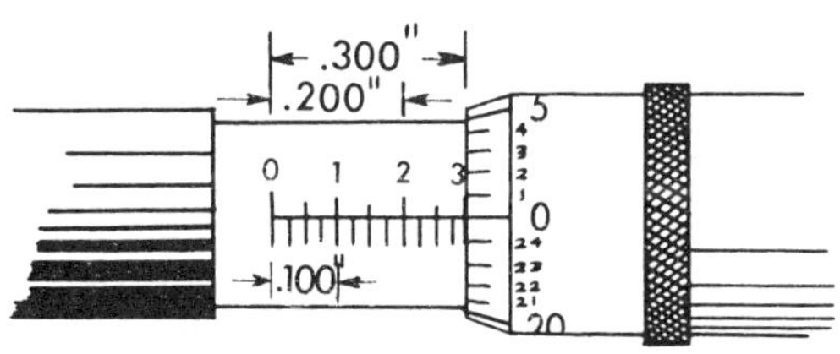

READING OF .100", .200", 300"

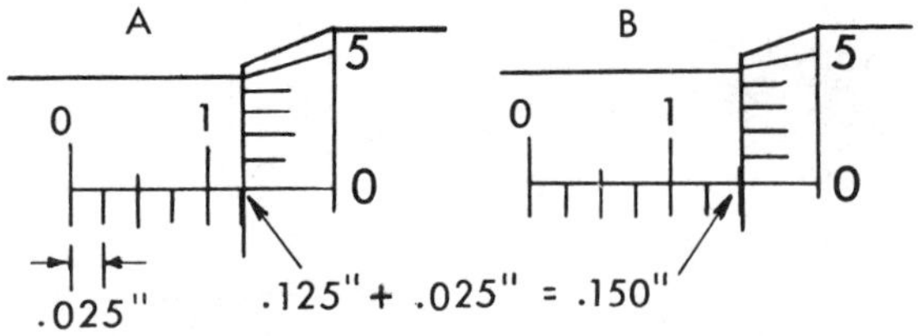

.025 DIVISIONS

Each of the sections between the 10th divisions (between 1, 2, 3, 4, etc.) is divided into four equal parts of .025 in. each (drawing A opposite). One complete revolution of the thimble (from the zero on the thimble around to the same zero) moves it one of these divisions (.025 in.) along the barrel (drawing B opposite).

The bevel edge of the thimble is divided into 25 equal parts. Each of these parts represents one 25th of the distance the thimble travels along the barrel in moving from one of the .025 in. divisions to another. Thus, each division on the thimble represents one thousandth (.001 in.) of an inch. These divisions are marked for convenience at every five spaces by 0, 5, 10, 15, and 20. When 25 of these graduations have passed the horizontal line on the barrel, the spindle, having made one revolution, has moved .025 in.

The micrometer is read by first noting the last visible figure on the horizontal line of the barrel, representing 10ths of an inch. Add to this the amount represented by the visible graduations beyond this figure (by multiplying the number of them by .025 in.), and the division on the bevel edge of the thimble that coincides with the line of graduation.

Add the following expressed in thousandths:

1. 10th on barrel (last visible figure)
2. Number of (.025) spaces on barrel x .025
3. Number of division on thimble ______ Total Reading

The drawings below show four sample readings of a micrometer.

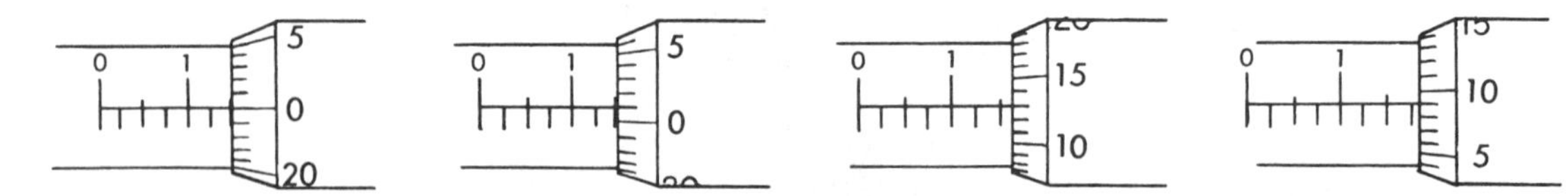

.100" + .050" = .150" .150" + .001" = .151" .150" + .013" = .163" .175" + .009" = .184"

CARE OF THE MICROMETER

Since the micrometer is an accurate tool, it must be handled carefully. If it is dropped, its accuracy may be permanently affected. Continual sliding of work between the anvil and spindle will wear the surfaces. If the spindle is tightened too much, the frame may be sprung permanently.

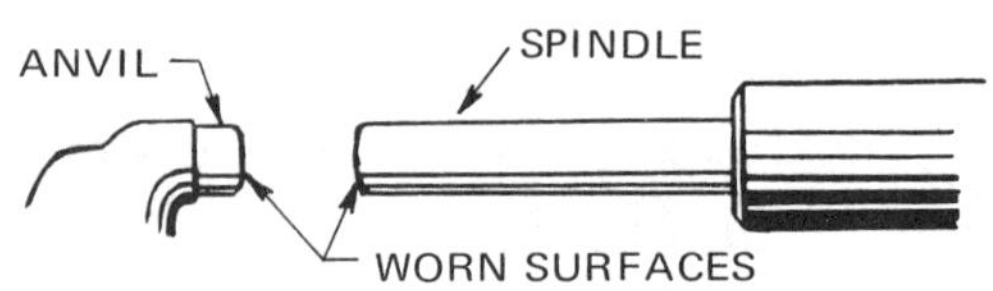

WEAR ON SPINDLE AND ANVIL

ASSIGNMENT

A. Set a micrometer for each of the following measurements.

1. .079 in. 2. .024 in. 3. .004 in. 4. .098 in. 5. .259 in. 6. .501 in.

B. Read each setting shown below:

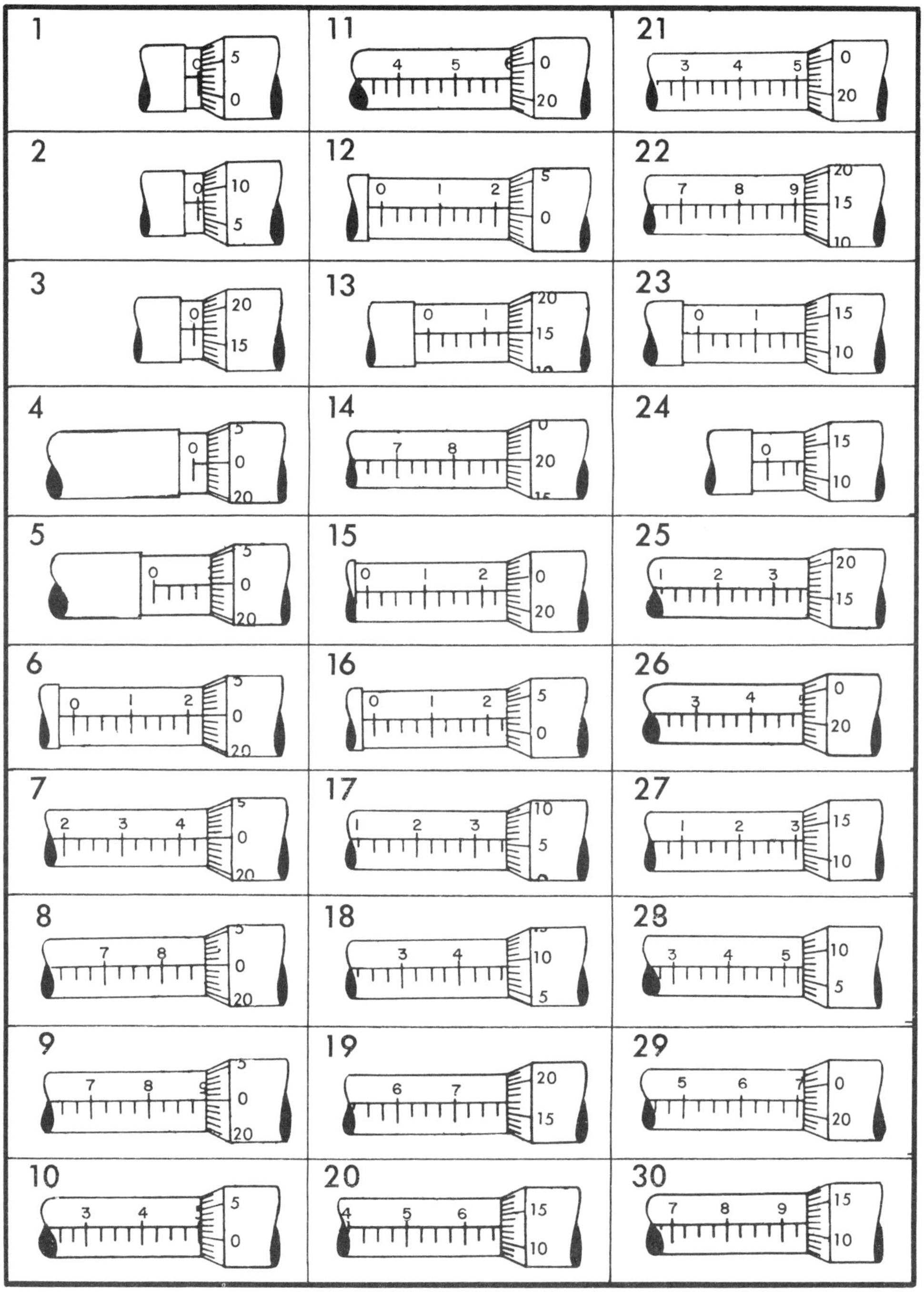

UNIT 20 DEGREE OF ACCURACY

Since the production cost of a job depends partly on the accuracy to which parts must be made, it is important that speed of production is not reduced because of unnecessary attention to accuracy. In many sheet metal processes, extreme accuracy is not required.

CONDITIONS WHICH DETERMINE THE DEGREE OF ACCURACY

The degree of accuracy required is dependent upon two conditions: (1) the type of job and (2) the use of the job. Some building jobs must be accurate to 1/4 in. or 1/8 in.; industrial jobs require an accuracy of 1/64 in. in some cases and .001 in. in others. When metal is cut to make a part, extra material is often allowed for trimming and finishing.

FACTORS WHICH DETERMINE THE DEGREE OF ACCURACY

Specified Limits

Specified limits indicate the degree of accuracy desired and are on the drawing. If a dimension on a drawing is 2″ ± 1/32″, the finished dimension of the job can vary from 1 31/32″ to 2 1/32″. These are the lower and the upper limits. Tolerance is the amount of variation permitted in the size of a part and is 1/32″ + 1/32″, or 1/16″ in this case. The basic dimension is 2″.

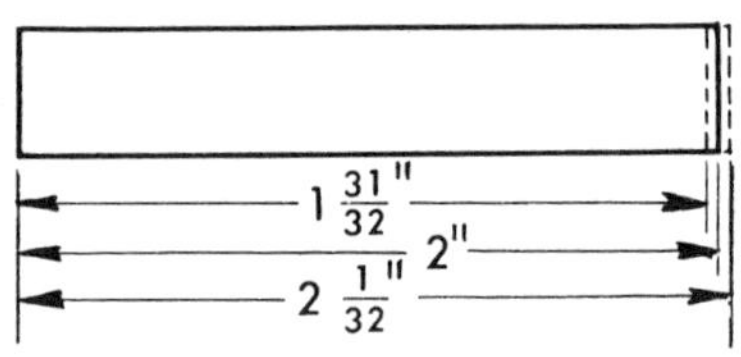

Limitations of the Measuring Instrument

- Steel Scales: 1/64″ to 1/100″
- Micrometers: .001″ to .0001″
- Precision Gage Blocks: .000002″ to .000008″
- Toolmakers' Microscopes: approximately .0001″ to .00001″

Errors in Measurement

- Measuring Tool
  1. Manufacturer's error (End divisions may be inaccurate.)
  2. Wear (Check ends of rules and micrometers periodically.)
  3. Temperature (This affects the 4th decimal place.)
- Human Element
  1. Eyesight (Have proper light, and look straight at part being measured.)
  2. Touch (Develop a sense of *feel* when using a micrometer.)
  3. Good Judgment (Select proper measuring instrument and closest division to suit job.)
- Machine and Material – In order that parts (jobs) will be interchangeable, the following factors must be considered:
  1. Variation in material
  2. Kind of machine and its rigidity
  3. Wear and support of tools

ASSIGNMENT

A. Allowances for Trim

Using the values given, compute the dimensions x and y for each problem.

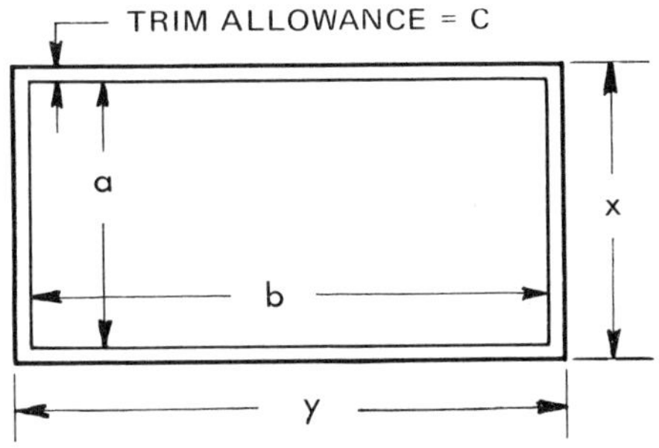

1. b = 8″   a = 10″   c = 1/8″
2. b = 4″   a = 6″   c = 1/16″
3. b = 8 1/2″   a = 10 1/2″   c = 1/8″
4. b = 15.75″   a = 20.5″   c = .25″

B. Tolerances and Dimensions

1. A dimension is 2″ ± 1/64″. What are the permissible dimensions?
2. What are the permissible dimensions of a job which has a basic dimension of 4″, if the tolerance is ± 1/32″?
3. A blueprint gives a dimension of 3.250″ ± .005″. What are the upper and lower limits? What is the amount of tolerance?
4. What are the permissible dimensions if the tolerance is $^{+.002}_{-.003}$ and the basic dimension is 3.625?
5. Compute the permissible dimensions for the rivet box below.

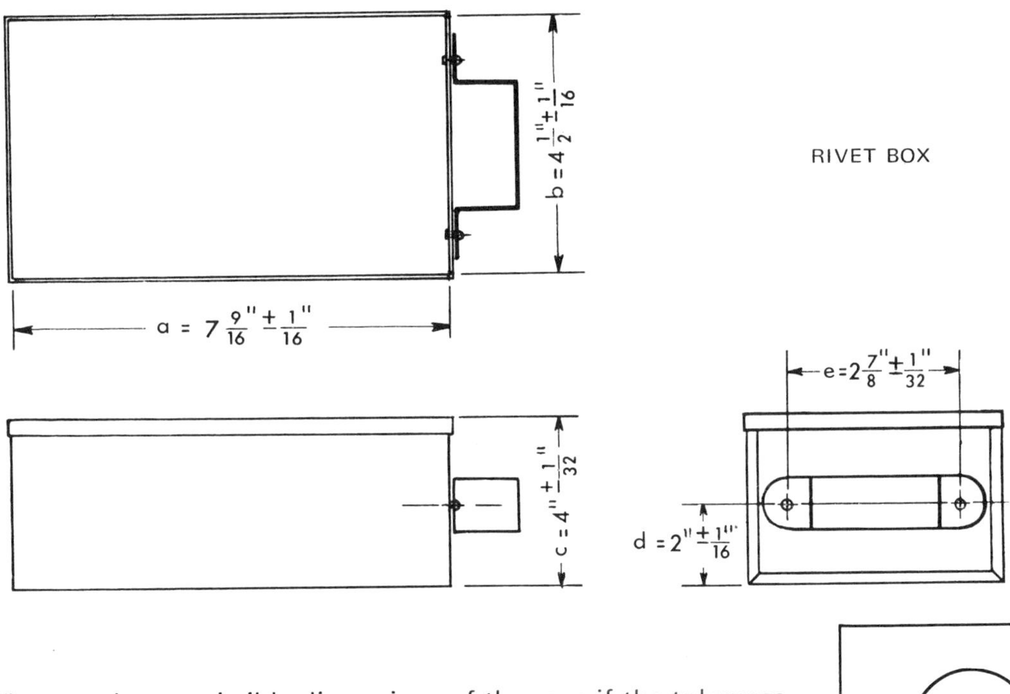

RIVET BOX

6. What are the permissible dimensions of the gage if the tolerance is $^{+.000}_{-.005}$?

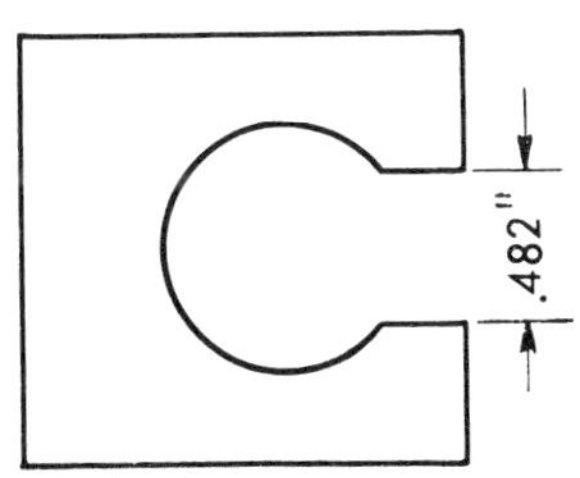

C. Summary of Decimal Fractions

1. Divide a line 12 9/16 in. long into 12 parts by construction and check the length of each part by computation and measurement.

2. Four pieces, 13.12 in. long, 14.62 in. long, 11.44 in. long, and 12.50 in. long respectively, are to be cut from a piece of wire 120 in. long. How much material will be left?

3. Three pieces of sheet metal, 12-gage, 20-gage, and 20-gage respectively, are to be riveted together. What will be the total thickness of the metal? What length of flathead rivet will be needed if the diameter of the rivet is 1/4 in.?

4. Three pieces, 5.1 in. long, 4.8 in. long, and 5.9 in. long respectively, are to be cut from a piece of band iron 30 in. long. If .1 in. is allowed for each saw cut, how much material will remain?

5. A bar 25.92 in. long is laid off in equal divisions of 1.62 in. How many spaces are there?

6. How many pieces of band iron, each 8.24 in. long, can be cut from a 120-in. piece? Allow .1 in. for each cut.

7. Find the distances x and y on the drawing below.

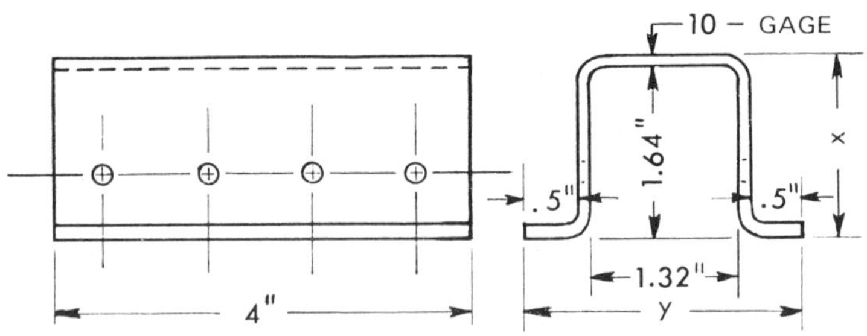

8. Find the missing dimensions (closest 64th of an inch) on the drawing below.

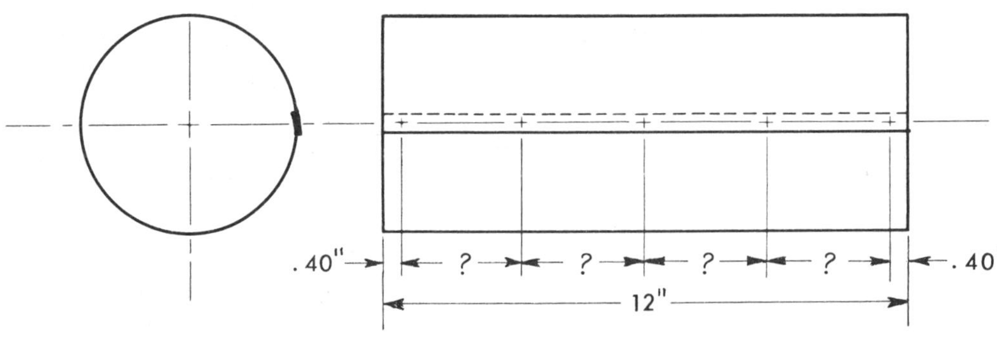

9. Arrange the following drills in order of size, starting with the smallest: 31/64; .0145; 1/8; .0312; .0310; .4040; .0135; 1/16; 7/64; 9/64.

SECTION 2 — LINEAR MEASURE

UNIT 21 THE UNITS OF LINEAR MEASURE

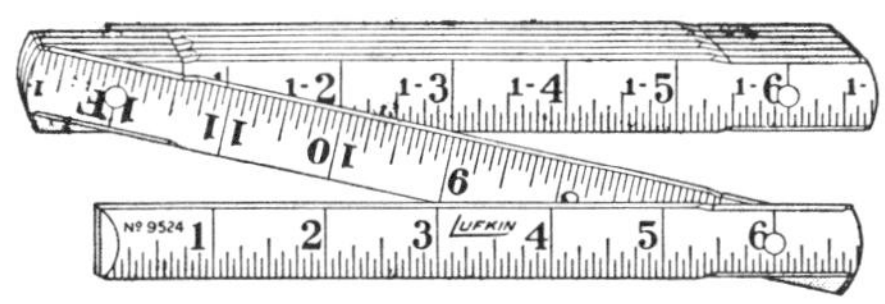

In some cases, as in the sheet metal work used in building, feet are used in measurements. It is, therefore, necessary to know the table of linear measure, to know how to change feet to inches and inches to feet, and how to add, subtract, multiply, and divide feet and inches.

When problems involving feet and inches are worked, much confusion and many errors will be avoided if the units (feet and inches) are correctly labeled at each stage of the computation. The method of checking problems of linear measurement is the same as the method used to check problems involving common fractions and decimals.

12 inches	=	1 foot (ft.)
3 feet	=	1 yard (yd.)
5280 feet	=	1 mile (mi.)

Inches are designated with a double prime (″). (3 inches = 3″)

Feet are designated with a single prime (′). (5 feet = 5′)

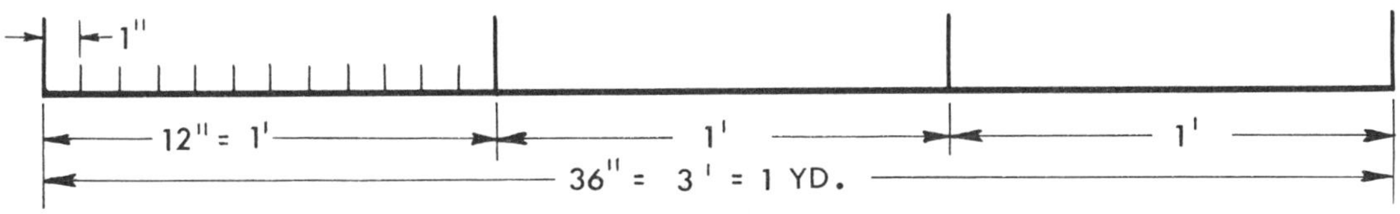

REDUCTION OF INCHES TO FEET AND INCHES

RULE: To change inches to feet and inches, divide the number of inches by 12, since 12 inches equal 1 foot.

Example: Change 29 3/4 inches to feet and inches.

1. Change 29 inches to feet by dividing by 12 (12 in. = 1 ft.).
2. Combine the remainder with the quotient.

```
      2 ft.
12 ) 29 in.
     24
      5 in.
```

2 ft. 5 in. + 3/4 in., or 2′ 5 3/4″ Ans.

Note: The remainder is always the same unit of measurement as the dividend.

REDUCTION OF FEET AND INCHES TO INCHES

RULE: To change feet and inches to inches, multiply the number of feet by 12, and then add the number of inches.

Example: Change 11 ft. 63/4 in. to inches.

1. Change 11 ft. to inches. 11 ft. x 12 = 132 in.
2. Add the inches. 132 in. + 6 3/4 in. = 138 3/4 in. Ans.

REDUCTION OF FEET TO INCHES (Decimal Form)

RULE: To change feet in decimal form to inches, multiply the number of feet by 12, and then change the decimal part of this result to a common fraction.

Example: Change .279 ft. to inches and 16ths of an inch.

1. Multiply .279 ft. by 12 to obtain inches in decimal form.

```
  .279
  x 12
  ----
   558
  279
  ----
 3.348 in.
```

2. Change .348 in. to 16ths of an inch. Add the result (3/8 in.) to the 3 in. The answer is 3 3/8 in.

```
  .348
  x 16
  ----
  2088
  348
  ----
 5.568 = 6/16 in., or 3/8 in.
```

REDUCTION OF INCHES TO FEET (Decimal Form)

RULE: To change inches to feet in decimal form, divide the number of inches by 12.

Example: Change 4 3/4 in. to a decimal part of a foot (Three decimal places).

1. Change 4 3/4 in. to decimal form. 4 3/4 = 4.75 in.
2. Divide by 12 and round off.

```
     .3958          = .396 ft. Ans.
12 ) 4.7500 in.
```

SUMMARY OF RULES

Whenever any length is measured in larger units, there are *fewer* of the larger units; whenever the same length is measured in smaller units, there are *more* of the smaller units. Since each foot has 12 in., two feet will have two times 12 in. or 24 in.; therefore, there are more of the smaller units.

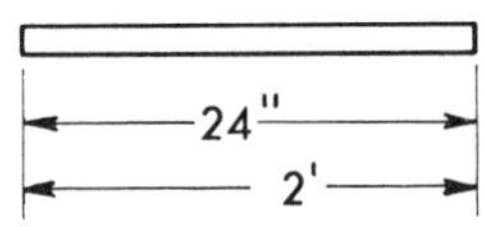

ASSIGNMENT

A. Reduction of Inches to Feet and Inches

1. Change 289 in. to ft. and in.
2. Change 125 3/4 in. to ft. and in.
3. Change 356 3/8 in. to ft. and in.
4. Change the dimensions on the drawings of the airplane to feet and inches.

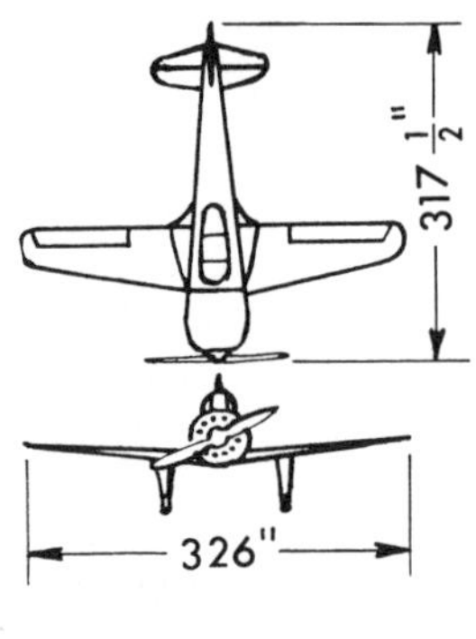

B. Reduction of Feet and Inches to Inches

1. Change 2′ 9 3/4″ to in.
2. Change 5′ 10 1/2″ to in.
3. Change 15′ 8 7/8″ to in.
4. Change 29′ 11 1/2″ to in.
5. Change both dimensions of the plate shown so that a rule divided in inches can be used to lay it out.

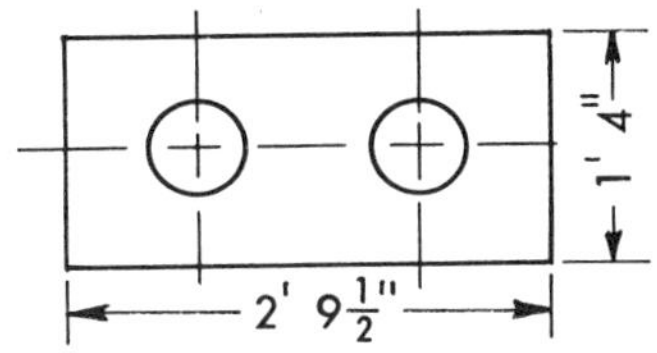

C. Reduction of Feet (Decimal Fractions) to Inches (Common Fractions)

The following measurements are taken from a blueprint: .250 ft.; .750 ft., .333 ft., .607 ft., .258 ft., .893 ft., .0125 ft., 1.203 ft.

1. What are the above measurements in inches to the nearest 1/16 of an inch?
2. What are the above measurements in inches to the nearest 1/64 of an inch?

D. Reduction of Inches to Feet (Decimal Form – 3 Decimal Places)

1. Change 6 in.; 3 in.; 9 in.; to feet.
2. Change 4 in.; 8 in.; 1 in.; 2 in.; 5 in.; 7 in.; 10 in.; 11 in.; to feet.
3. Change 1 1/2 in.; 1 1/4 in.; 2 1/2 in.; 2 3/4 in.; to feet.
4. Change 4 1/8 in.; 5 7/8 in.; 2 7/16 in.; 3 5/32 in.; to feet.
5. Change the dimensions of the bracket shown to feet in decimal form.

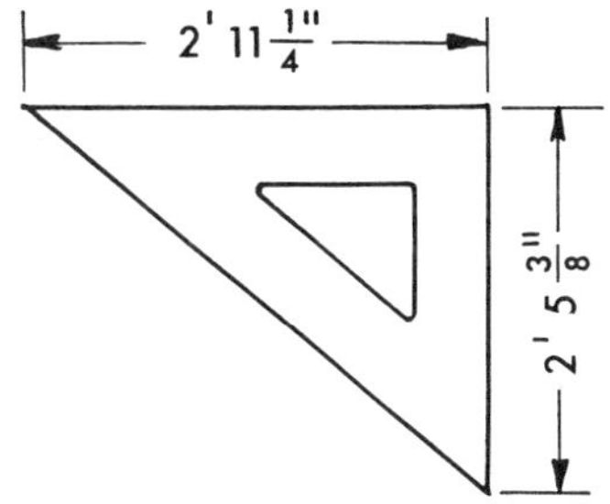

E. Miscellaneous Problems

1. Change 5 yards to inches.
2. Change 205,836 inches to yards, feet, and inches.

UNIT 22 ADDITION AND SUBTRACTION OF UNITS OF LINEAR MEASURE

When it is necessary to find unknown dimensions on larger jobs, feet and inches must be added or subtracted. This occurs in building construction when the length of pipe, gutter, etc. is needed.

RULE: When adding or subtracting units of linear measure, arrange the units in like columns, add or subtract each column separately, and express the answer in its lowest terms.

ADDITION

Example: Add 10 ft. 3 1/2 in. and 8 ft. 9 3/4 in.

1. Arrange the numbers in columns: feet under feet, inches under inches. Add each column separately, changing the fractions to the same denominator.

10 ft. 3 1/2 in.	=	10 ft. 3 2/4 in.
+ 8 ft. 9 3/4 in.	=	+ 8 ft. 9 3/4 in.
		18 ft. 12 5/4 in.

2. Reduce the inches to feet and inches.

12 5/4 in. = 1 ft. 1 1/4 in.

3. Add this result to 18 ft.

	18 ft.
	+ 1 ft. 1 1/4 in.
Answer:	19 ft. 1 1/4 in.

SUBTRACTION

Example: Subtract 3 ft. 8 1/2 in. from 10 ft. 2 in.

1. Arrange the numbers in columns: feet under feet, and inches under inches.

10 ft. 2 in.
− 3 ft. 8 1/2 in.

2. Change the inches into a number which can be subtracted by borrowing 1 ft. or 12 in., from 10 ft.

10 ft. 2 in.	=	9 ft. 14 in.
		− 3 ft. 8 1/2 in.

3. Change one inch into a fraction by borrowing 1 in., or 2/2 in., from 14 in.

9 ft. 14 in.	=	9 ft. 13 2/2 in.
		− 3 ft. 8 1/2 in.
Answer:		6 ft. 5 1/2 in.

4. Subtract each column.

ASSIGNMENT

Solve and check each of the following problems.

1. Add 3′ 6″, 2′ 3″
2. Add 6′ 8″, 9′ 10″
3. Add 7′ 5″, 10′ 11″
4. Add 3′ 5 1/2″, 11′ 10 1/2″
5. Add 14′ 9 1/4 ″, 3′ 9 1/2″
6. Add 5′ 8 3/4″, 24′ 11 7/8 ″
7. Add 36′ 9 7/8″, 5′ 10 9/16″
8. Subtract 3′ 8″ from 4′ 10″

9. Subtract 5′ 10″ from 8′ 2″
10. Subtract 1′ 11″ from 3′ 1″
11. Subtract 18′ 9 1/2″ from 25′ 6 1/2″
12. Subtract 6′ 8 1/4″ from 9′ 6 3/4″
13. Subtract 12′ 10 3/4″ from 14′ 2″
14. Subtract 1′ 8 7/8″ from 2′ 10 1/4″
15. 3′ 1 3/4″ + 2′ 10 1/8″ + 5′ 9 1/16″ − 1′ 11 7/8″ = ?
16. Compute the unknown dimension on the drawing below.

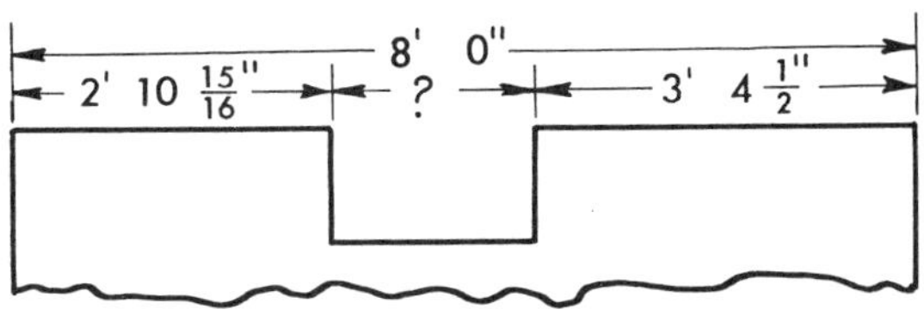

17. Find the unknown dimension.

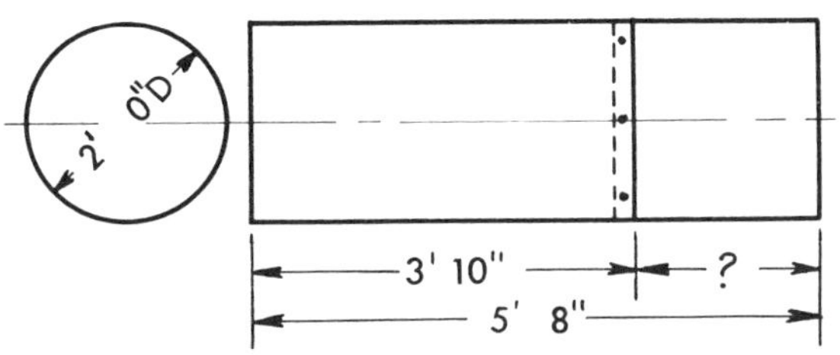

18. Find the unknown dimension.

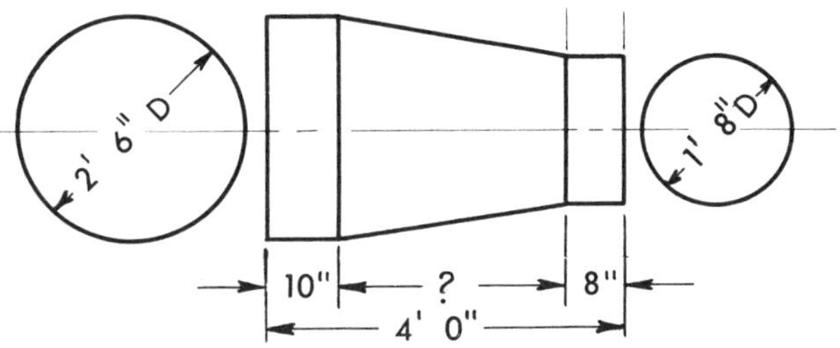

19. Compute the unknown dimension on the drawing of the duct below.

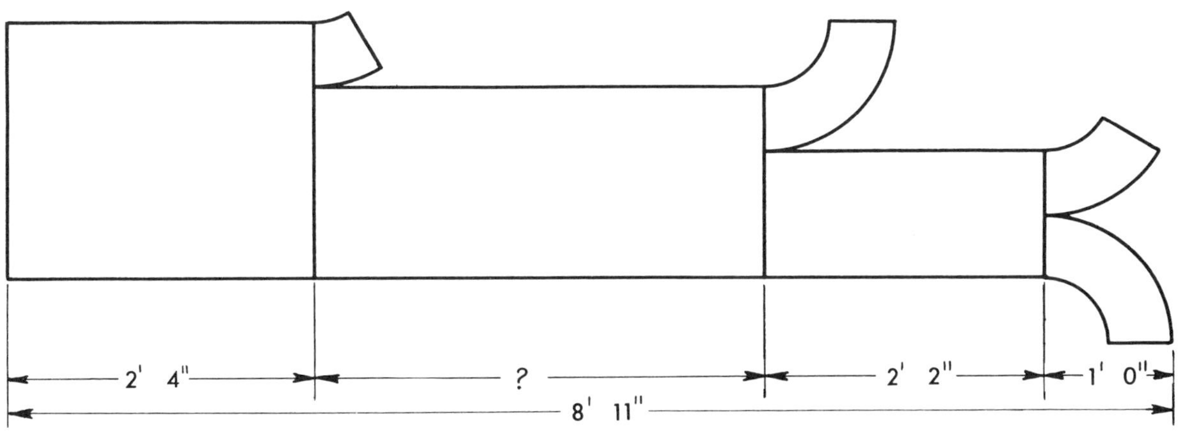

UNIT 23 MULTIPLICATION OF UNITS OF LINEAR MEASURE

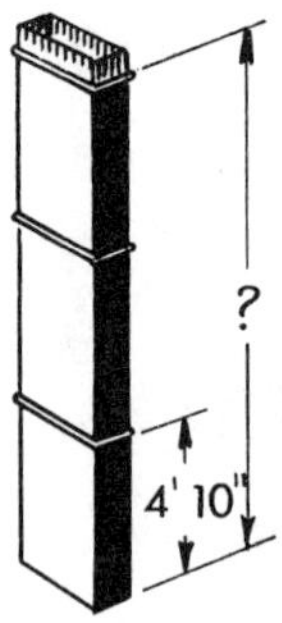

When long pieces of the same length (wire, pipe, band iron, etc.) are needed, it may be desirable to use multiplication of the units of linear measure (generally feet and inches) to determine quickly the total length of material required.

In some cases, it is more convenient to convert the dimensions given into inches before the problem is worked and then back again into feet and inches.

RULE: Arrange the units of measurement in columns, multiply each column by the number required, and express the answer in its lowest terms.

Example: How much wire is required for eight pieces, each 10 ft. 7 3/4 in. long?

1. Multiply each column by the number of pieces.

   10 ft. 7 in. $\frac{3}{4}$ in. $\quad \frac{3}{\cancel{4}_{1}} \times \frac{\cancel{8}^{2}}{1} = 6$ in.

   x 8

   80 ft. 56 in. + 6 in.

2. Add the inches. — 80 ft. 56 in. + 6 in. = 80 ft. 62 in.

3. Change the inches to feet. — 62 in. = 5 ft. 2 in.

4. Add this result to 80 ft. — 80 ft. + 5 ft. 2 in. = 85 ft. 2 in. Ans.

Note: This problem can be solved by reducing the dimension to inches.

1. Change the feet to inches. — 10 ft. x 12 = 120 in.

2. Add the inches. — 120 in. + 7 3/4 in. = 127 3/4 in.

3. Multiply the inches by the number of pieces.

   127 in. $\frac{3}{4}$ in.

   x 8

   1016 in. + 6 in. = 1022 in.

4. Change the result to feet and inches. — 1022 in. = 85 ft. 2 in. Ans.

ASSIGNMENT

Solve and check each of the following problems.

1. Multiply 1′ 6″ by 2.
2. Multiply 2′ 6″ by 3.
3. Multiply 2′ 9″ by 6.
4. Multiply 5′ 3″ by 9.
5. Multiply 1′ 8 1/2″ by 2.
6. Multiply 2′ 9 1/2″ by 4.
7. Multiply 13′ 7 3/4″ by 2.
8. Multiply 14′ 10 1/8″ by 5.
9. Multiply 1′ 9 7/8″ by 7.
10. Multiply 12′ 10 15/16″ by 10.

11. Find the total length of 10 pieces of wire, each 15′ 9″ long.
12. Find the total length of 15 pieces of wire, each 2′ 9 3/4″ long.
13. a. Compute the total length of 8 pieces of band iron cut from 20′ of stock, each 1′ 8 7/8″ long. Allow 1/16″ for each cut.

    b. How much will remain?
14. Compute the height of the cabinet shown on the drawing below.

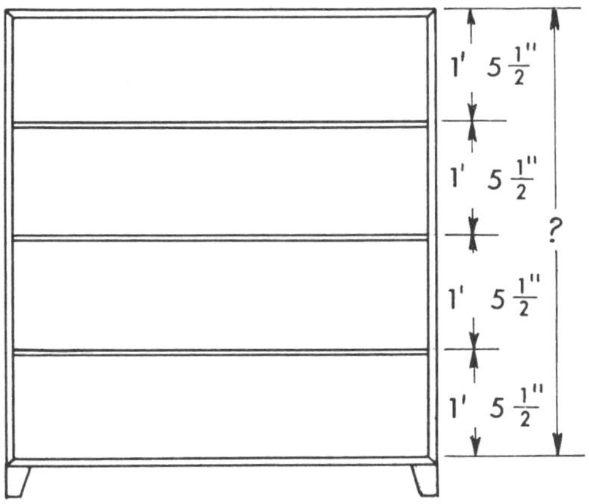

15. What is the total length of 10 sections of smoke pipe, if each section measures 2′ 4 1/2″ from the end of pipe to the bead?

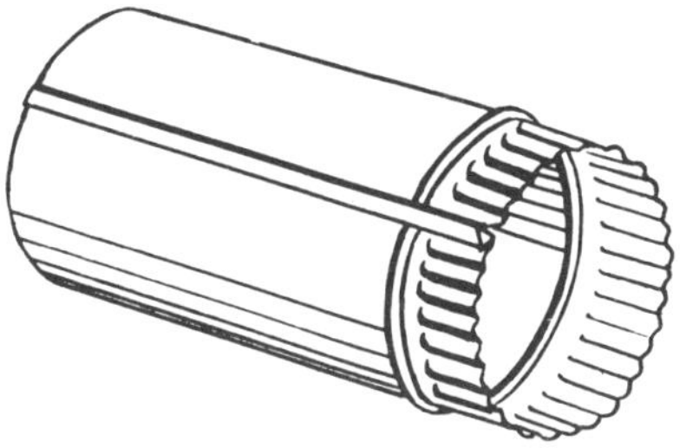

16. Find the total length of 5 pieces of sheet metal, each 2′ 3″ long.
17. Compute the total length of 15 sections of rectangular duct (pipe). The length of each section is shown on the drawing.

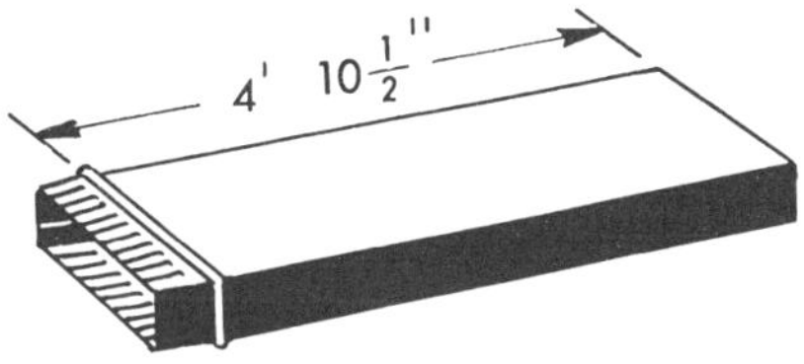

UNIT 24 DIVISION OF UNITS OF LINEAR MEASURE

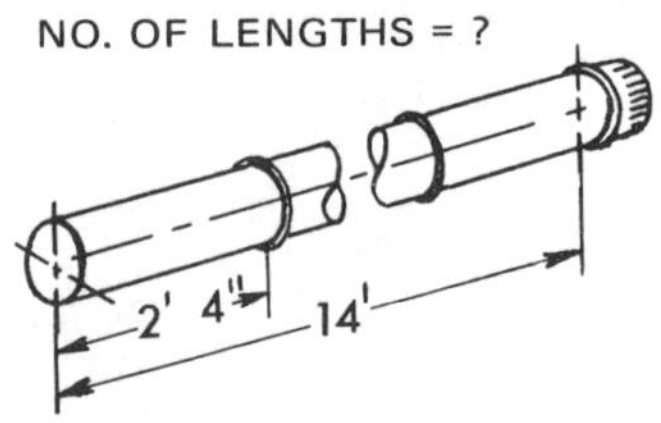

Division of the units of linear measure (generally feet and inches) has two applications: (1) dividing a piece of stock or a length into equal parts, and (2) finding the number of pieces of a given length which can be obtained from a length of stock. These problems can be worked either in feet or in inches, whichever is the more convenient method.

Example 1: Divide 60 ft. 8 1/2 in. into 4 equal parts.

1. Arrange the measurement in columns. — 60 ft. 8 in. 1/2 in.
2. Divide each column by the number of parts. — 4)60 ft. 8 in. 1/2 in. = 15 ft. 2 in. 1/8 in. Ans.

Note: 60 ft. 8 1/2 in. could be changed into inches, and then divided by 4.

1. 60 ft. 8 1/2 in. = 728 1/2 in.
2. 728 1/2 in. ÷ 4 = 182 1/8 in., or 15 ft. 2 1/8 in. Ans.

Example 2: How many pieces of wire, each 10 ft. 8 1/2 in. long, can be obtained from 100 ft. of wire? How much wire will be left?

1. Change the inches to a decimal part of a foot by dividing 8 1/2 in. by 12. Then add the result to 10 ft.

   12) 8.5000 = .7083 ft.

   10 ft. 8 1/2 in. = 10.7083 ft.

2. Divide 10.7083 ft. into 100 ft. The answer is 9 pieces and 3.625 ft. waste.

   10.7083∧) 100.0000∧ ft. = 9. pieces
   96 3747
   3.6253 ft. waste

Note: This problem can also be solved as follows.

1. Change both dimensions to inches.

   10 ft. 8 1/2 in. = 128.5 in.
   100 ft. x 12 = 1200 in.

2. Divide 1200 in. by 128.5 in. The answer is 9 pieces and 43.5 in. waste

   128.5∧) 1200.0∧ in. = 9. pieces
   1156 5
   43.5 in. waste

ASSIGNMENT

Solve and check the following problems.

1. Divide 2' 6" by 2.
2. Divide 3' 9" by 2.
3. Divide 6' 8 3/4" by 2.
4. Divide 10' 6" by 3.
5. Divide 6' 9 3/4" by 4.
6. Divide 10' 9 5/8" by 8.

7. A 100-ft. coil of wire is to be divided into 16 equal parts. How long (ft. and in.) will each part be? Work the problem by both the *foot method* and *inch method.*
8. Divide a 25-ft. coil of wire into 12 equal parts.
9. If 4 shelves are to be spaced equally between the ceiling and the floor, height of ceiling 8′ 4″, what will be the spacing of the shelves?
10. How many 2′ pieces of wire can be cut from a 100′ coil?
11. How many 2′ 9″ pieces of wire can be cut from a 100-ft. coil of wire? How much waste will there be?
12. How many sections of the pipe shown below will be required?

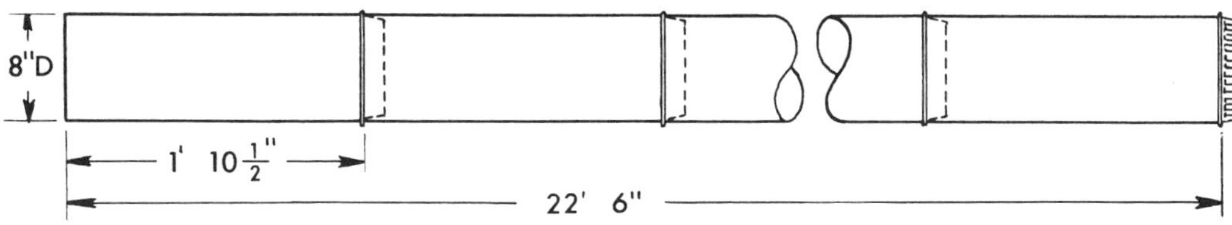

13. How many sections of pipe are required for a duct 40′ long, if each section is 2′ 8″ long?
14. Compute the number of 8′ sheets required for 24 pieces, each 1′ 4″ long.
15. Compute the approximate number of lengths of 10′ gutter required for the house. (Regard a fractional part of the answer as a whole length.)

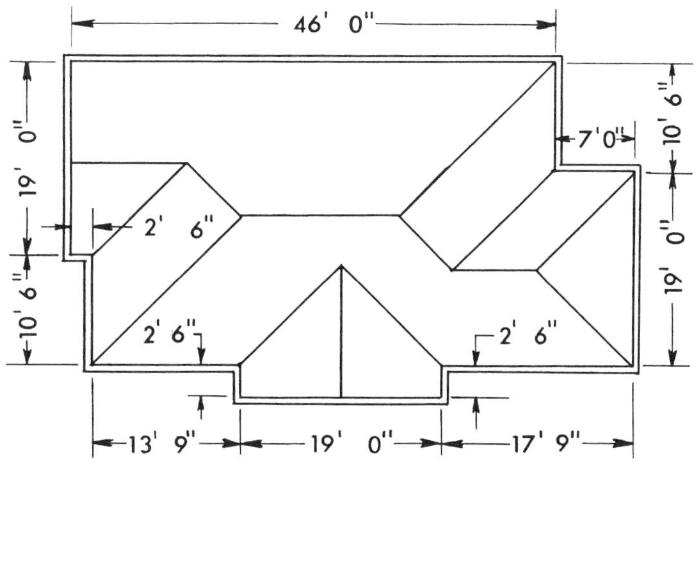

SECTION 3 AVERAGES PERCENTAGES COSTS AND WAGES

UNIT 25 AVERAGES

Material varies in weight, thickness, etc., and so the measurement of material must often be expressed in terms of average weight and average thickness. In order to explain the meaning of averages, the average height of several lines is shown.

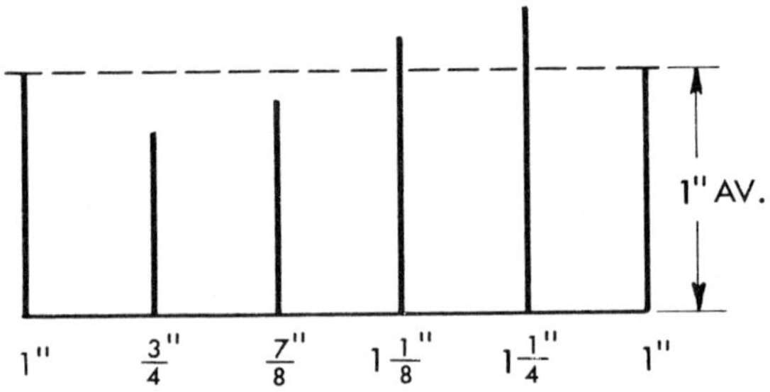

AVERAGING SEVERAL QUANTITIES

To find the average of several numbers, divide their sum by the number of quantities (numbers).

Example: Find the average of the following numbers: 39.1, 40.8, 39.4, 40.1, 39.9

1. Add the 5 given quantities (numbers).
2. Divide by the number of quantities, 5, to obtain the average, 39.9. Ans.

$$\begin{array}{r} 39.2 \\ 40.8 \\ 39.4 \\ 40.1 \\ \underline{39.9} \\ 199.4 \end{array}$$

$$5\overline{)199.40}$$
$$39.88 \quad \text{Ans.}$$

Note: If the numbers are in fractional form, it is generally best to change them to decimals before averaging.

FINDING A MISSING QUANTITY FOR A KNOWN AVERAGE

To find the value of one number which, when combined with several numbers, will result in a given average,multiply the given average by the number of quantities and then subtract the sum of the other numbers.

Example: Find the fourth number which, when combined with 92, 90, and 95, will give an average of 93.

1. Multiply the required average by the number of quantities. $93 \times 4 = 372$
2. Add the three given numbers. $92 + 90 + 95 = 277$
3. Subtract $372 - 277 = 95$ Ans.

Check: Average the four numbers. $\dfrac{92 + 90 + 95 + 95}{4} = 93$

ASSIGNMENT

A. Averaging Several Quantities

1. Find the average thickness of a sheet of metal which measures at four points as follows: .059″, .058″, .060″, .059″.
2. Find the average wage per day if a workman received $30.80 on Monday; $28.85 on Tuesday; $42.35 on Wednesday; $30.80 on Thursday; $30.80 on Friday; and $34.65 on Saturday.
3. Find the average weight of five sheets of metal which weigh as follows: 12 1/4 lb., 11 3/4 lb., 12 lb., 12 3/8 lb., 12 7/8 lb.
4. Find the average wage per day if a sheet metal mechanic apprentice received $21.60 on Monday; $18.90 on Tuesday; $21.60 on Wednesday; $17.55 on Thursday; and $33.75 on Friday.
5. The weekly earnings of six (6) men were: $154.00; $173.75; $100.10; $130.90; $147.50; and $183.50. What was their average weekly wage?
6. Find the average length of the following rods: 6′ 11″; 7′ 1″; 7′; 6′ 11 1/2″; 7′ 1/2″. **Hint:** Change all dimensions to inches, average, and then change back to feet and inches.
7. If the total weight of 8 sheets of metal is 108.75 lb., what is the average weight per sheet?
8. If 10 sheets of metal weigh an average of 15 1/4 lb. each, what is the total weight of the sheets?
9. Find the average breaking strength in pounds of five samples of steel wire if the wires break under the following loads: 278 lb.; 276 lb.; 285 lb.; 270 lb.; and 281 lb.

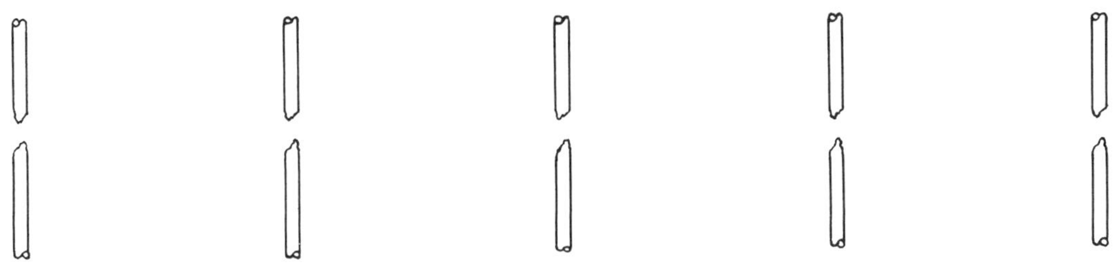

278 LB. 276 LB. 285 LB. 270 LB. 281 LB.

B. Finding a Missing Quantity for a Known Average

1. If a man made 13 elbows on Monday, 15 on Tuesday, 18 on Wednesday, 14 on Thursday, 16 on Friday, how many must he make on Saturday to average 16 per day?
2. If a man earned $21.60 on Monday, $30.30 on Tuesday and $28.75 on Wednesday, how much must he earn on Thursday to average $27.00 a day for the four days?

UNIT 26 PERCENTAGE

Percentage is another way of expressing parts of a whole. It is useful in making comparisons. It is easier to see that 50% is larger than 37 1/2% than to compare 1/2 with 3/8, or .5 with .375. Percent (the symbol of which is %) simply means hundredths parts of the whole.

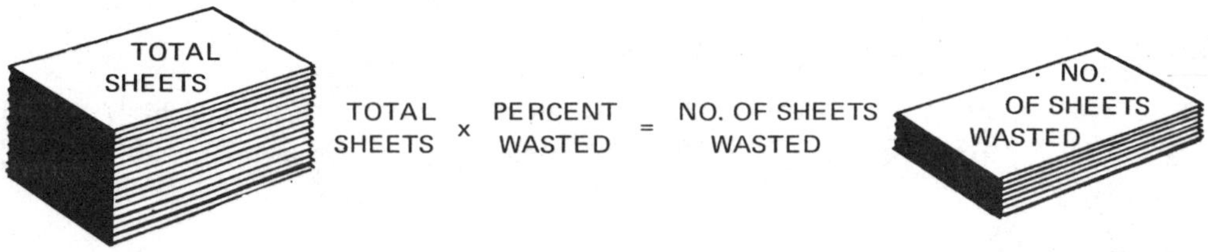

The chart below shows the relationship between fractions, decimals, and percents. In sheet metal work, percents are used in connection with waste, solders, costs, wages, etc.

EQUIVALENTS

Fractions in Lowest Terms	Fractional Equivalents of Decimals	Decimal Fractions	Percent
1/8	125/1000	.125	12.5
1/4	25/100	.25	25
3/8	375/1000	.375	37.5
1/2	5/10	.5	50
5/8	625/1000	.625	62.5
3/4	75/100	.75	75
7/8	875/1000	.875	87.5
8/8	10/10	1.00	100

RULE 1: To change from percent to decimal form, remove the percent sign and move the decimal point two places to the left.

Example: Change 8.5% to a decimal. 8.5% = 8.5/100, or .085 Ans.

Example: Change 1/2% to a decimal. 1/2% = .5%, or .005 Ans.

RULE 2: Part equals whole times percent. Part = Whole x Percent.

Example: Find 75% of 2 inches.

1. Percent = 75%, or .75. Whole = 2 in.
2. Multiply the whole (2 in.) by the percent (.75) in decimal form. 2 x .75 = 1.50 in. Ans.

The diagram below represents a section of heating duct which was damaged in a fire. We could report (1) 3/4 of the heating duct was damaged, (2) 75 parts per hundred were damaged, (3) .75 parts were damaged, (4) 75% of the duct was damaged.

Consider how you would report this damage and the reasons for your decision.

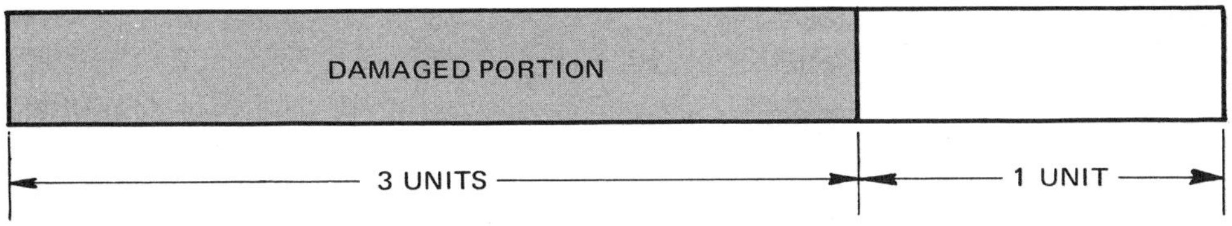

ASSIGNMENT

A. Change the following percents to decimal form:

1. 1%	5. 6%	9. 62 1/2%	13. 250%
2. 2%	6. 8%	10. 38 1/2%	14. .1%
3. 4%	7. 10%	11. 59 1/4%	15. .2%
4. 5%	8. 60%	12. 189%	16. .3 1/2%

B. Solve the following problems:

1. 25%	of	120	=	?	5. 34 1/2%	of	249	=	?
2. 37 1/2%	of	88	=	?	6. 19 1/4%	of	110	=	?
3. 60%	of	160	=	?	7. 40 3/4%	of	300	=	?
4. 55%	of	200	=	?	8. .1 1/2%	of	261.25	=	?

9. Find 1/10 of 58; .1 of 58; 10% of 58.
10. Find 1/8 of 64; .125 of 64; 12 1/2% of 64.
11. Find the weight of tin and of lead in a bar of solder weighing 1 1/2 lb. if it is composed of 50% tin and 50% lead. Check the answer by adding the weight of tin and of lead.
12. Find the weight of tin and of lead in a bar of solder, 60% tin and 40% lead, weighing 1 1/2 lb. Check the answer.
13. Find the weight of tin and of lead in an alloy of 10% tin and 90% lead weighing 1 1/2 lb. Check the answer.
14. Find the number of sheets of metal wasted if 5% of a total of 120 sheets are wasted.
15. Find the number of rivets wasted if 2 1/2% of a total of 5 boxes of 1000 each are wasted.
16. A man's salary was $7,700 per year. He receives a 7 1/2% raise. How much increase does he receive? What is his new yearly salary?
17. A man's rate of pay was $4.15 per hour. He receives an 8% raise. How much increase does he receive and what is his new pay rate?
18. There is a turnover of 3% of men a month in an industrial plant employing 2200 men. How many men change jobs per month?
19. In a plant of 25,000 employees, 2% were absent in one day. How many employees were absent?

C. Study the chart on page 60, and complete the charts which follow by filling in the missing equivalents. Remember to round off the percents to the nearest tenth.

	Fractions in Lowest Terms	Fractional Equivalents of Decimals	Decimal Fractions	Percent
1	1/16	$\frac{625}{10000}$	.0625	?
2.	3/16	$\frac{1875}{10000}$	?	18.8
3.	5/16	?	.3125	31.3
4.	7/16	$\frac{4375}{10000}$	.4375	?
5.	9/16	$\frac{5625}{10000}$	.5625	?
6.	11/16	?	.6875	68.8
7.	13/16	$\frac{8125}{10000}$	.8125	?
8.	15/16	?	.9375	93.8
9.	17/32	?	.53125	53.1
10.	$\frac{1}{64}$	$\frac{15625}{1000000}$	.015625	?
11.	$\frac{15}{64}$	?	.234375	23.4
12.	$\frac{5}{32}$	$\frac{15625}{100000}$	?	15.6
13.	$\frac{9}{32}$	$\frac{28125}{100000}$	.28125	?
14.	$\frac{1}{32}$	?	.03125	3.1
15.	$\frac{5}{64}$	$\frac{78125}{1000000}$	.078125	?
16.	$\frac{3}{32}$	$\frac{9375}{100000}$	?	9.4
17.	$\frac{7}{32}$	?	.21875	21.9
18.	$\frac{17}{64}$	$\frac{26562}{100000}$	?	26.6
19.	$\frac{21}{32}$	$\frac{62625}{100000}$	.62625	?

20.	$\frac{3}{16}$	?	.1875	18.8
21.	$\frac{13}{16}$	$\frac{8125}{10000}$	?	81.3
22.	$\frac{23}{64}$	$\frac{35937}{100000}$	.35937	?
23.	$\frac{19}{32}$	?	.59375	59.4
24.	$\frac{11}{32}$	$\frac{34375}{100000}$	?	34
25.	?	$\frac{46875}{1000000}$	.046875	4.7
26.	$\frac{9}{16}$	?	.5625	56.2
27.	$\frac{27}{64}$	$\frac{421875}{1000000}$	.421875	?
28.	$\frac{15}{16}$	?	.9375	93.8
29.	$\frac{31}{64}$	$\frac{484375}{1000000}$	?	48.4
30.	$\frac{15}{32}$	?	.46875	46.9
31.	$\frac{25}{32}$	$\frac{78125}{100000}$	.78125	?
32.	$\frac{11}{64}$	$\frac{171875}{1000000}$	?	17.1
33.	?	$\frac{40625}{100000}$	.40625	40.6
34.	$\frac{3}{64}$	$\frac{46875}{1000000}$	.046875	?
35.	$\frac{23}{32}$	?	.71875	71.9
36.	?	$\frac{84375}{100000}$	.84375	84.4
37.	$\frac{13}{64}$	$\frac{203125}{1000000}$	.203125	?
38.	?	$\frac{96875}{100000}$	.96875	96.9
39.	$\frac{9}{64}$	$\frac{140625}{1000000}$	.140625	?
40.	?	$\frac{984375}{1000000}$	.984375	98.4

UNIT 27 PERCENTAGE – ADDITIONAL APPLICATIONS

The following drawings show additional applications of percentage.

% WASTED = ?

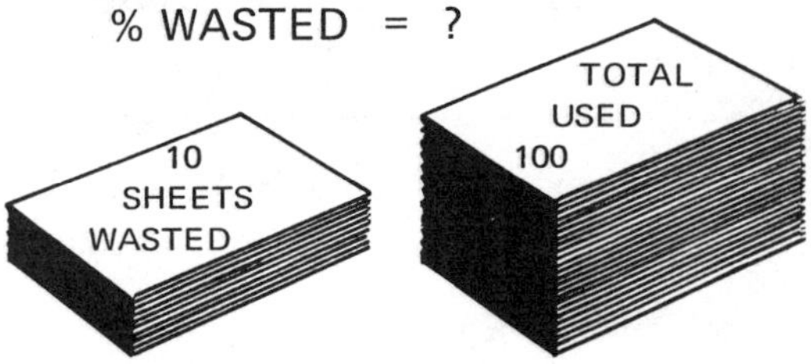

TOTAL USED = ?

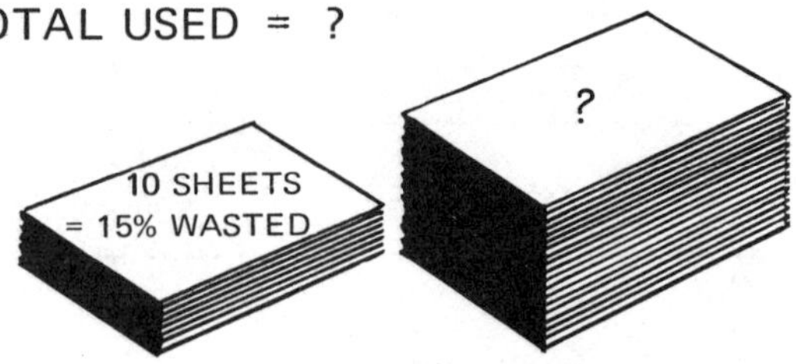

CONVERTING TO PERCENTAGE

RULE: To change a decimal to a percent, move the decimal point two places to the right and add the percent sign.

Example 1: Change .125 to a percent .125 = 12.5% Ans.

RULE 2: To change a fraction to a percent, change the fraction to a decimal and then to a percent.

Example 2: Change 3/8 to a percent.

a. Divide the numerator by the denominator. $8\overline{)3.000}$ = .375

b. Move the decimal point two places to the right and add the percent sign. .375 = 37.5% Ans.

FINDING THE PERCENT

RULE: To find the percent, divide the part by the whole and change to percent. Percent = Part ÷ Whole.

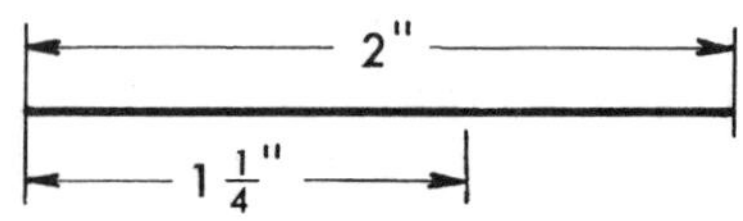

Example: What percent of a 2-in. line is a line 1 1/4 in. long?

1. Divide the part (1¼") by the whole (2") 1.25/2 = .625.
2. Change the result to a percent. .625 = 62.5%, or 62½% Ans.

FINDING THE WHOLE

RULE: To find the whole, divide the part by the percent expressed as a decimal. Whole = Part ÷ Percent.

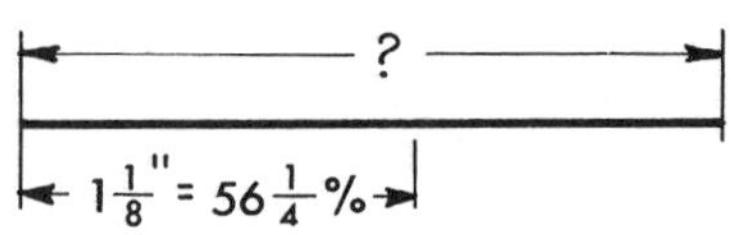

Example: What is the length of a line if 56¼% of the line is 1 1/8 in. long?

1. Change the percent to a decimal. 56¼% = .5625
2. Divide the part (1 ⅛") by the decimal (.5625) to obtain the whole. $.5625\overline{)1.1250}$ = 2 in. Ans.

ASSIGNMENT

A. Change the following to percent:

1. .165	5. .0925	9. 1/4	13. 3/4
2. .283	6. .001	10. 3/8	14. 7/8
3. .37 1/2	7. .00025	11. 1/2	15. 3/20
4. .015	8. 1/8	12. 5/8	16. 1/40

B. Solve the following problems to find the percent.

	Part	Whole	Percent		Part	Whole	Percent
1.	4	16	?	4.	18	270	?
2.	9	27	?	5.	61 1/2	184.5	?
3.	7	8	?	6.	199	200	?

7. If 50 rivets of a total of 1000 rivets are spoiled, what percent is spoiled?
8. One sheet of metal is wasted of a total of 20 sheets. What percent is wasted?
9. A firm made a $105 profit on an air-conditioning system costing $2500. What was the percent of profit?
10. When a government bond, costing $18.75 is later redeemed for $25.00, how much has it increased in value and what has been the percent of increase over the original investment?

C. Solve the following problems to find the whole:

	Part	Percent	Whole		Part	Percent	Whole
1.	16	2	?	4.	66	16 1/2	?
2.	21	3	?	5.	82.5	8 1/4	?
3.	118	59	?	6.	51.5	25 3/4	?

7. The interest on a sum of money for a year is $202. If the rate of interest was 7%, how much was the sum of money?

UNIT 28 WAGES

The sheet metal worker must be able to keep account of the time he spends on various pieces of work, the total time he works each day, and he must be able to check the computation of his weekly wage. Such problems involve time and cost computations.

TABLE OF TIME

60 seconds (sec.)	=	1 minute (min.)
60 minutes	=	1 hour (hr.)
24 hours	=	1 day (da.)
7 days	=	1 week (wk.)
52 weeks	=	1 year (yr.)
365 days	=	1 year (calendar)

Example 1: Change 2 hr. 45 min. to hours in decimal form.

$$2\ 45/60 = 2\ 3/4, \text{ or } 2.75 \text{ hr. Ans.}$$

Example 2: Compute the time for the following workday: 8:15 A.M. - 4:45 P.M.; lunch, 12:00 noon - 12:30 P.M.

A.M.	Noon	P.M.
12:00 noon = 12.00 8:15 A.M. = -8.25 3.75	1:00 P.M. 12:30 P.M. = .50 hr.	4:45 P.M. = 4.75 1:00 P.M. = -1.00 3.75 hr.

Total Time: 3.75 + .50 + 3.75 = 8.00 hr. Ans.

PAYROLLS – HOURLY RATE

One method of computing a weekly wage is to find the total hours worked for the week, including the overtime and the double time, and then to multiply this sum by the hourly rate.

Example: Find the week's wages of a man whose rate is $4.15 per hour for an 8-hour day. The overtime rate is time and a half for Saturdays and for all time over 8 hours per day; the double time rate applies to Sundays.

	Straight Time	Overtime	Double Time
Tuesday	8 hr.		
Wednesday	8 hr.	2 hr. 30 min.	
Thursday	8 hr.	3 hr.	
Friday	8 hr.		
Saturday		8 hr.	
Sunday			4 hr.
Monday	8 hr.		

Computation of Weekly Wage

1. Multiply the straight time by 1, the overtime by 1.5, and the double time by 2.

Day	Computations	Totals
Tuesday	8 x 1	8.00
Wednesday	8 x 1 + 2.50 x 1.5	11.75
Thursday	8 x 1 + 3.00 x 1.5	12.50
Friday	8 x 1	8.00
Saturday	8 x 1.5	12.00
Sunday	4 x 2	8.00
Monday	8 x 1	8.00
		68.25 hr.

2. Computation of weekly wage. 68.25 hr. x \$4.15 = \$283.2375 or \$283.24 Ans.

PAYROLLS – HOURLY RATE PLUS BONUS PLAN

The hourly rate plus bonus plan is used as an incentive to increase production. This bonus rate may apply to individual workmen, to groups or departments, or to the whole firm.

Example: A man produced 200 sheet metal parts in a 40-hour week. The standard is 120 parts per week and the wage is \$3.85 per hour, with a bonus of 35 cents for exceeding the rate per hour by one part. Compute his wage for the week.

1. Average number produced per hour 200/40 = 5.00
2. Standard rate per hour 120/40 = 3.00
3. Excess produced per hour 5.00 - 3.00 = 2.00
4. Bonus rate per hour \$0.35 x 2 = \$0.70
5. Total rate per hour \$3.85 + \$0.70 = \$4.55
6. Wage for week 40 x \$4.55 = \$182.00

Note: There are various wage problems which depend on percentage methods for their solution. For instance, if the pay of a man is raised from \$3.80 to \$4.20 per hour the percent of increase in pay is .40/3.80 x 100 = .1052 or 10 1/2%.

ASSIGNMENT

A. Time Problems

1. If a man started a job at 9:30 A.M. and finished it at 11:25 A.M., how much time did he take to do the job?
2. If a man started a job at 1:45 P.M. and finished it at 5:15 P.M., how much time did he take to do the job?
3. If a man started a job at 8:10 A.M., ate lunch from 12:00 noon to 12:30 P.M., and finished the job at 4:40 P.M., how much time should be charged to the job?

B. Payrolls – Hourly Rate

1. Using the information on the timecard below, make the following computations:
   a. The total number of hours worked each day
   b. The total numbers of hours worked on each job
   c. The weekly wage at a rate of $4.25 per hour for a 40-hour week; time and a half for Saturday; double time for Sunday and time and a half for overtime (time in excess of 8 hours a day.)

TIMECARD

For week beginning ______________________ **and ending** ______________________

No.	Description	Mon.	Tues.	Wed.	Thurs.	Fri.	Sat.	Sun.	Total
34583	Fittings	8	10	12	7				
37844	Tees				1	8	6		
37895	Elbows						2	4	
Totals									

2. What is the rate per hour of a man who earned $122.40 in 40 hours?
3. Find each period wage for an apprentice, if the journeyman's wage is $4.35 per hour. The chart below gives the rate of an apprentice as a percent of the prevailing journeyman's rate.

First	six months – 25%		Sixth	six months – 42%
Second	six months – 27%		Seventh	six months – 48%
Third	six months – 30%		Eighth	six months – 58%
Fourth	six months – 33%		Ninth	six months – 68%
Fifth	six months – 37%		Tenth	six months – 88%

C. Payrolls – Hourly Rate Plus Bonus Plan

1. A workman is paid on an hourly wage plus bonus plan. His standard per week is 160 pieces; his production for the week is 240 pieces. He worked 40 hours in a week at a rate of $4.15 per hour with a bonus of 35 cents for every piece produced over his standard. Compute his total wage for this week.
2. A worker is paid on an hourly wage plus bonus plan. His standard per hour is 12 pieces. His production for the week is 600 pieces. He worked 40 hours in a week at a rate of $3.95 per hour with a bonus of 20 cents for every piece produced over his standard. What are his earnings for the week?

D. Additional Problems

1. If a helper earns $3.15 per hour and a machinist $4.20 per hour, what percent of the machinist's wage is the helper's wage?

2. If a man's pay is raised from $3.65 per hour to $4.10 per hour, what is the percent of increase?
3. Compute the daily, the weekly and the yearly wage for a journeyman who receives $4.27 per hour and works 8 hours per day, 5 days a week and 50 weeks a year.
4. A contractor employs eight (8) journeymen at $4.15 per hour and two apprentices at $2.70 per hour. Each man works 8 hours a day, 5 days a week. What is the contractor's weekly payroll?
5. A mechanic earns $29.20 in 8 hours. What is his hourly rate?
6. Find the total contribution to the Social Security Fund made by an employee for 1 year if 1 1/2% is deducted each year and he earns $8,169.50 that year.
7. Find the amount contributed to the Unemployment Insurance Fund for one year by an employer if his yearly payroll is $129,462 and the yearly rate is 3%.
8. If a workman received 65% of his weekly wage as disability insurance, and his average weekly wage is $166.00, how much will he receive per week if he is disabled?
9. If a workman has $1,859.28 in a savings account at the bank and the interest is 6%, what is the annual and the semiannual interest?
10. There is a first mortgage of $18,000 on a house. If the interest rate is 9 1/4% per year, what is the yearly interest charge? What is the quarterly interest charge?

UNIT 29 WEIGHTS AND COSTS

Mechanics may be confronted with the problem of computing costs of materials and tools. This involves the use of tables of weights of materials and list prices, and the computation of trade discount. A typical table is shown at the bottom of this page.

Trade discount is a percentage reduction given to retailers by wholesalers. This discount enables the retailer to meet the costs of his overhead expenses, including rent, heat, light, insurance, etc. and to make a profit on his investment. After the discount is deducted from the *list price* given in the catalog, the result is called the *net price*. Either the net price or the trade discount is often given on a separate sheet.

WEIGHT TABLES

TABLE 29-1 AVOIRDUPOIS WEIGHT (Commercial)

16 ounces (oz.)	=	1 pound (lb.)
100 pounds	=	1 hundredweight (cwt.)
2000 pounds	=	1 ton (t.)

Problems involving weight can be handled in a manner similar to those concerning the units of linear measure.

Example: Change 2.375 lb. to oz. Multiply by 16 (oz. in 1 lb.). 2.375 x 16 = 38 oz., or 2 lb. 6 oz. Ans.

Example: Change 40 oz. to lb. Divide by 16 (oz. in 1 lb.). 40 ÷ 16 = 2.5 lb., or 2 lb. 8 oz. Ans.

TABLE 29-2 WEIGHTS AND LIST PRICES OF BLACK BAND IRON

Size	Wt. per Lineal Ft. (lb.)	Length (ft.)	List Price Per 100 lb.
1/2 in. x 12-gage	.185	12	$ 20.70
3/4 in. x 12-gage	.278	12	19.70
1/2 in. x 1/8 in.	.212	20	20.50
3/4 in. x 1/8 in.	.318	20	19.50
1 in. x 1/8 in.	.425	20	19.10
1 ½ in. x 1/8 in.	.6375	20	19.00
3/4 in. x 3/16 in.	.478	20	19.40
1 ½ in. x 3/16 in.	.956	20	18.90

Example: Find the weight of fifty 12-foot lengths of black band iron, 1/2 in. x 12 gage.

Solution: Multiply .185 lb. (weight of one foot) by 12 and then by 50. The total weight is 111 lb. Ans.

COST PROBLEMS

Unit Prices

Prices may be listed per 100 lb., per box, per 100 lineal feet, per 100 pieces, per dozen, etc. If the quantity ordered is less than a lot (quantities similar to the above), the price may be given *per unit* (pound, piece, foot).

Counting Table

	12 units	=	1 dozen (doz.)
	12 dozen	=	1 gross

Example: If the price is given as $18.60 per cwt., then the price per pound will be $18.60 divided by 100 or $.186 per lb.

Example: If the price is given as 16 3/4 cents per lb., then the price per lb. will be 16.75 cents or $.1675.

Single Discounts

To find the net price, the amount of discount is computed and then subtracted from the list price.

Example: Find the net price of 200 feet of 1/2-in. x 12-gage band iron if a trade discount of 40% is given.

1. Find the weight of 200 ft. 200 x .185 lb. = 37 lb.
2. Divide $20.70 (price per 100 lb. from table) by 100 lb. and multiply by 37 lb. $\frac{20.70}{100} \times 37 = \$7.659$ or 7.66
3. Find the trade discount. $7.66 x .40 = $3.064 = $3.06
4. Subtract the trade discount from the list price. 7.66 - 3.06 = $4.60
5. If the discount is 40%, then the net price must be 100 - 40 = 60%. Check above example 7.66 x 60 = $4.60 (to nearest cent).

Successive Discounts

If more than one discount is given, the first discount applies to the list price; the second discount applies to the amount remaining after the first discount has been deducted, etc.

Example: In the example above, a discount of 4% is given if the bill is paid within 10 days. Find the net cash price.

1. Cash discount 4.60 x .04 = $ .184 (.18 nearest cent)
2. Net cash price 4.60 - .18 = $4.42

Alternate solution to complete problem:

7.66 x .60 = $4.60 x .96 = 4.416 = $4.42 (nearest cent)

ASSIGNMENT

A. Weight Problems

1. Find the weight in pounds of one length of 1" x 1/8" black band iron.
2. Find the weight in pounds and ounces of one length of 1/2" x 12-gage black band iron.
3. Find the total weight in pounds of sixty-five 12-foot lengths of band iron (black), size 3/4" x 12-gage.
4. Find the total weight in pounds and ounces of three 20-foot lengths of 3/4" x 1/8" band iron.

B. Cost Problems

1. Find the net price of 5 lengths of 1/2" x 1/8" black band iron if a discount of 52% is given. (See chart on page 70.)
2. Find the net price of 25 lengths of 3/4" x 1/8" black band iron if the discount is 49%. (See chart on page 70.)
3. Find the net price of one 100-lb. coil of round black iron wire, list price 18 cents per lb. The trade discount is 40%. (See chart on page 70.)
4. Find the net cost of two 100-lb. coils of round black iron wire. The list price is 19 1/2 cents per lb. The trade discount is 45%.
5. If the list price per box of 1-lb. tinners' rivets is $1.40 and the trade discount is 40%, what is the net price per box?
6. If a dozen pairs of 12-in. combination snips with a 2 3/4-in. cut are listed at $140.40 with a 49% discount, find the net cost per pair of snips.
7. The list price per box (112 sheets of 20" x 28") of 107 lb. bright coke tin plate is $78.00. (a) If the discount is 48%, what is the net cost per box? (b) What is the net cost per sheet?
8. The list price of a standard handbrake 5' 1" long, 16-gage capacity, is $725.00. Find the net cost price if discounts of 53% and 2% (for cash) are given.
9. A 36-in., 16-gage squaring shear is listed at $810.00, and discounts of 21% and 19% (for cash) are given. What is the net price?
10. A 30-in. bar folder is listed at $295.00 and the discounts are 45% and 2 1/2% (cash discount if paid within 10 days of billing). (a) What is the net price? (b) If the bill is not paid in 10 days, how much will the bar folder cost?
11. Mill bastard files (10 in.) are listed at $10.20 per dozen or 95 cents each. If the discount is 45% in each case, what is the saving per file when a dozen is purchased?

12. Find the net price per drill if 1/4-in. high-speed steel drills are $11.20 per dozen and carbon steel drills are $5.80 per dozen, each with a 50% discount. How much more will one high-speed steel drill cost than one carbon steel drill? If the high-speed drill lasts three times as long, which is the more economical?

13. The list price of 10-in. dia., 1 C tinplate pipe (not locked) is $0.72 per foot. Find the net price of 25 two-foot lengths if the trade discount is 45%. Compare this with 28-gage galvanized pipe at $0.58 per foot and the same trade discount. What would be the total amount saved if the galvanized iron pipe were used?

14. The list price of a squaring shear is $610.00. One jobber offers a discount of 50%, another offers trade discounts of 25% and 25%. Which is the better offer and what is the difference in price?

15. The net price of a burring machine is $67.37 and the list price is $122.50. What is the trade discount in percent? **Hint:** Subtract the net price from the list price, divide the result by the list price, and then change this result to a percent.

16. The net price of a drill press is $640.25 and the list price is $985.00. What is the trade discount in percent?

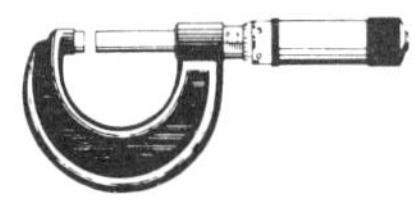

17. The net price of a micrometer caliper is $11.13 and the list price is $15.90. What is the trade discount in percent?

18. If a trade discount of 40% is offered on a slip roll former and the dollar value of this discount is $140.25, what is the list price of the machine? **Hint:** Divide the dollar value of the discount by the trade discount (percent) expressed in decimal form.

19. If a trade discount of 45% is given on a ring and circle shear and the value of this discount is $195.80, (a) what is the list price of the machine? (b) What is the net price of the machine?

UNIT 30 FACTORY COST AND SELLING PRICE

A manufacturer cannot profitably sell an article at the actual factory cost of labor and material. There are two other items which must be added: (1) the overhead; (2) the profit. The illustration below shows the relationship of these items with numerical values given to each. Problems involving these items, are, like discount, an application of percentage.

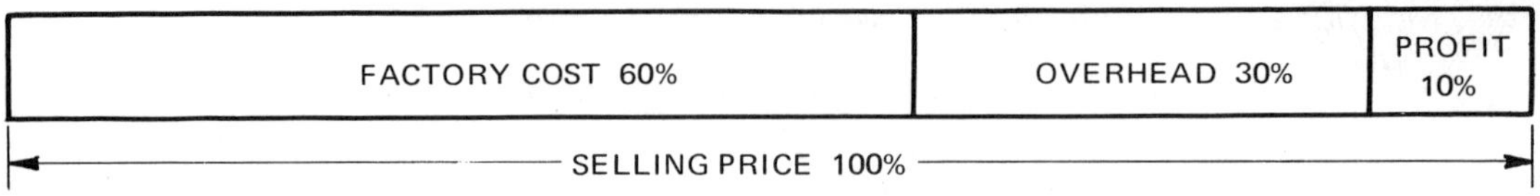

FACTORY COST

Overhead = Percent Overhead x Selling Price
Profit = Percent Profit x Selling Price
Factory Cost = Selling Price - Overhead - Profit

Example: If the selling price of a job is $200, the overhead 30%, and the profit 10%, what is the amount of the overhead, the amount of the profit, and the factory cost?

1. Find the overhead. $200 x .30 = $60 Ans.
2. Find the profit. $200 x .10 = $20 Ans.
3. Subtract the overhead and then the profit from the selling price to obtain the factory cost.

$$\begin{array}{r} \$200 \\ -\ \$\ 60 \\ \hline \$140 \end{array} \qquad \begin{array}{r} \$140 \\ -\ \$\ 20 \\ \hline \$120 \end{array} \text{ Ans.}$$

SELLING PRICE

$$\text{Selling Price} = \frac{\text{Factory Cost}}{100\% - \%\text{ Overhead} - \%\text{ Profit}}$$

Example: If the factory cost is $130, the overhead 25%, and the profit 10%, find the selling price.

1. Subtract the % overhead from 100%. 100% - 25% = 75%
2. Subtract the % profit from this result. 75% - 10% = 65%
3. Divide the factory cost by .65 (from step 2) to obtain the selling price.

$$\begin{array}{r} \$\,2\,0\,0.\ \text{Ans.} \\ .65_{\wedge}\overline{)\,1\,3\,0.0\,0_{\wedge}} \\ \underline{1\,3\,0} \end{array}$$

ADDITIONAL RULES

1. $\%\text{ Profit} = \dfrac{\text{Profit}}{\text{Selling Price}}$

2. $\%\text{ Overhead} = \dfrac{\text{Overhead}}{\text{Selling Price}}$

3. $\text{Selling Price} = \dfrac{\text{Profit}}{\%\text{ Profit}}$

ASSIGNMENT

A. Factory Cost

1. The selling price of a squaring shear is $1,200, the overhead 30%, and the profit 10%. What is the amount of overhead, the amount of profit, and the factory cost?
2. If a lot quantity of safety guards is sold at $850, the overhead 35%, and the profit 9%, what is the factory cost?

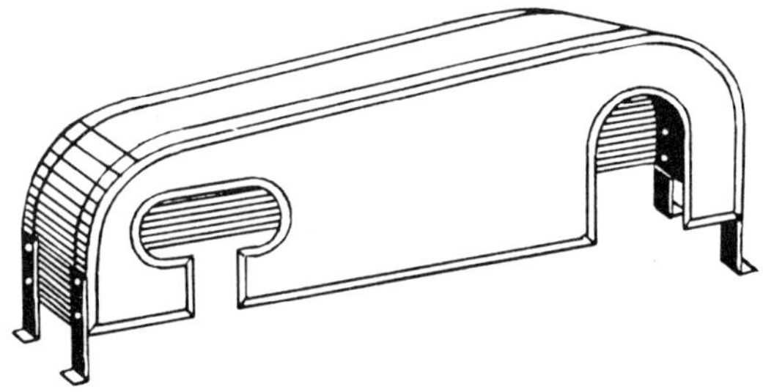

3. The selling price of an oil tank is $479.50, the overhead 39%, and the profit 11%. What is the amount of overhead, the amount of profit, and the factory cost?

B. Selling Price

1. The factory cost of a job is $300, the overhead 30%, and the profit 10%. Find the selling price of the job.
2. If the factory cost of a milk tank is $156.00, the overhead 35%, and the profit 9%, what is the selling price of the tank?
3. If the factory cost of 5 gross of pulley guards is $825.00, the overhead 38%, and the profit 7%, find the selling price of the entire lot of pulley guards. What is the selling price of each pulley guard?

C. Additional Problems

1. If a profit of $2000 is made on an air-conditioning system, and the contract price is $50,000, what is the percent of profit?
2. A contractor bid $1000 on the sheet metal work for a building. Labor, material, and overhead together were estimated to be $900. What will be his percent of profit?
3. A sheet metal firm sold a tool cabinet at a profit of $75. This profit was 5% of the selling price. Compute the selling price of the tool cabinet.
4. The factory cost of a gross of 14-in. elbows is $200, the overhead is $60, and the profit desired is $25. What should be the selling price, the percent overhead, and the percent profit?

SECTION 4 GEOMETRY

UNIT 31 THE CIRCLE

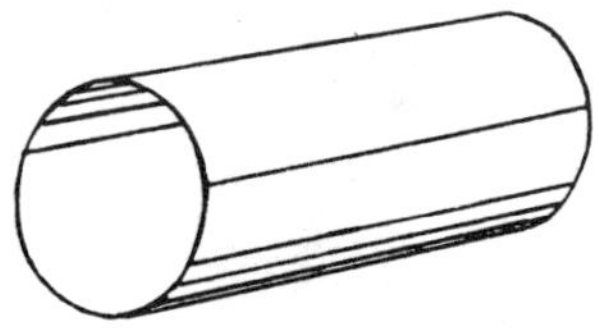

Sheet metal work requires the construction of pipes, containers, etc. which are circular in shape. Consequently the workman must understand the circle and the relationship of its parts to one another.

DEFINITIONS

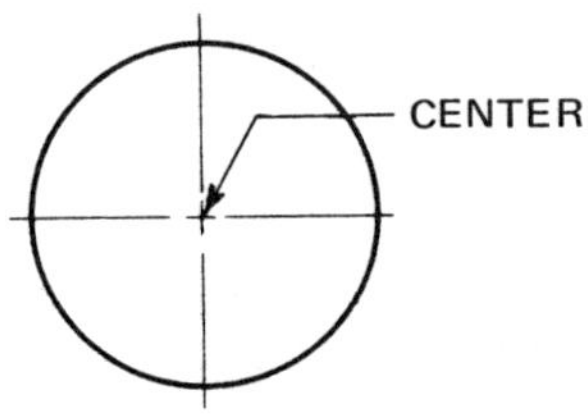

A *circle* is a plane curve, all parts of which are equally distant from a point called the center. The symbol for a circle is ⊙ .

The *circumference* of a circle is the distance around the circle measured in linear units.

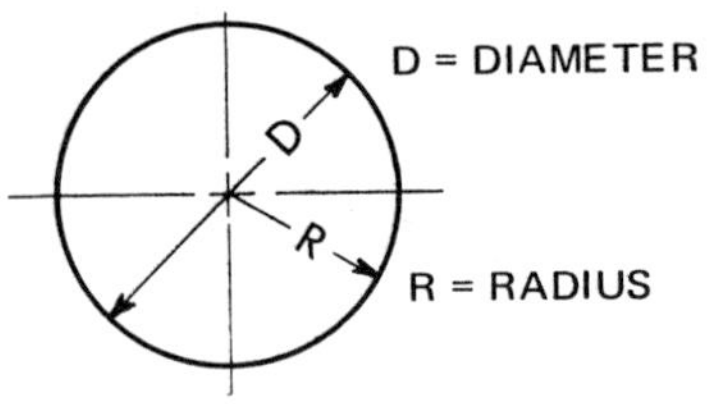

The *radius* (R or r) is a straight line from the center to the circumference.

The *diameter* (D or d) is a straight line through the center terminating in the circumference. It divides a circle into half circles (semicircles).

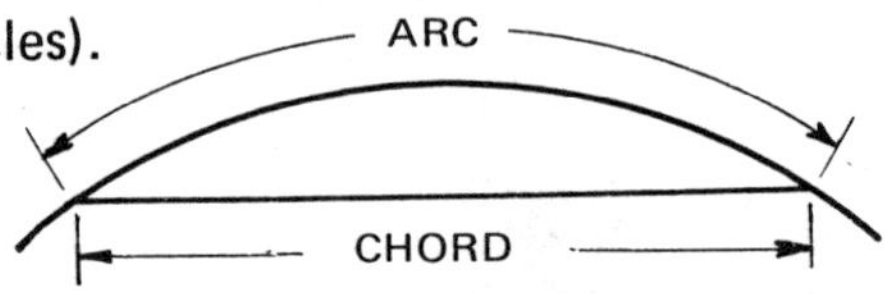

The *arc* of a circle is a part of the curved line.
A *chord* is a straight line between ends of an arc.

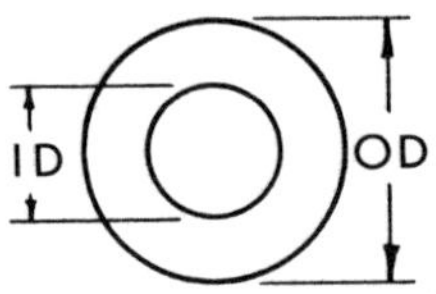

Concentric circles are circles having the same center. The diameter of the outer circle is sometimes designated by the letters OD and the inner circle by ID.

RELATIONSHIP OF THE RADIUS AND THE DIAMETER

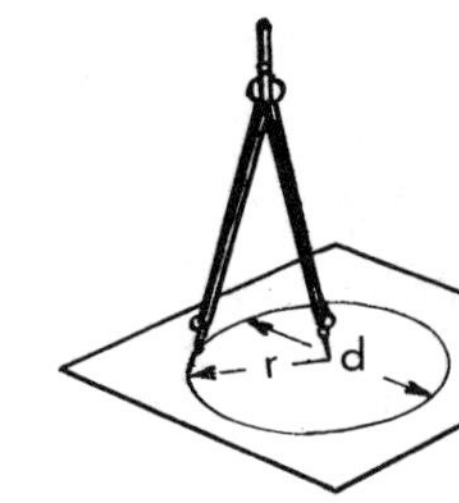

RULE 1: The radius is equal to one-half the diameter; the diameter is equal to twice the radius. The compass is set a distance equal to the radius.

Example 1: If the diameter of a circle is 2 3/16 in., find radius.

2 3/16 in. ÷ 2, or 35/16 x 1/2 = 35/32, or 1 3/32 in. Ans.

Example 2: If the radius of a circle is 1 7/8 in., find diameter.

1 7/8 in. x 2, or 15/8 x 2/1 = 15/4, or 3 3/4 in. Ans.

ASSIGNMENT

1. Compute the missing quantity in each of the following problems:

	Radius	Diameter		Radius	Diameter
a.	2 1/2″	?	f.	?	9 3/16″
b.	3 7/8″	?	g.	2.6″	?
c.	4 9/10″	?	h.	2 63/64″	?
d.	?	2 1/4″	i.	1′ 4″	?
e.	?	5 3/4″	j.	?	5′ 9″

2. Using the diameter given below, compute the radius, and draw the circle for each problem. Check the accuracy of each circle by measuring the diameter.

   a. 3″ b. 3 1/2″ c. 2 3/8″ d. 2 15/32″

3. Compute the unknown dimensions on each of the drawings below. Check the answers of problems *a* to *d* by measurement.

a.

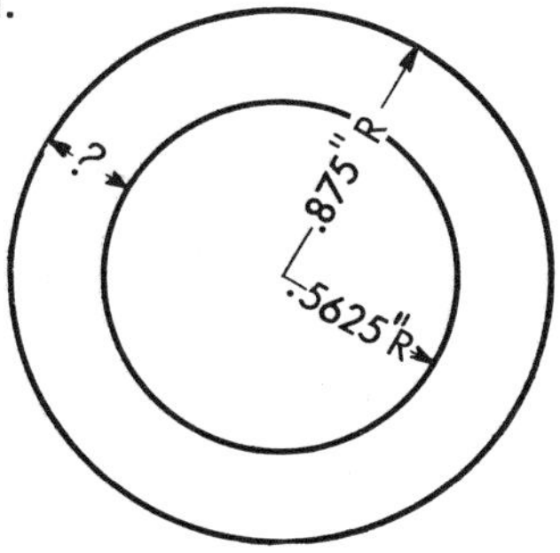

b.

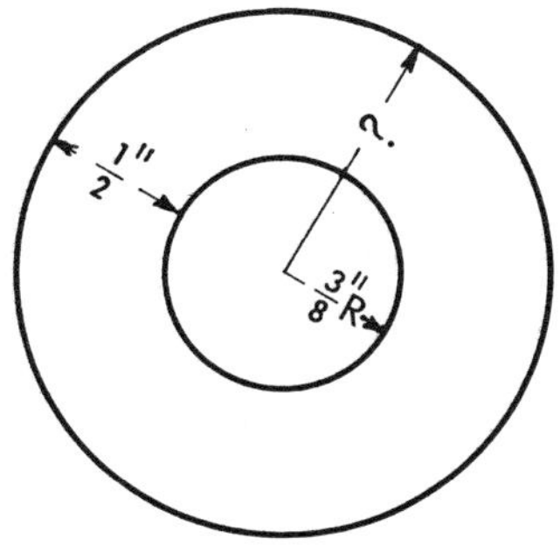

c.

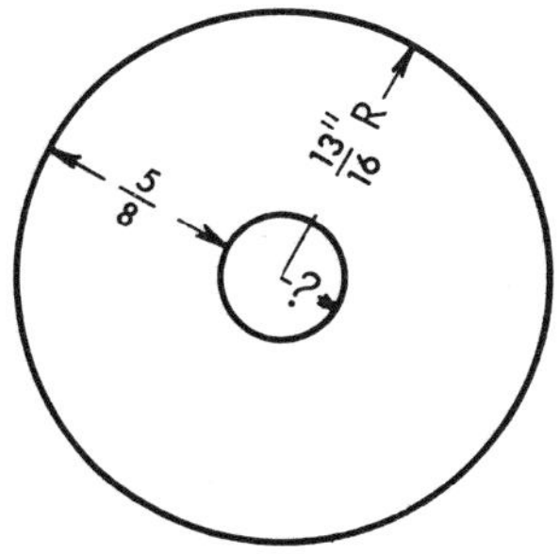

d.

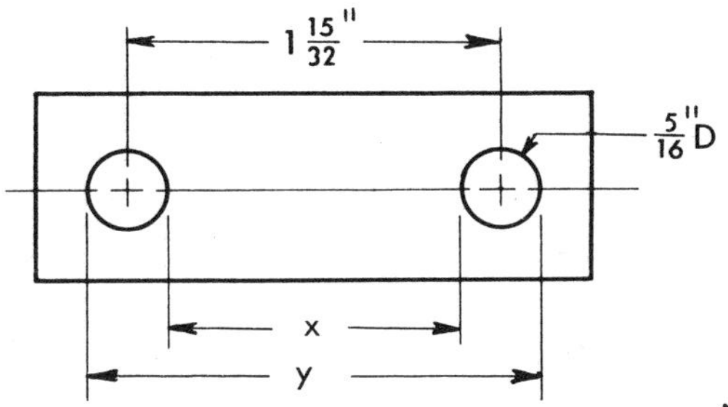

e.

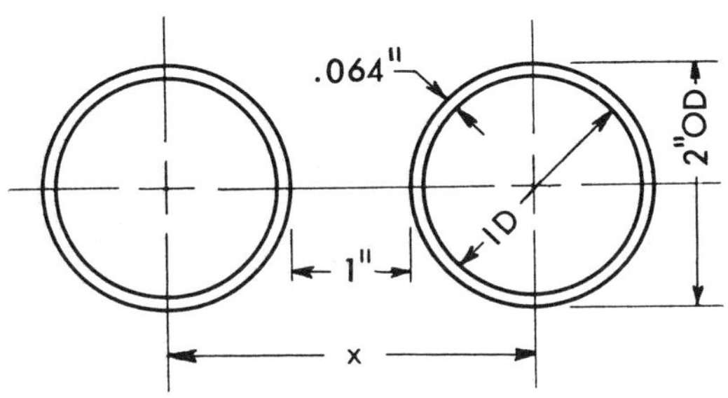

f.

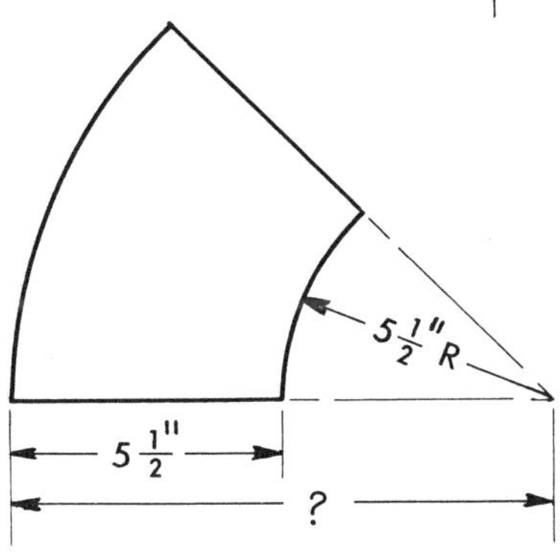

UNIT 32 THE UNITS OF ANGULAR MEASURE

An *angle* is formed by two straight lines intersecting at a point called the vertex. Angles are measured in degrees, minutes, and seconds. Angles appear in practically all jobs made by sheet metal workers. Roof flanges, miters, skylights, etc. involve angles.

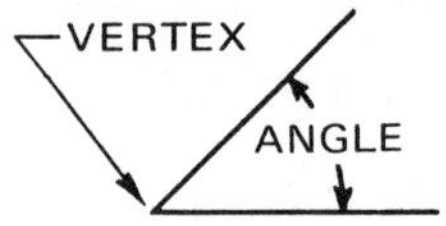

Degrees, minutes, and seconds can be added, subtracted, multiplied, and divided in a manner similar to that used for units of linear measure.

TABLE OF ANGULAR MEASURE

One circle	=	360 degrees (deg.)
One degree	=	60 minutes (min.)
One minute	=	60 seconds (sec.)

SYMBOLS USED IN ANGULAR MEASURE

- *Degrees* are designated by a small zero as 2 deg. = 2°.
- *Minutes* are designated by a single prime as 5 min. = 5′.
- *Seconds* are designated by a double prime as 30 sec. = 30″.
- *Angles* are designated by the symbol ∠.

The word "therefore" is designated by the symbol ∴ when it is used in solving problems.

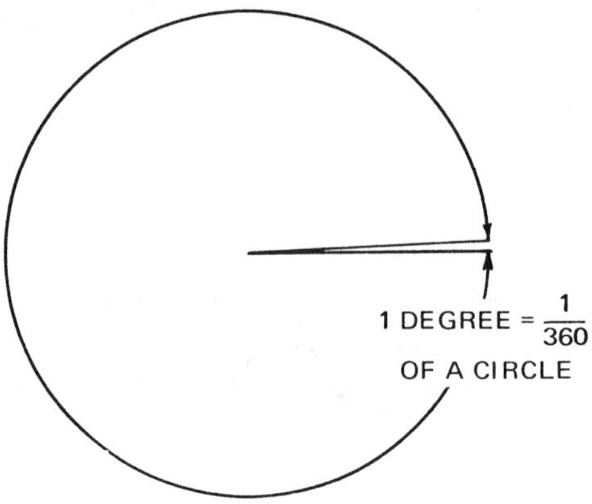

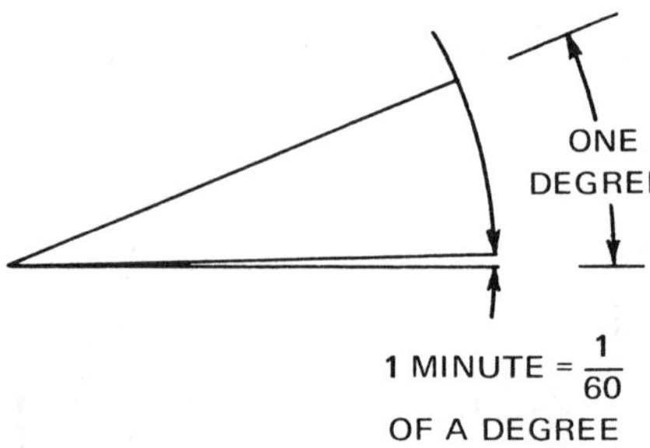

Note: The value of the angle does not change regardless of how far the sides of the angle are extended.

RULES AND EXAMPLES

1. There are 90 degrees in a quarter circle: $360 \times 1/4 = 90°$

   There are 180 degrees in a half circle: $360 \times 1/2 = 180°$

   There are 270 degrees in a three-quarter circle: $360 \times 3/4 = 270°$

2. To change 60 degrees to a part of a circle, divide 60° by 360.

   $\therefore$ 60/360 = 1/6 of a circle

3. To change 15 min. to deg., divide 15′ by 60: 15/60 = 1/4°, or .25°.
4. To change .5 deg. to min., multiply .5° by 60: .5 x 60 = 30 min.

ASSIGNMENT

A. Parts of a Circle

1. How many degrees are there in 1/8 of a circle?
2. How many degrees are there in 5/12 of a circle?
3. How many degrees are there in 11/30 of a circle?
4. How many degrees are there in 1/5 of a circle?
5. What part of a circle is 40 degrees?
6. What part of a circle is 45 degrees?
7. What part of a circle is 72 degrees?
8. What part of a circle is 120 degrees?

B. Reduction Problems

1. Change 45 min. to deg.
2. Change 42 min. to deg.
3. Change 12 min. to deg.
4. Change 3 deg. to min.
5. Change 12 deg. to min.
6. Change 16 deg. to min.

C. Calculation Problems

1. Find the value of each of the angles marked x.

a.

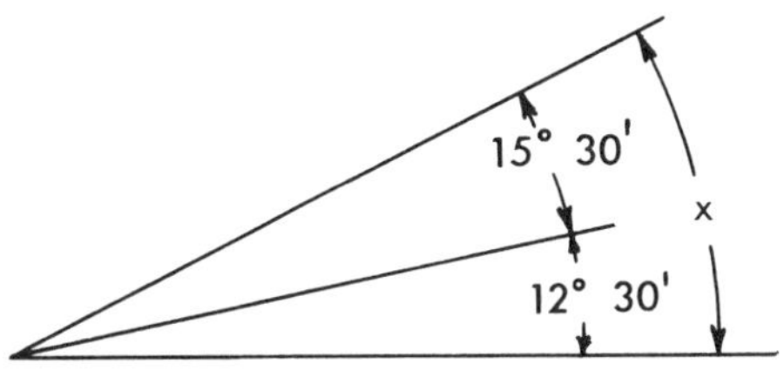

b.

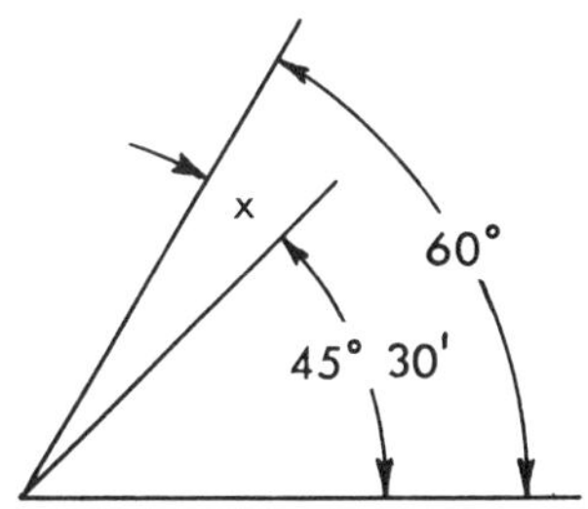

2. Multiply 5 1/2° by 8.
3. How many degrees will there be in seven angles, each 8° 30′?
4. Divide 72 1/2° by 5.
5. If an angle of 90° 30′ is divided into 4 angles, what is the value of each angle?
6. What is the value of the angles between the equally spaced holes in the reinforcing plate shown?

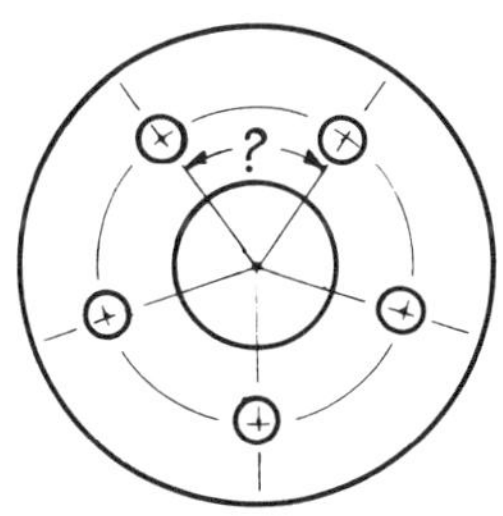

UNIT 33 THE PROTRACTOR

Angles are measured with a device called a protractor. There are many different kinds of protractors for use in drawing rooms and shops, such as the semicircular protractor, the vernier protractor, and the swinging-blade protractor.

The semicircular protractor shown has two scales, each of which is divided into 180 degrees. (Some scales are further divided into one-half degrees.) Every ten degrees is labeled with two numbers; the inner number gives the angle starting at the right, and the outer number gives the angle starting at the left. The sum of the two readings at any point is 180 degrees.

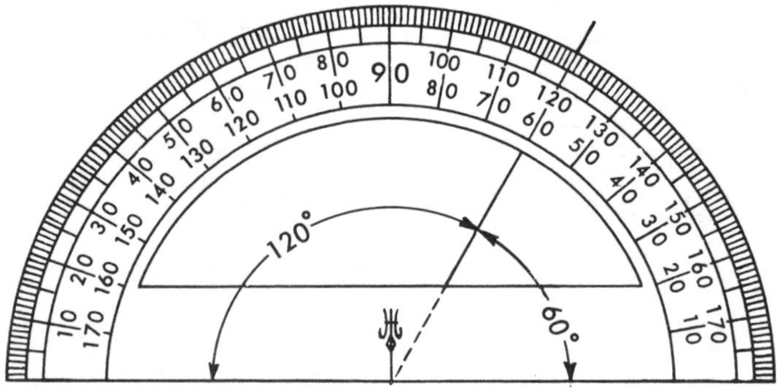

MEASURING ANGLES WITH A PROTRACTOR

1. Place the lower edge of the protractor along one edge of the angle with the center point of the protractor at the vertex of the angle.
2. Read the graduation (38°) where the side of the angle crosses the graduated edge of the protractor.

Note: When the angle extends to the right, use the inner scale; and when the angle extends to the left, use the outer scale.

Note: If the sides of the angle do not meet, extend them until they meet at the vertex.

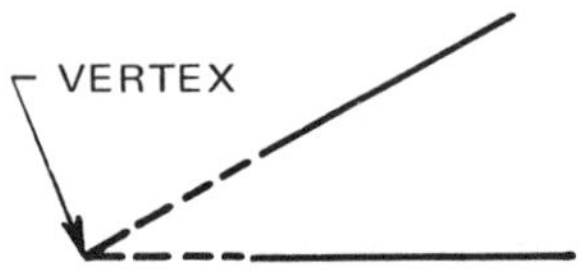

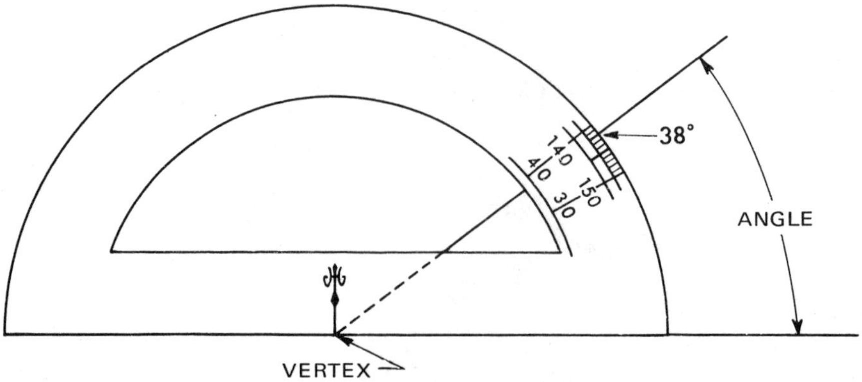

ASSIGNMENT

1. Read the values of the angles indicated by each letter:

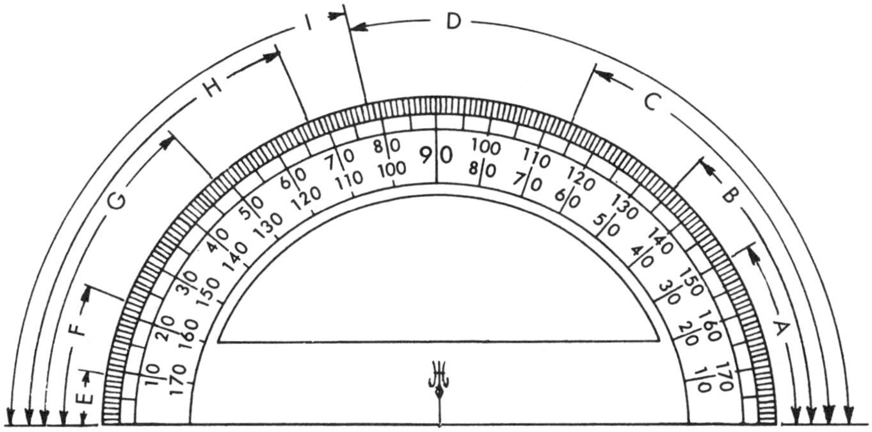

2. Measure the angles below with a protractor:

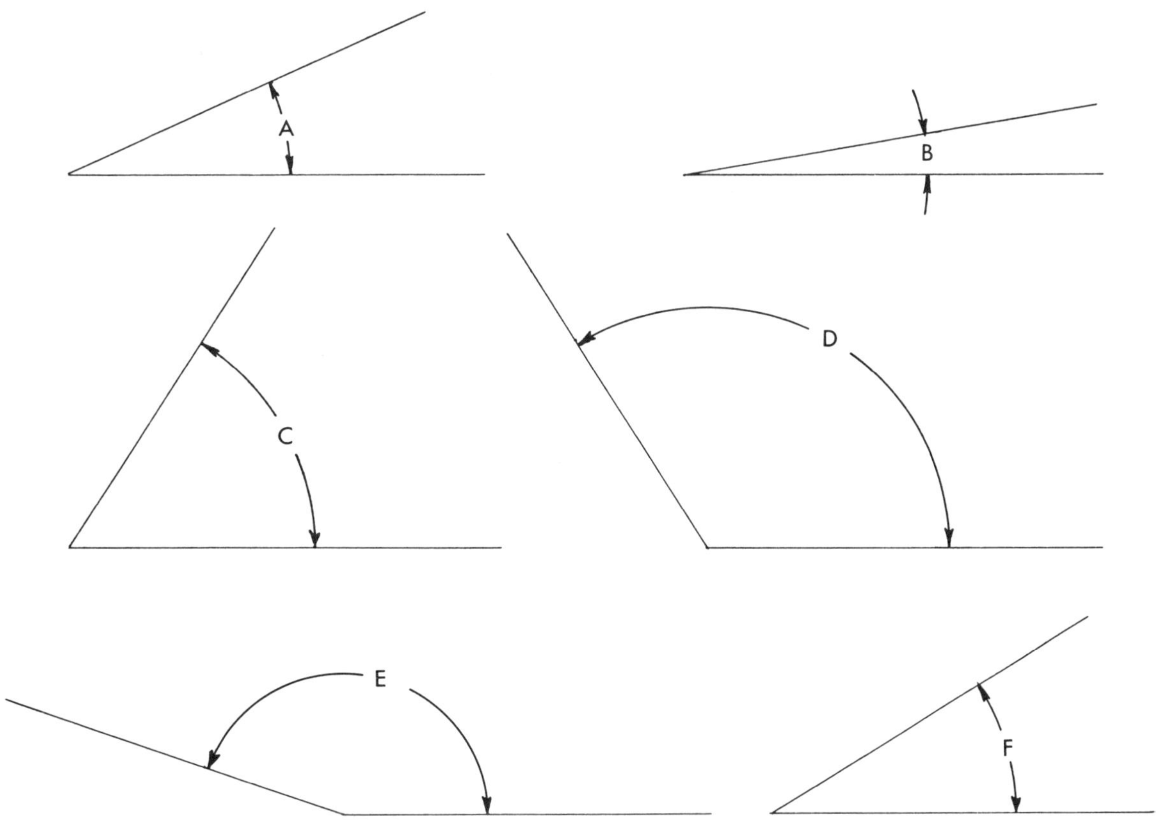

3. Lay out the following angles:

a. 21°	c. 3°	e. 109°
b. 49°	d. 91 1/2°	f. 167 1/2°

UNIT 34 TRIANGLES

Polygons are plane figures formed by three or more straight lines. The number and relationship of the sides governs the name of the plane figure.

A *triangle* is a plane figure with three straight sides. These sides may or may not be equal. A triangle also has three angles. The triangle symbol is a small triangle △.

Figures of triangular shape occur in many sheet metal jobs. In the drawing of the chimney cap shown at the right, the front view of the upper part is triangular in shape.

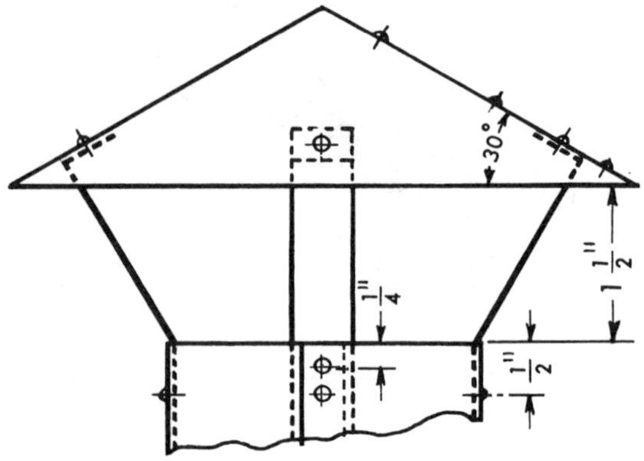

Triangles are divided into groups according to the lengths of their sides. The altitude of a triangle is the perpendicular distance from the vertex to the base. Altitude is abbreviated alt.

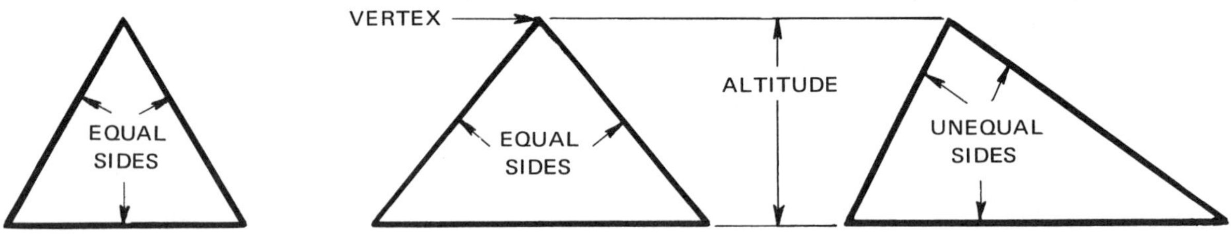

EQUILATERAL TRIANGLE (ALL SIDES EQUAL) — ISOSCELES TRIANGLE (TWO SIDES EQUAL) — SCALENE TRIANGLE (NO SIDES EQUAL)

Triangles are also divided into groups according to the sizes of their angles.

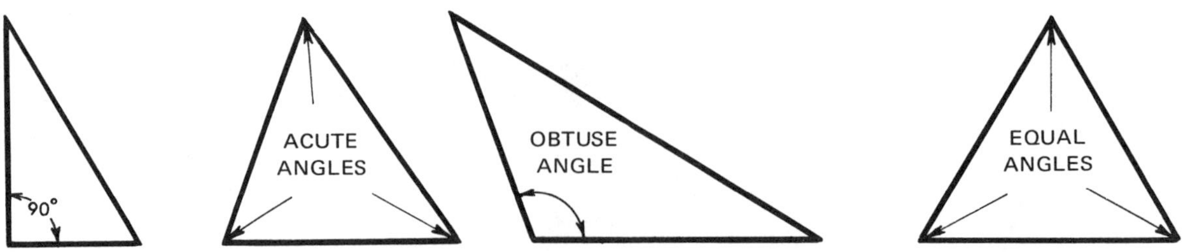

RIGHT TRIANGLE (ONE RIGHT ANGLE) — ACUTE TRIANGLE (ALL ANGLES ACUTE) — OBTUSE TRIANGLE (ONE ANGLE OBTUSE) — EQUIANGULAR TRIANGLE (ALL ANGLES EQUAL)

Note. Observe that an equilateral triangle is always equiangular.

CONSTRUCTION OF A TRIANGLE: (Three Sides Given)

Example: Draw a triangle: base = 2 3/4″, sides = 2 1/4″ and 2″.

STEP 1

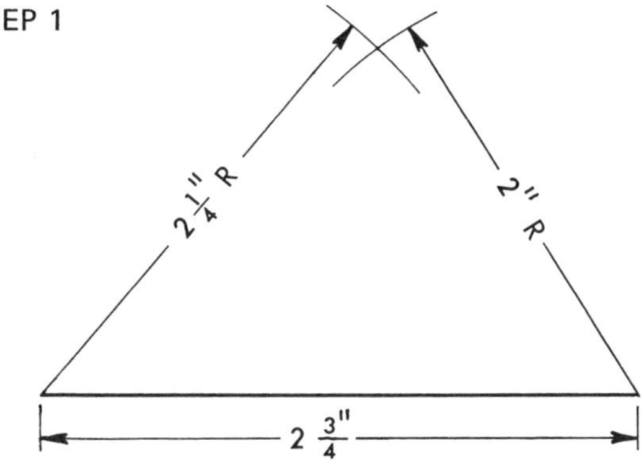

STEP 2

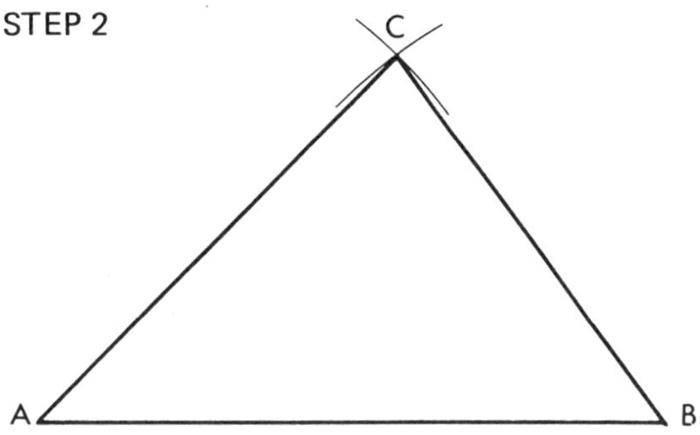

1. Draw a 2 3/4-in. line. Using the ends of this line as centers, draw arcs with 2 1/4-in. and 2-in. radii.
2. Complete the triangle. Draw a straight line from point at ends of line to point where arcs cross, or intersect.
3. Measure the angles of the triangle at A, B, and C. The sum of the angles should equal 180 deg.

PRINCIPLE: Two triangles are equal if the three sides of the one triangle equal three sides of the other triangle.

CONSTRUCTION OF A TRIANGLE: (Three Sides and Included Angle Given)

Example: Draw a triangle: sides 2 11/16″ and 2 7/16″, ∠ = 38°.

STEP 1

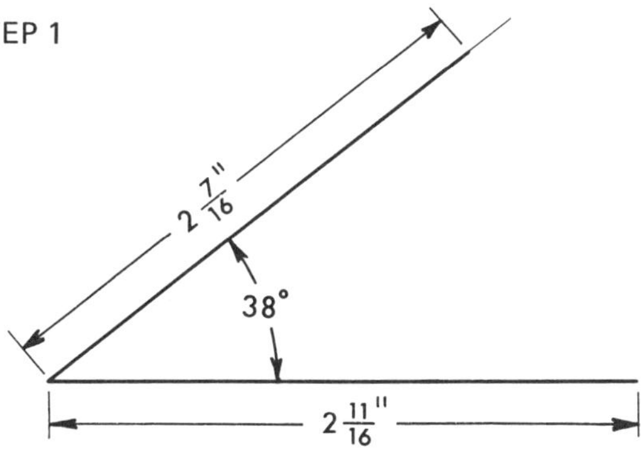

STEP 2

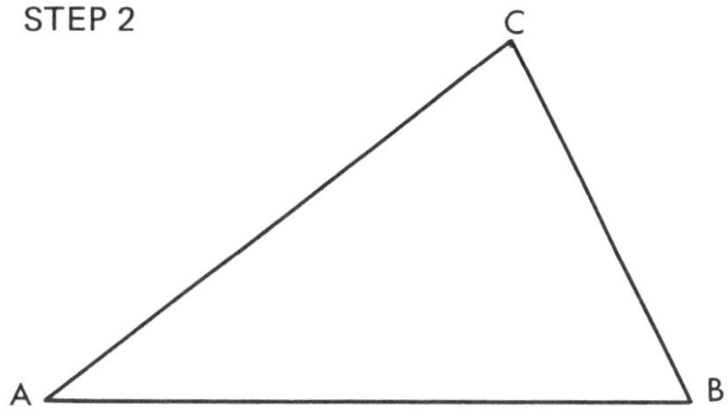

1. Draw a 2 11/16-in. line. Lay out a 38-deg. angle at one end of this line. Measure 2 7/16 in. on the top side of this angle.
2. Complete the triangle.
3. Measure the angles of the triangle at A, B, and C. The sum of the angles should equal 180 deg.

PRINCIPLE: Two triangles are equal if two sides and the included angle of one triangle equal two sides and the included angle of the other triangle.

CONSTRUCTION OF A TRIANGLE: (Two Angles and Included Side Given)

Example: Draw a triangle: base = 2 3/4″, base angles = 42°.

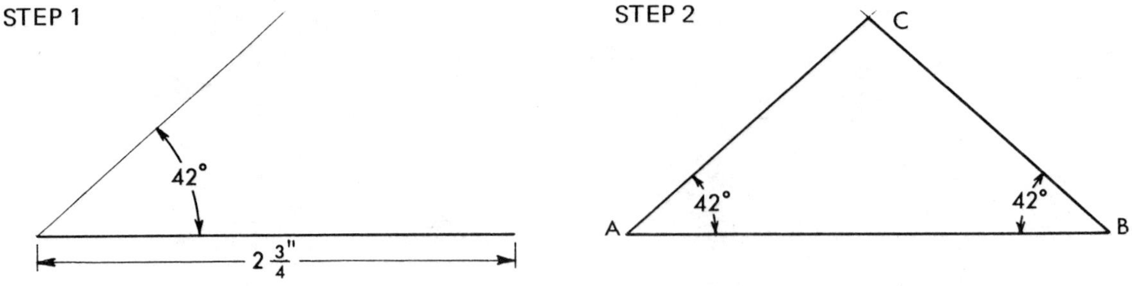

1. Draw a 2 3/4-in. line and lay out a 42-deg. angle at one end.
2. Lay out a 42-deg. angle at the other end of the line.
3. Measure the angles of the triangle at A, B, and C. The sum of the angles should equal 180 deg.
4. Measure the sides AC and CB. These should be equal and so make an isosceles triangle.

PRINCIPLE 1: Two triangles are equal if two angles and the included side of one triangle equal two angles and the included side of the other triangle.

PRINCIPLE 2: If two angles of a triangle are equal, the sides opposite these angles are equal.

PRINCIPLE 3: If two sides of a triangle are equal, the angles opposite these sides are equal.

APPLICATION OF PRINCIPLES

The Sum of the Angles of a Triangle Equals 180°.

Example: If angle A = 40° and side AC = side CB, compute the values of the angles at B and at C.

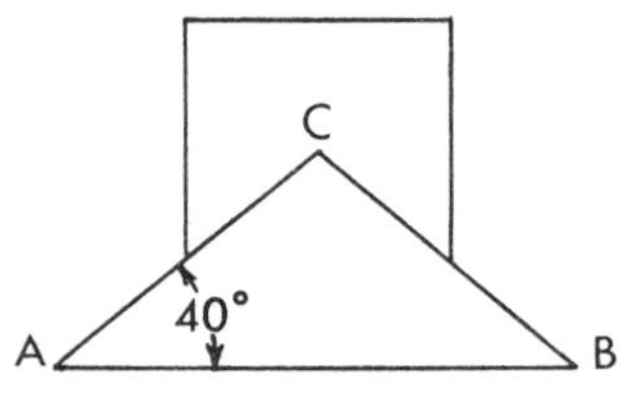

1. Since side AC = side CB, then ∠ A = ∠ B.
2. If ∠ A = 40°, then ∠ B = 40° (isosceles triangle)
3. ∠ C = 180° − 2 x 40°, or 100° Ans.

IF TWO TRIANGLES ARE EQUAL, THEIR CORRESPONDING PARTS ARE EQUAL.

Example: If the angles AOB and BOC of the elbows are equal, are the lines AB and BC equal? Give the reasons.

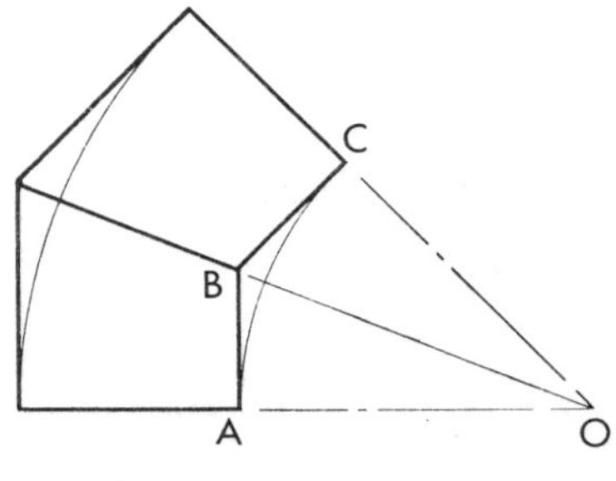

1. ∠ AOB = ∠ BOC, lines OA = OC and OB = OB
2. △ AOB = △ BOC (Two triangles are equal if two sides and the included angle of each are equal.)
3. ∴ line AB = line BC (Corresponding parts of equal triangles are equal.)

ASSIGNMENT

1. Draw an equilateral triangle with sides each 4 in. long. Compute each angle, and check by measuring with a protractor.
2. Draw a right triangle with a base of 2 3/8 in. and an altitude of 4 in. Measure each angle.
3. Draw an isosceles triangle with sides of 3 in. and an included angle of 32 deg. Compute and measure the values of the remaining angles.
4. Draw an isosceles triangle with a base of 3 1/8 in. and base angles of 53 deg. Compute the value of the remaining angle, and check by measuring with a protractor.
5. Draw an isosceles triangle with a base of 2 9/16 in. and a vertex angle of 36 deg. (The value of each of the remaining angles must be computed before laying out the triangle.)
6. Draw a right triangle with a base of 2 3/4 in. and a base angle of 50 deg. Compute and measure the value of the remaining angle.
7. Compute the values of the unknown angles in the triangles below.

a.

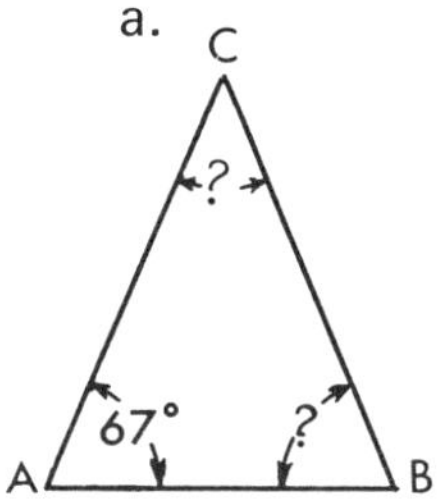

b.

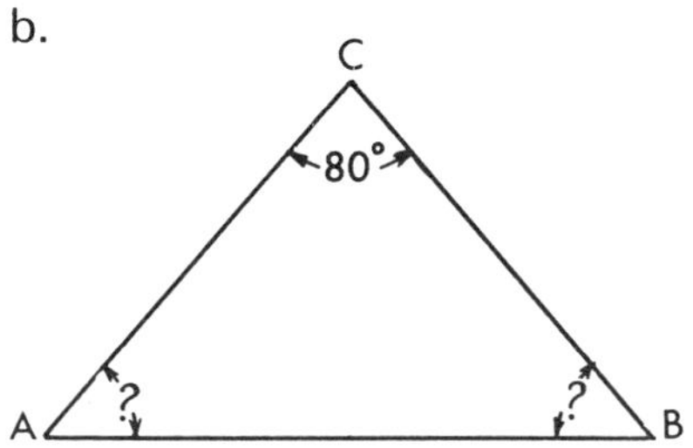

c.

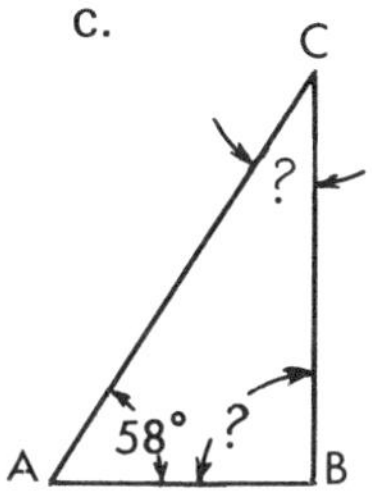

d.

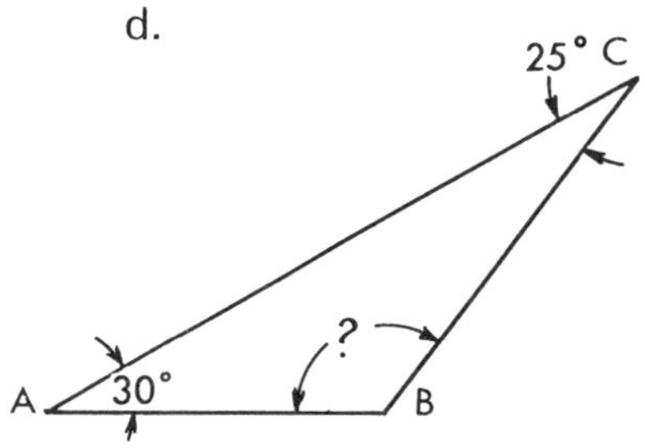

8. Compute the values of the unknown angles of the jobs below.

a.

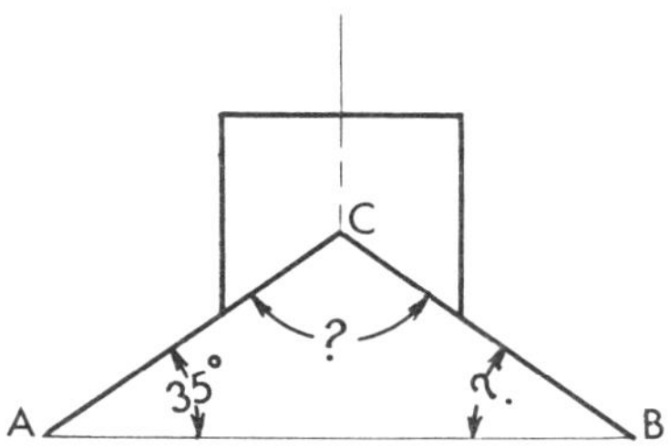

b.

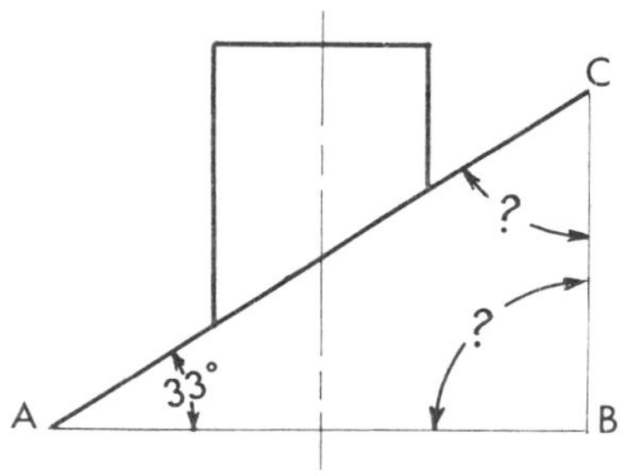

c.

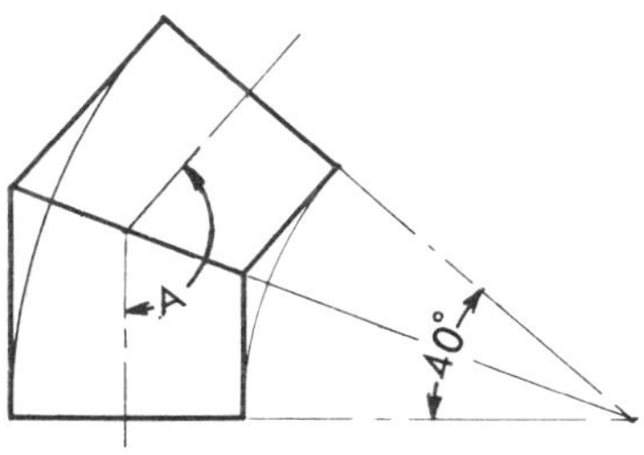

d

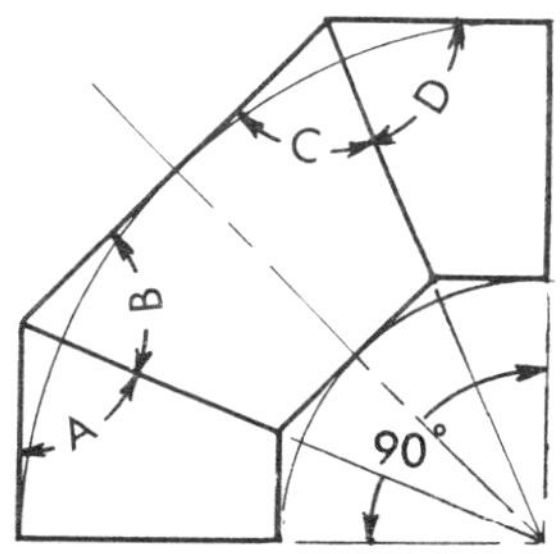

UNIT 35 INTERSECTING LINES

An understanding of lines is very important to the sheet metal worker because every job consists of curved or straight lines, or a combination of both.

The straight lines may be at various angles to each other. If they are at 90 deg. to each other, they are said to be *perpendicular.* If they are at other degrees except 90 deg. or 180 deg. they are said to be *oblique.* The drawing shows a job with perpendicular and oblique lines. The sides of the 60 deg. angle are also oblique to each other.

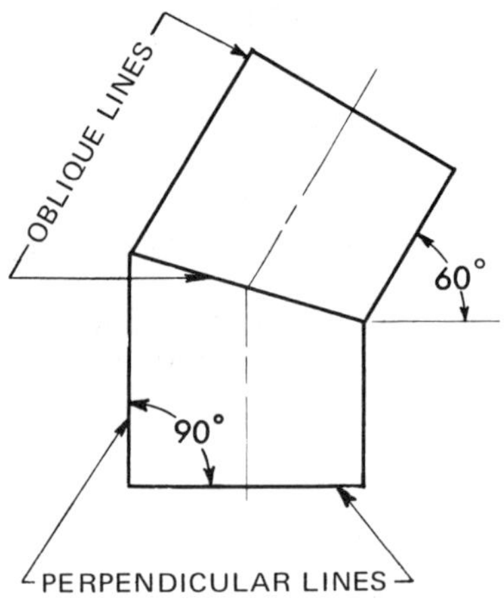

PERPENDICULAR LINES

Perpendicular lines are lines which intersect at 90 deg., or which, when extended, intersect and form angles of 90 deg. The symbol for perpendicular is ⊥. Lines that are perpendicular can be in any position. Angles of 90 deg. are called *right angles.*

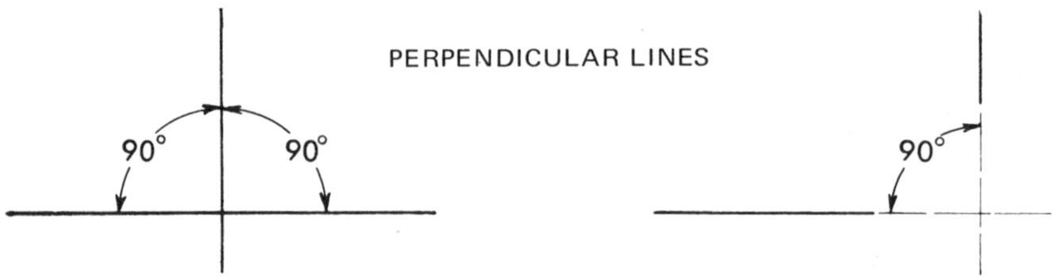

Construction of a Right Angle From a Given Point in a Line

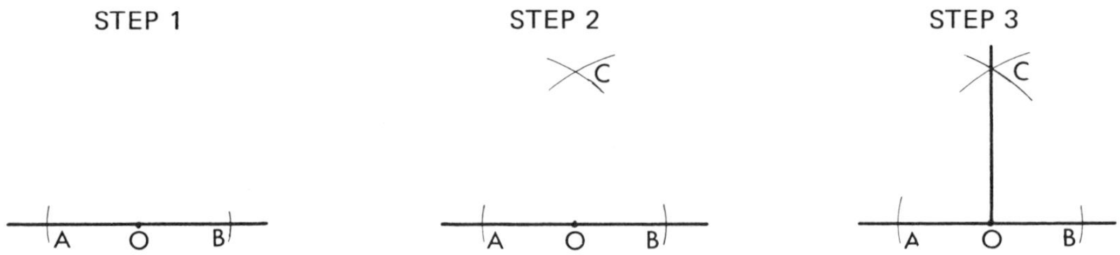

1. Using given point O as a center, draw two arcs with the same radius.
2. Using points A and B as centers, draw arcs with the same radius intersecting at C.
3. Connect point C with given point O.

Note: Each angle formed is 1/2 of 180 deg., or 90 deg. (right angle).

Construction of a Right Angle From a Point Outside the Line

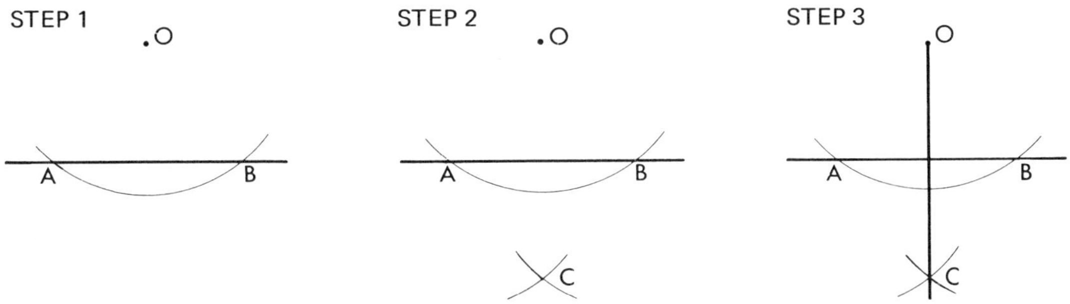

1. Using given point O as a center, draw arc AB.
2. Using points A and B as centers, draw arcs with the same radius intersecting at C.
3. Connect the point C with given point O.

Note: Each angle formed is 1/2 of 180 deg., or 90 deg.

Construction of a Right Angle at the End of a Line

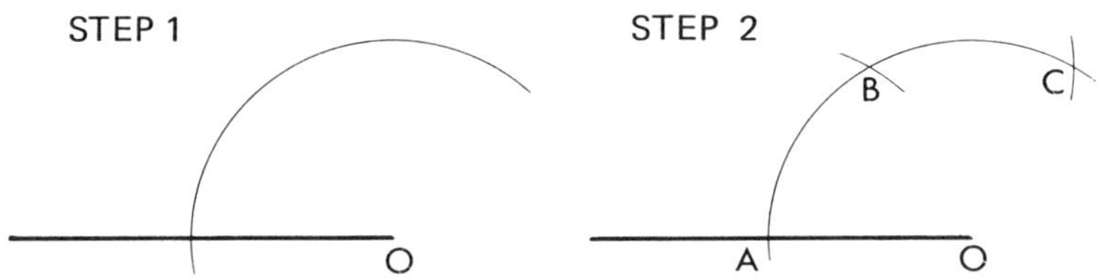

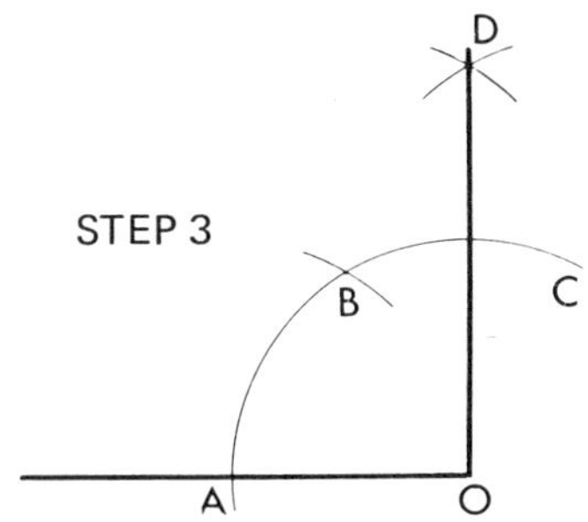

1. Using given point O at the end of the line as a center, draw an arc.
2. Using the same radius and point A as a center, mark two points B and C on this arc with the compass.
3. Using points B and C as centers, draw arcs with the same radius intersecting at D. Connect points D and O.

Note: Since $\triangle$ AOB and $\triangle$ BOC are equilateral, the value of $\angle$ AOB, and also of $\angle$ BOC is $180°/3$, or $60°$.

a. $\angle$ BOD = $\angle$ BOC/2, or $30°$ ($\angle$ BOC was bisected)

b. $\angle$ AOD = $\angle$ AOB + $\angle$ BOD

$\therefore$ $\angle$ AOD = $60° + 30°$, or $90°$ Ans.

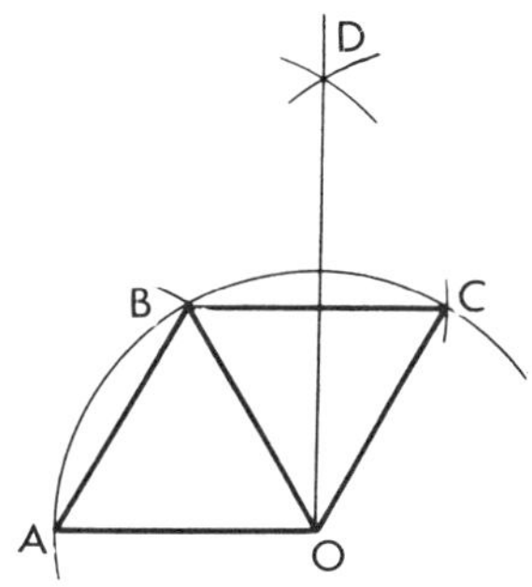

OBLIQUE LINES

OBTUSE ANGLE

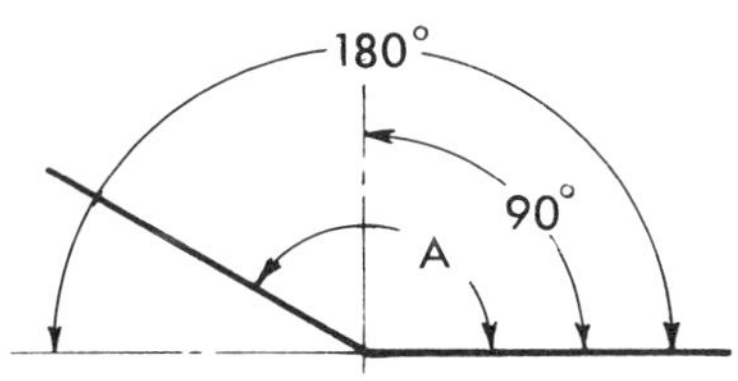

ACUTE ANGLE

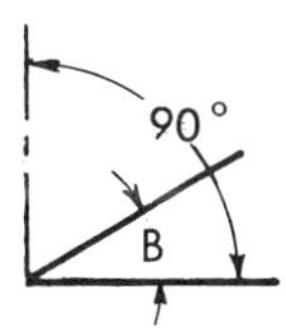

The *obtuse* or open angle A is an angle greater than 90 deg., but less than 180 deg.

The *acute* or closed angle B is an angle less than 90 deg., but more than 0 deg.

Application to Sheet Metal Work

The drawing at the right shows the various kinds of angles on a typical sheet metal job.

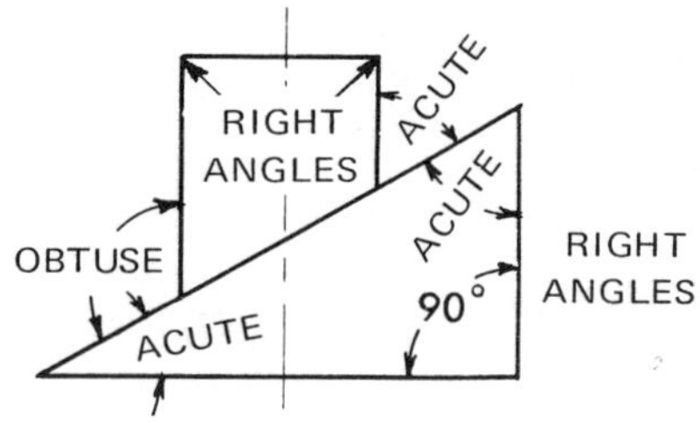

Dividing an Angle into Equal Parts

Divide ∠ MON into 4 equal parts.

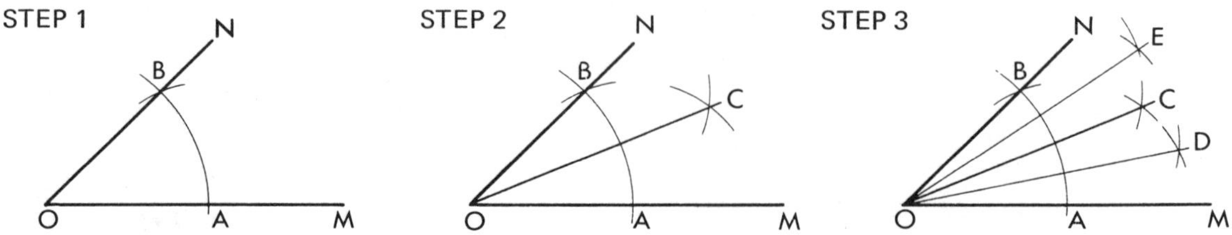

1. Draw arc AB using vertex O as a center of the given ∠ MON.
2. Draw arcs with same radius using points A and B as centers. Connect C and O.
3. Repeat to find points E and D. The ∠ MON is divided into four equal parts.

Application to Sheet Metal Work

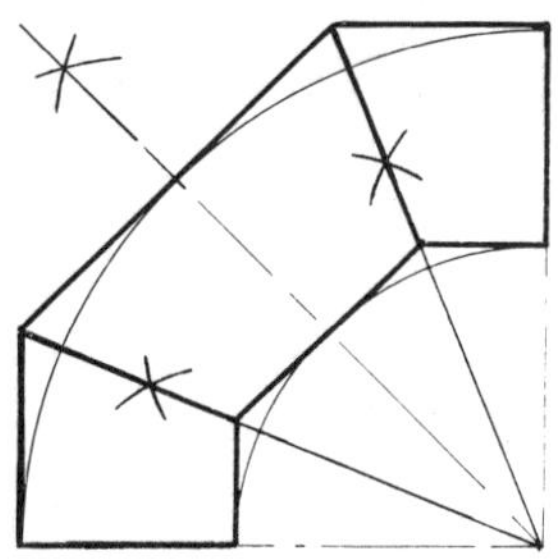

The drawing shows the application of the above construction to the layout of an elbow. If the angle of the elbow is 90 deg., each small angle will be 90°/4, or 22 1/2°. Check this answer with a protractor.

Transferring an Angle

Transfer *given angle* at *left* to *line NM* on *right* using point N as a vertex.

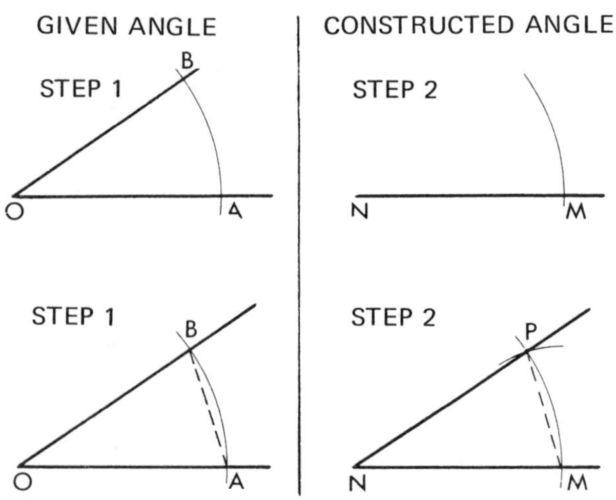

1. Using the vertex O of the *given angle* as center, draw arc AB. With the same radius and point N as a center, draw an arc intersecting the line NM.

2. With a radius equal to chord AB of the *given angle* and point M as a center, draw an arc intersecting arc PM. Draw a line from point N to point P to complete the angle being constructed.

PRINCIPLE: The angles are equal because the two triangles, NMP and OAB, are equal. The two triangles are equal because the three sides of the one were made equal to the three sides of the other.

ANGLE PROBLEMS

Total Degrees Around a Point

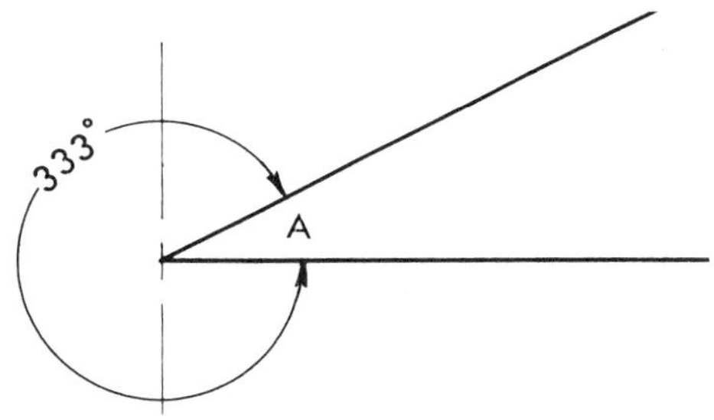

∠ A is 360° - 333°, or 27°. Check this answer by measuring the angle on the drawing with a protractor.

Supplementary Angles

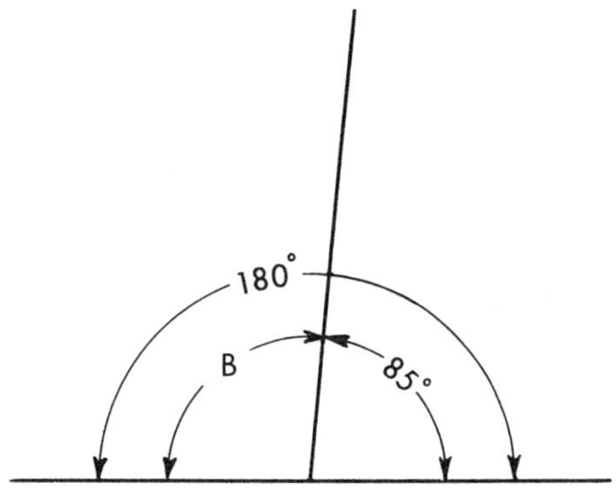

If the sum of two angles is 180°, the angles are supplementary. When an angle is 180°, it is said to be a straight angle.

∠ B is 180° - 85°, or 95°. Check this answer with a protractor.

Complementary angles

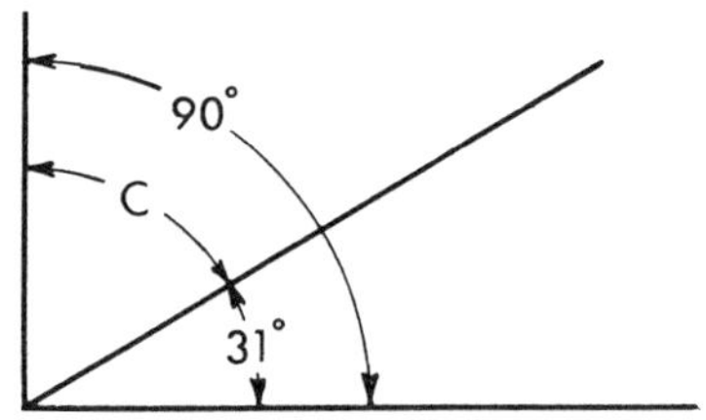

When the sum of two angles is 90°, the angles are complementary.

∠ C is 90° - 31°, or 59°. Check Ans.

Vertical Angles

When two straight lines intersect, the opposite angles formed are called vertical angles. On the drawing below, angles A and C, and angles B and D are vertical angles.

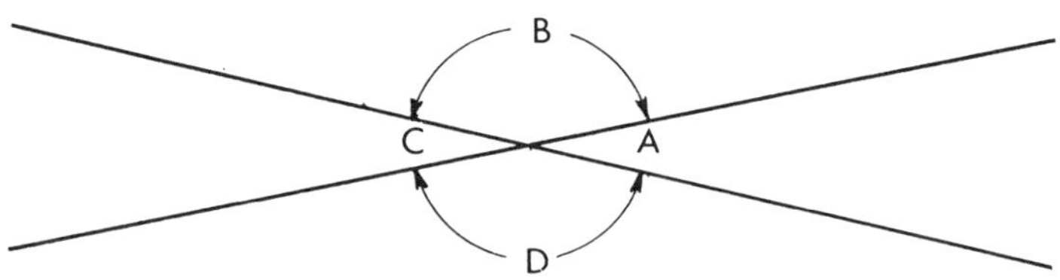

By measurement:

∠ A = 25° ∠ B = 155°
∠ C = 25° ∠ D = 155°
∴ ∠ A = ∠ C; ∠ B = ∠ D

PRINCIPLE: If two straight lines intersect, the vertical angles are equal.

Summary Problem

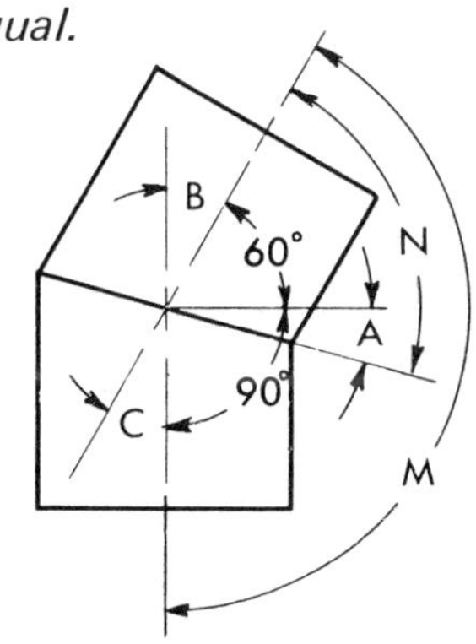

Compute the values of angles A, B, and C of the elbow shown.

1. ∠ M = 60° + 90°, or 150°.
2. ∠ N = ∠ M/2, or 75° (∠ M was bisected when the elbow was laid out.)
3. ∠ A = 75° - 60°, or 15° Ans.
   ∠ B = 90° - 60°, or 30° Ans.
   ∠ C = 30° Ans.

ASSIGNMENT

A. Perpendicular Lines

Check the right angles with a protractor.

1. Erect a perpendicular to a line 4 1/2 in. long from a point at the center of this line.
2. Erect a perpendicular to a 5-in. line from a point 3 in. above this line.
3. Erect a perpendicular 3 1/2 in. long from the right-hand end of a line 4 3/4 in. long.
4. Draw an isosceles triangle, base 3 in., sides 4 in. Draw a perpendicular from the vertex to the base. How is the base divided?

B. Oblique Lines

Check the angles with a protractor.

1. Divide an acute angle of 60 degrees into 4 equal angles.
2. Draw an isosceles triangle, base 4 in., sides 5 in. Bisect the angle at the vertex. How is the base divided?
3. Lay out an acute angle of 42 degrees with a protractor and transfer this angle to another part of the paper.

C. Angle Problems

Compute the values of the unknown angles.

1.

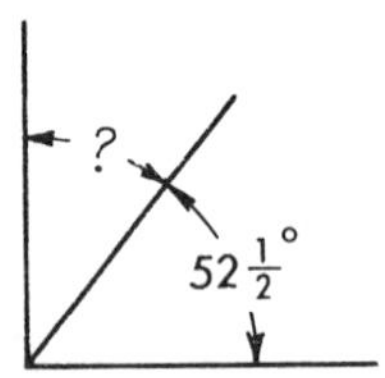

2.

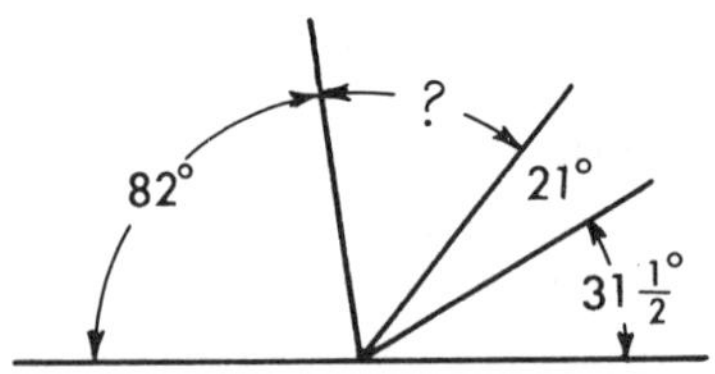

3.

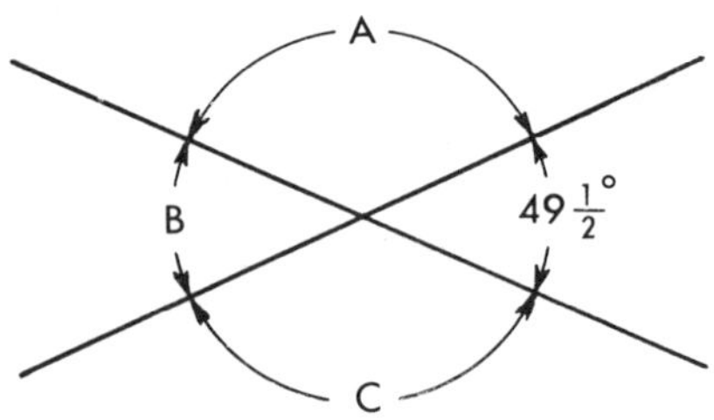

4.

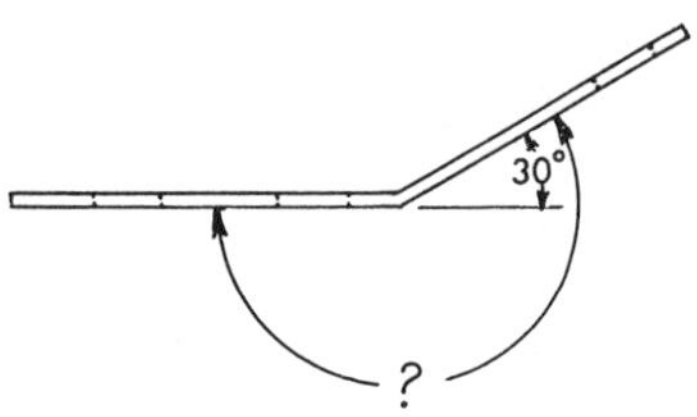

5.

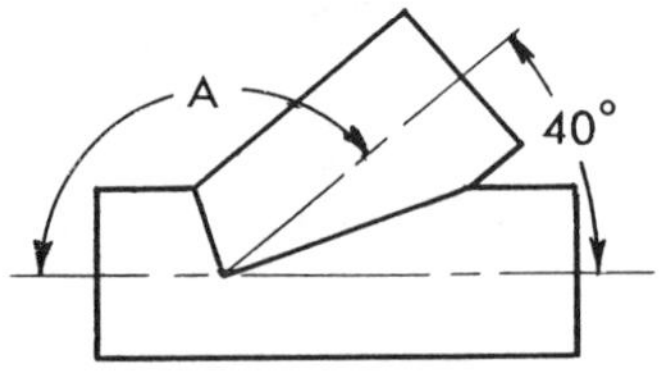

6.

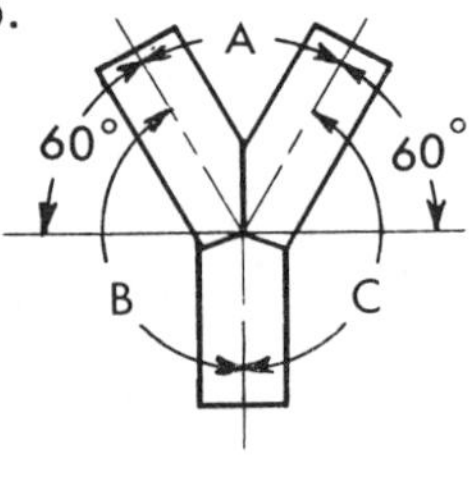

7.

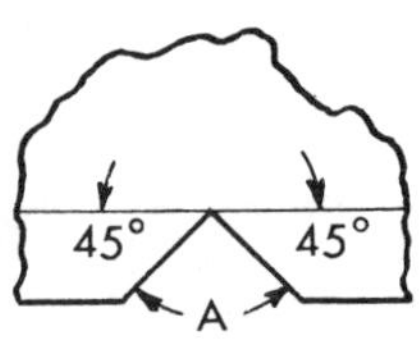

8.

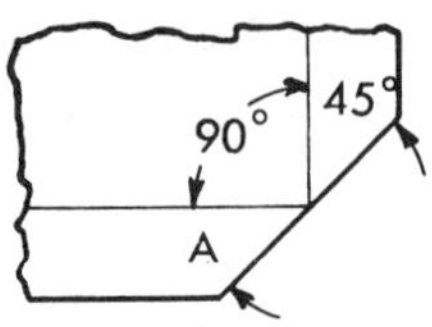

9.

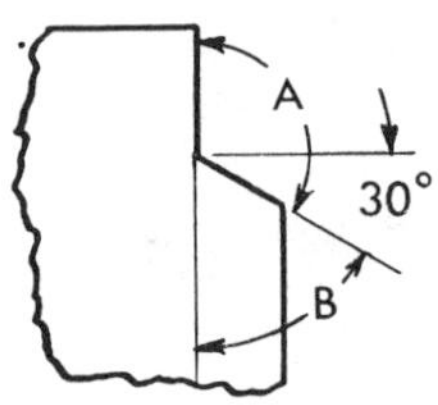

10.

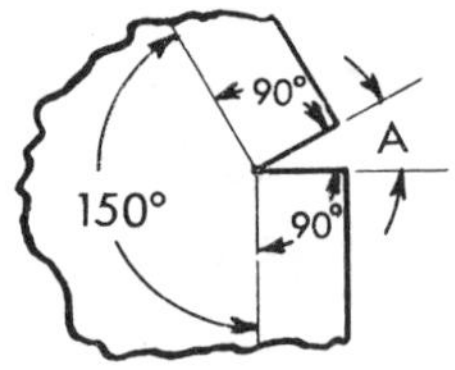

UNIT 36 PARALLEL LINES

Parallel lines are lines which are equidistant (equally distant) at all points. The symbol for the word parallel is ‖ .

Parallel lines appear in many sheet metal jobs and are used in the layout of such jobs as square, rectangular and circular pipe, pans, etc.

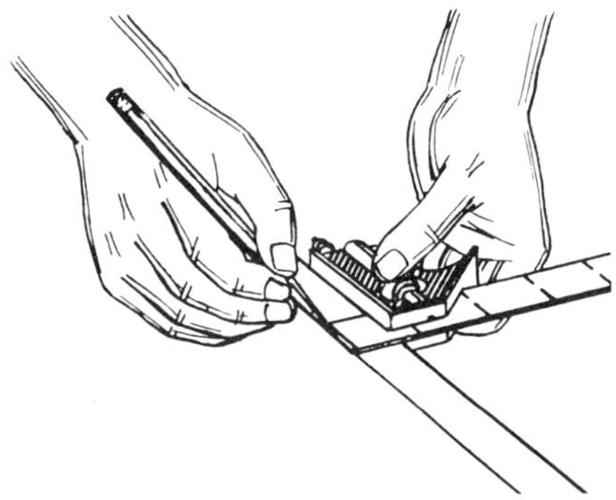

CONSTRUCTION OF PARALLEL LINES

By Perpendiculars

Draw two parallel lines 2 in. apart.

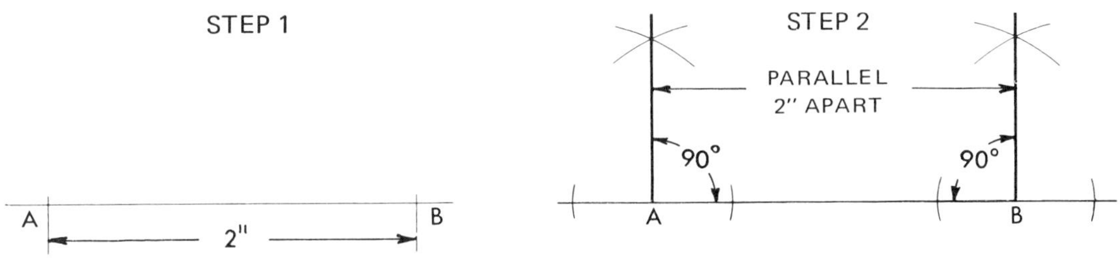

1. Draw a baseline. Lay out two points 2 in. apart on this line.
2. Erect perpendiculars to the baseline at A and at B.

PRINCIPLE: Two lines perpendicular to the same line are parallel. Check by measurement to prove the lines are parallel.

By Arcs of Circles

Draw two parallel lines 1 in. apart.

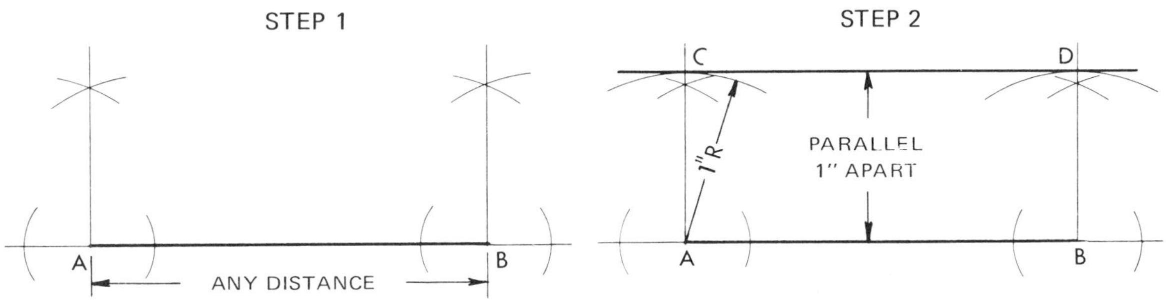

1. Draw a baseline. Erect perpendiculars to this line near the ends.
2. Using points A and B as centers, draw arcs of 1 in. radius. Connect points C and D.

PRINCIPLE: The lines AB and CD are parallel because they are the same distance apart at all points.

PARALLEL LINES INTERSECTED BY ANOTHER STRAIGHT LINE

Angles Formed

When a straight line crosses two parallel lines, the following pairs of angles are formed:

Alternate interior angles:	∠C and ∠E; ∠D and ∠F.
Exterior-interior angles:	∠A and ∠E; ∠B and ∠F; ∠G and ∠C; ∠H and ∠D.

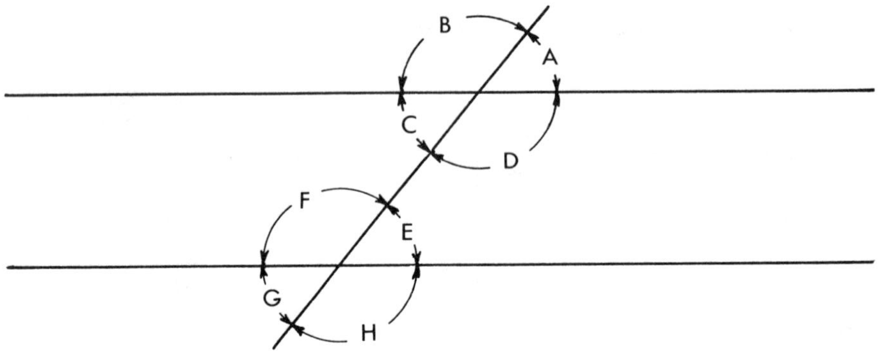

If each angle is measured, it will be found that:

∠A = 52°	∠B = 128°
∠C = 52°	∠D = 128°
∠E = 52°	∠F = 128°
∠G = 52°	∠H = 128°

The pairs of alternate interior angles which are equal are:

∠C and ∠E ∠D and ∠F

The pairs of exterior-interior angles which are equal are:

∠A and ∠E ∠G and ∠C

∠F and ∠B ∠H and ∠D

PRINCIPLE: If two parallel lines are cut by another line, the alternate interior angles formed are equal, and the exterior-interior angles formed are equal.

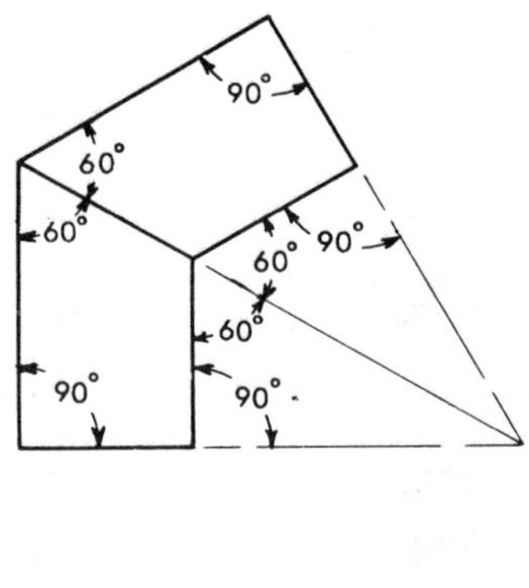

Application to Sheet Metal Work

The drawing of the elbow shows the application of the above principle. The sides of the elbow are parallel and the angles are equal as shown on the drawing. The pairs of 60-deg. angles and the pairs of 90-deg. angles are exterior-interior angles.

Construction of Parallels

By making the exterior-interior angles equal

STEP 1

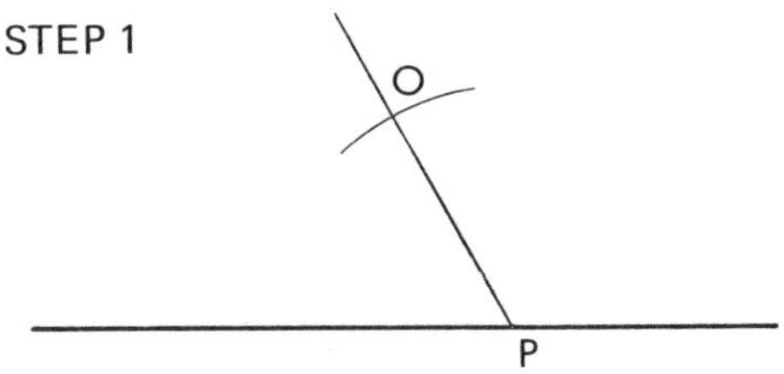

STEP 2

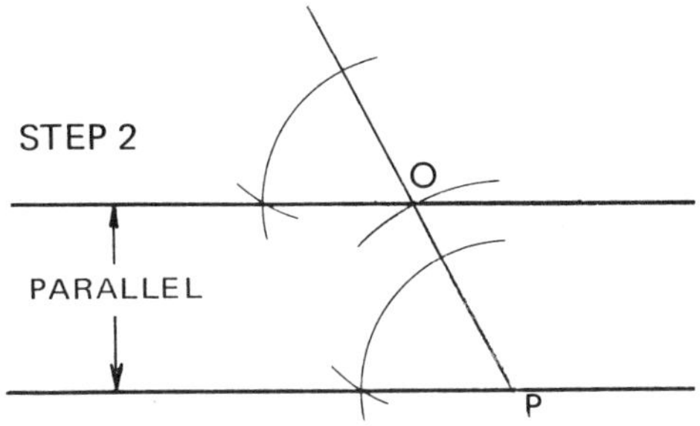

1. Draw line OP through the given point O at any angle to the given line.
2. Transfer acute angle at P to O. The upper line is parallel to the given line. Check by measurement.

Note: When a square or a triangle is used to draw parallel lines, these lines are parallel because the exterior-interior angles are equal.

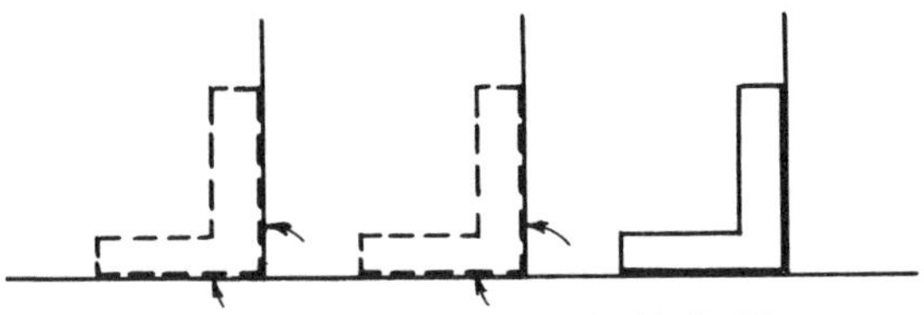

EXTERIOR - INTERIOR ANGLES

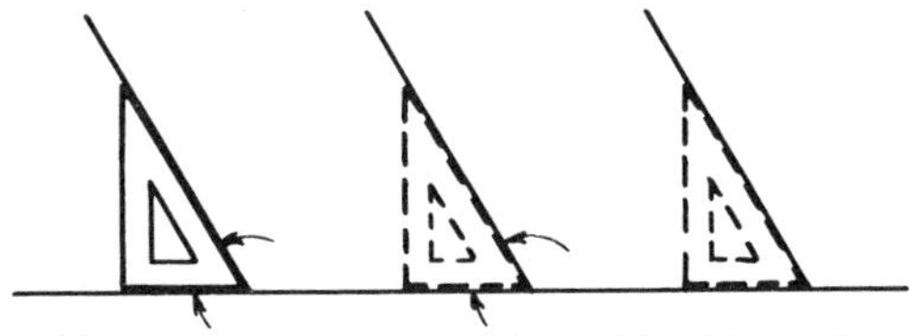

EXTERIOR - INTERIOR ANGLES

DIVIDING LINES INTO EQUAL PARTS

Lines may be divided into equal parts by using one of several construction methods. These methods all depend on the principle that three or more equally spaced parallel lines intercept equal parts on any straight line passing through them.

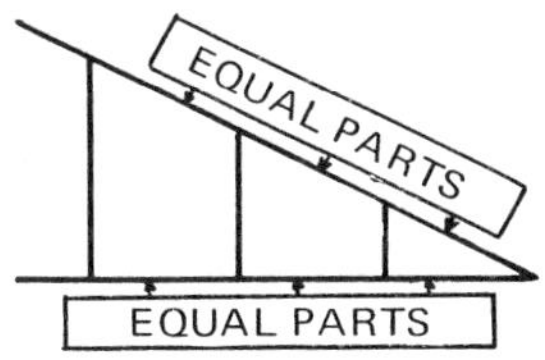

By Using Parallel Lines

Divide a 2 3/4-in. line into 5 equal parts.

STEP 1

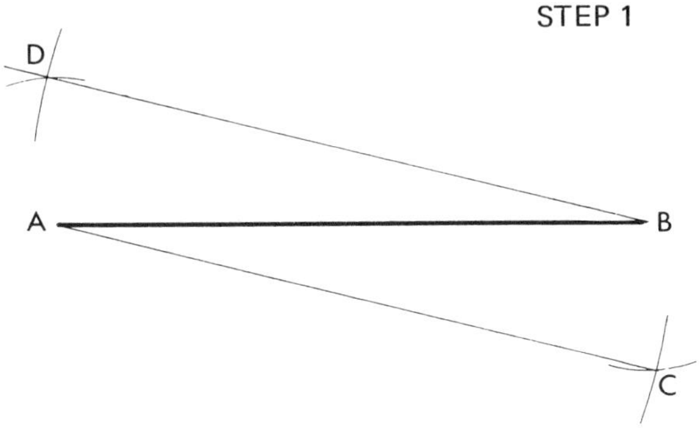

STEP 2

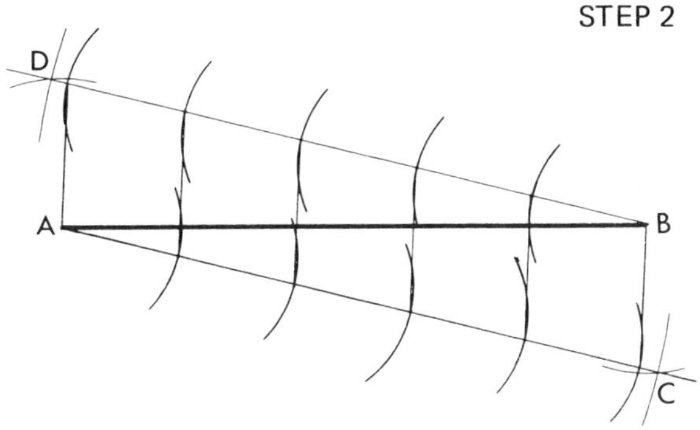

1. Draw a line AB, length 2 3/4 in. Draw two lines, AC and BD, from the ends of line AB parallel to each other by making ∠ CAB = ∠ DBA.
2. Starting from point A and point B, lay out 5 equal parts on lines AC and BD with a compass. Connect these points with lines (parallel) to divide given line AB into 5 equal parts.

By Transferring Angles

Divide a 2 1/2 in. line into 5 equal parts.

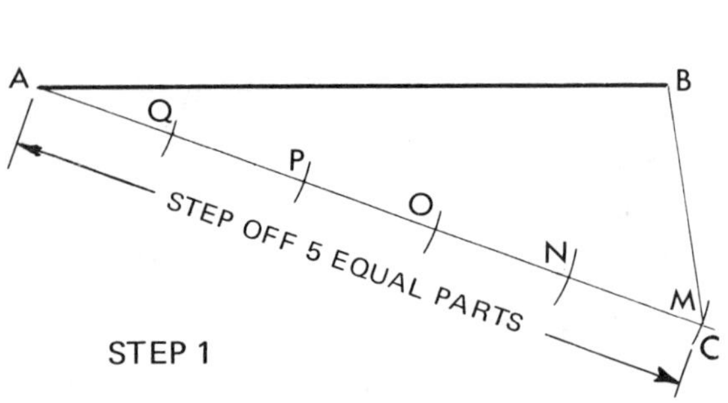

STEP 1

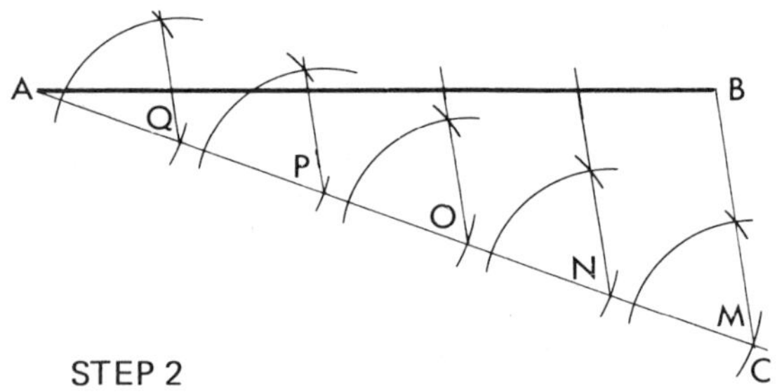

STEP 2

1. Draw line AB, length 2 1/2 in. Draw line AC at an angle to line AB. Using a compass, lay out 5 equal parts on line AC. Draw another line from point M to B.
2. Transfer the acute angle at M to N, O, P, and Q. The intersection of the sides of these angles with the given line divides that line into 5 equal parts.

Note: The sides of the angles are parallel because the exterior-interior angles are equal.

With a Scale

To divide a given line into equal parts, draw a line at right angles to one end of that line. Place a scale at such an angle to the given line so that the number of parts required will divide evenly into the distance covered by the scale.

On the drawing, the given line is 2 7/16 in. and is to be divided into six parts. Place the scale so that 3 in. will cover 2 7/16 in. on the given line. Draw lines from the 1/2-in. divisions (3″/6 = 1/2″) on the scale and perpendicular to the given line. **Note:** The perpendicular lines are parallel to each other because lines perpendicular to the same line are parallel.

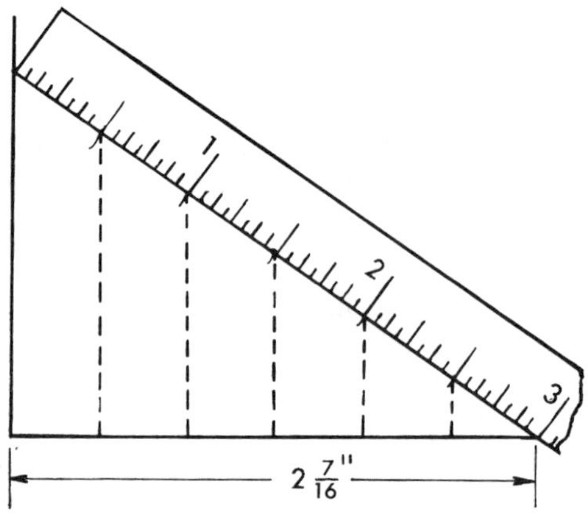

Explanation for the Division of the Lines into Equal Parts

1. Lines EC, FB, GA were constructed || and parts DC, CB, BA were constructed equal. ∴ lines DC, EH, FI are equal.
2. △ GIF = △ FHE = △ ECD because two angles and the included side of each are equal.
3. Then, GF = FE = ED because corresponding parts of equal triangles are equal.

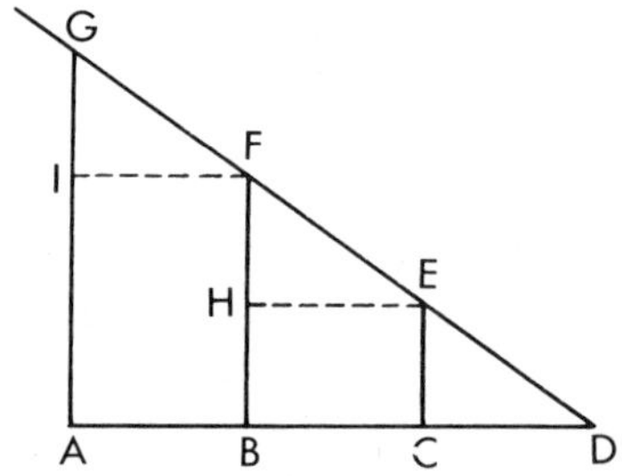

ASSIGNMENT

A. Parallel Lines

1. Draw two parallel lines 1 1/2 in. apart and 1 7/8 in. long (by perpendiculars).
2. Draw two parallel lines 1 3/4 in. apart and 4 in. long (by arcs of circles).
3. Draw two parallel lines 2 1/8 in. apart and 4 3/16 in. long (by measuring the distances apart on the perpendiculars).

B. Parallel Lines Intersected by Another Straight Line

1. Draw two parallel lines 1 1/4 in. apart and 3 in. long; cross these parallel lines with a line drawn at an angle of 35 deg. Measure and compute all the angles formed.
2. Compute the values of the unknown angles in each problem below:

a.

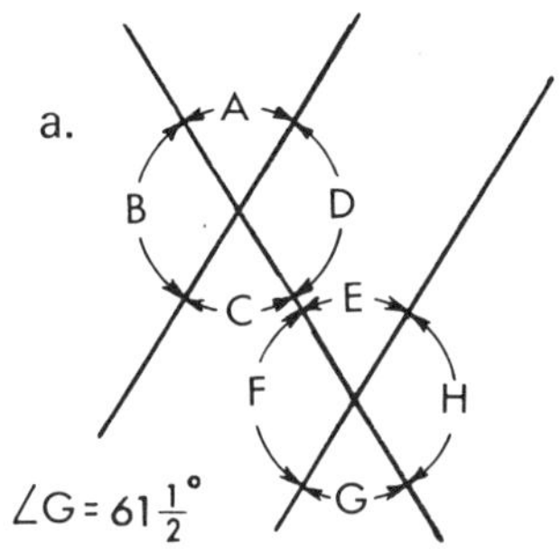

b.

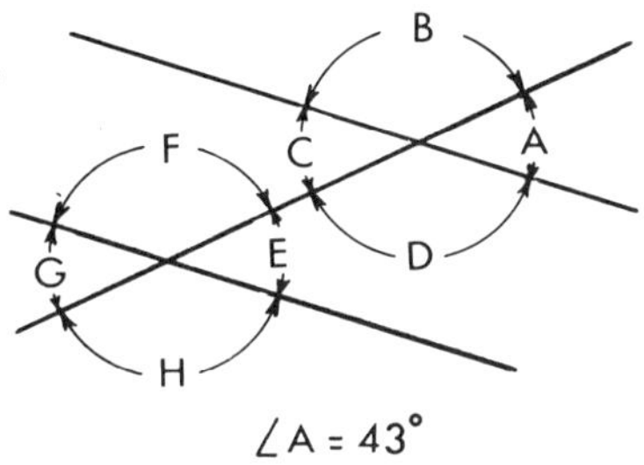

c.

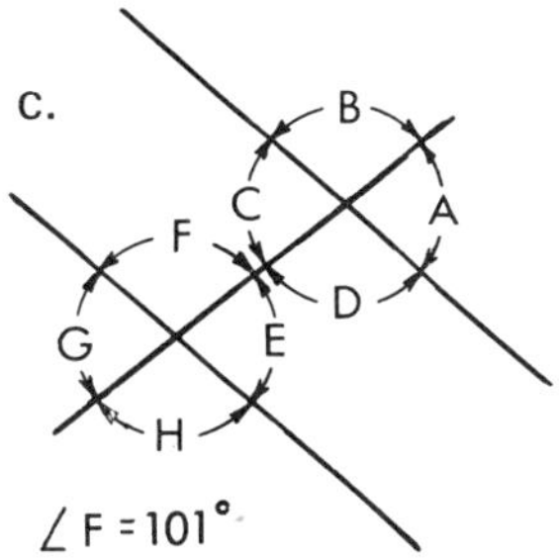

d.

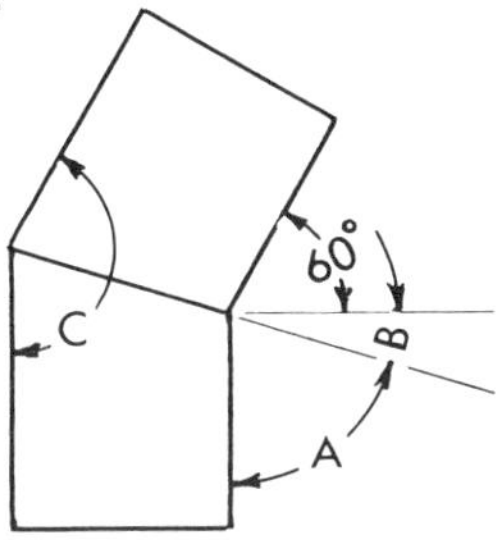

e.

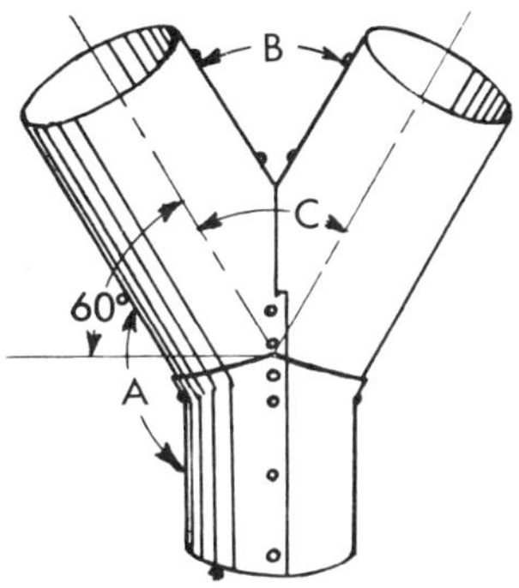

f.

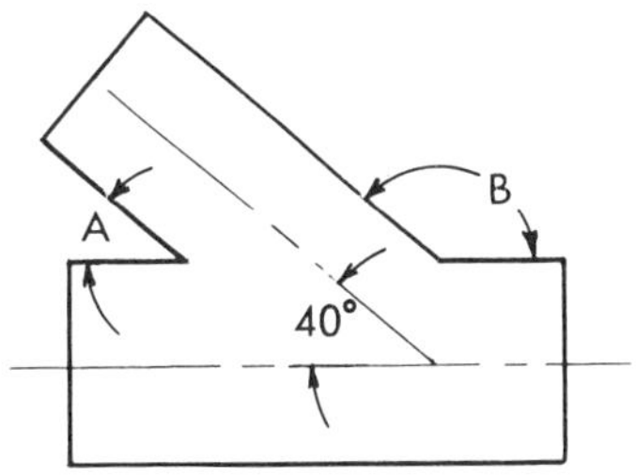

C. Dividing Lines

1. Divide a line 5 in. long into 7 equal parts by using parallel lines. Check by computation and measurement.
2. Divide a line 4 5/8 in. long into 3 equal parts by transferring angles. Check by computation and measurement.
3. Divide a line 7 in. long into 9 equal parts by using a scale. Check by computation and measurement.
4. Divide a line 3 9/16 in. long into 5 equal parts by any of the above methods. Check by computation and measurement.

UNIT 37 SQUARE, RECTANGLE, AND TRAPEZOID

Squares, rectangles, and trapezoids are four-sided plane figures (polygons). Such figures appear in many sheet metal jobs; sheets of black iron are of rectangular shape.

Squares and rectangles belong to a type of four-sided plane figures called *parallelograms.* Parallelograms are plane figures whose opposite sides are parallel. If the adjacent sides are at an acute or obtuse angle to each other, the figure is called an oblique parallelogram.

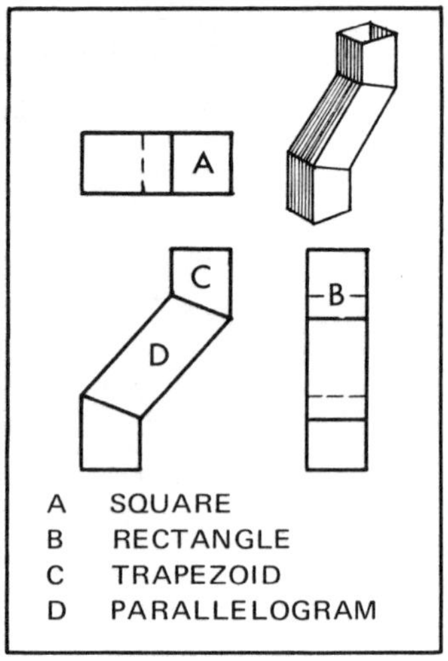

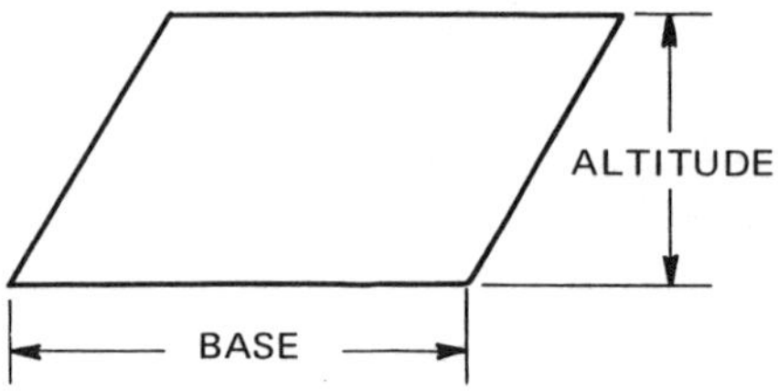

Any side of a parallelogram may be considered the base. The altitude is the perpendicular distance between the base and opposite side.

SQUARE

A square is a four-sided plane figure having the sides equal, and each angle equal to 90°. The opposite sides of a square are parallel. The symbol for a square is a small square □ . The letter s is used to designate the length of one side.

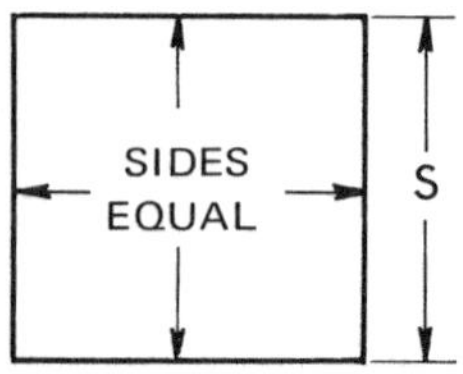

A square is constructed as follows:

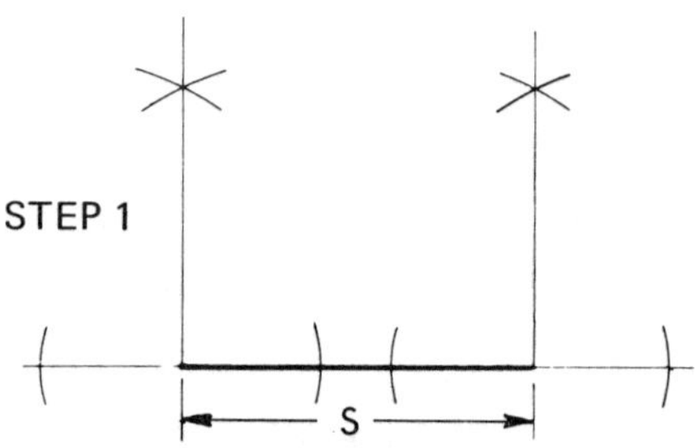

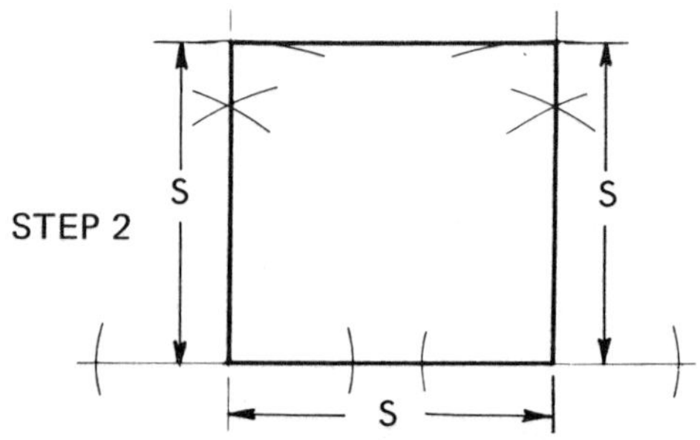

1. Lay out the length of one side on a baseline. Erect perpendiculars to this line at the ends of this side.
2. Lay out the length of the side on each perpendicular, and draw the upper side.

Note: The diagonals of a square are the lines connecting the opposite corners. They should be equal.

RECTANGLE

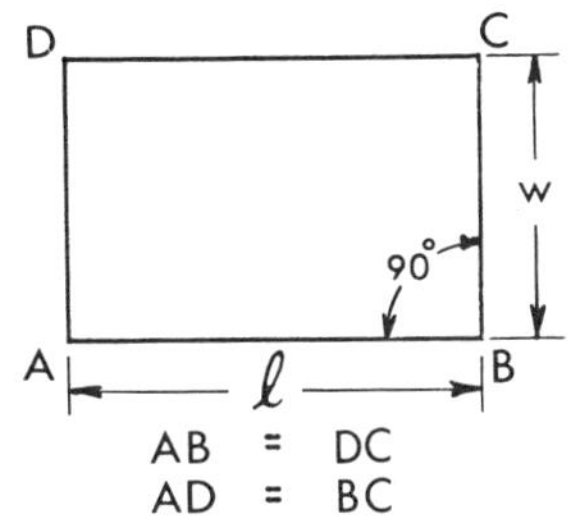

A rectangle is a four-sided plane figure having the opposite sides equal and each angle equal to 90 degrees. The length of the longer side of the rectangle is designated by the letter ℓ, the length of the shorter side by w. The symbol for a rectangle is a small rectangle ▭ . The opposite sides of a rectangle are parallel.

A rectangle is constructed in a manner similar to a square, using the length and width given.

Note: The diagonals of a rectangle are equal in length.

TRAPEZOID

Shapes

A trapezoid is a four-sided plane figure with only two sides parallel. The two parallel sides are called bases. Trapezoids have many shapes, as shown below. In the illustration, B = length of large base; b = length of small base; a = alt.

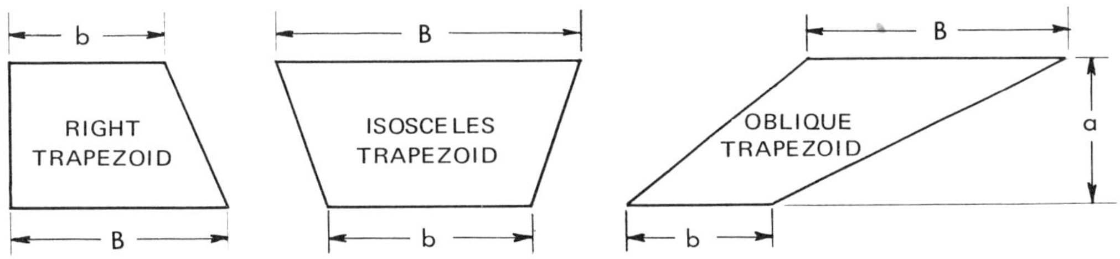

Construction of a Trapezoid

Draw an isosceles trapezoid with bases of 3 in. and 2 in., and altitude of 1 1/2 in.

STEP 1

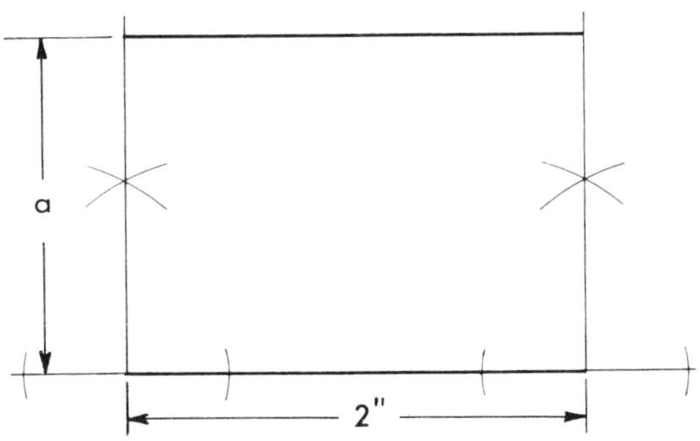

STEP 2

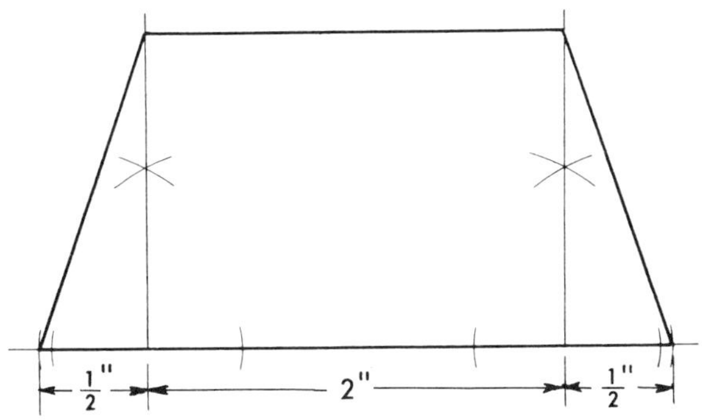

1. Draw a line. Erect perpendiculars to this line 2 in. apart. Lay out the altitude of 1 1/2 in. on each perpendicular. Draw the small base.
2. Lay out 1/2 in. on the lower line from the end of each perpendicular. (1/2 the difference of the bases: 3 in. - 2 in. = 1 in.; 1/2 of 1 in. = 1/2 in.) Draw the inclined sides.

   Note: The large and small bases are parallel because they are equidistant.

ASSIGNMENT

A. Square

Construct squares of the dimensions given, and check constructions by measurement.

1. s = 2″ 2. s = 3″ 3. s = 2 3/8″

B. Rectangle

Construct rectangles of the dimensions given, and check construction by measurement.

1. ℓ = 3″, w = 2″ 2. ℓ = 4 3/16″, w = 2 1/4″

C. Trapezoid

Construct trapezoids of the dimensions and the shapes shown, and check the construction by measurement.

1. B = 3″ b = 2″, a = 2″

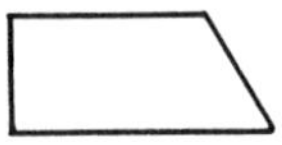

2. B = 4″, b = 3″, a = 2 1/2″

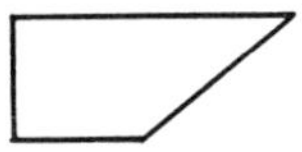

3. B = 5″, b = 4″, a = 3 3/4″

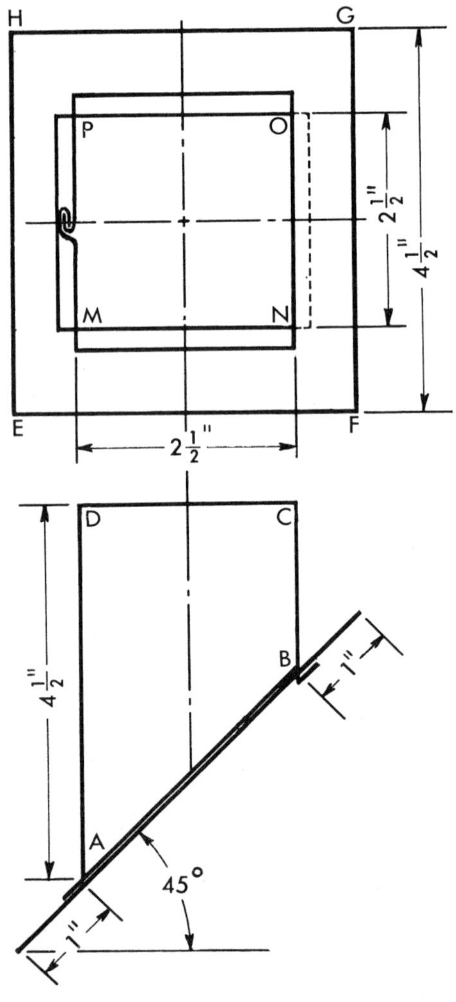

D. Identification Problem

On the drawing at the right, name the figures indicated by the following letters:

1. ABCD
2. EFGH
3. MNOP

Note: The illustration below shows a pictorial view of the same job.

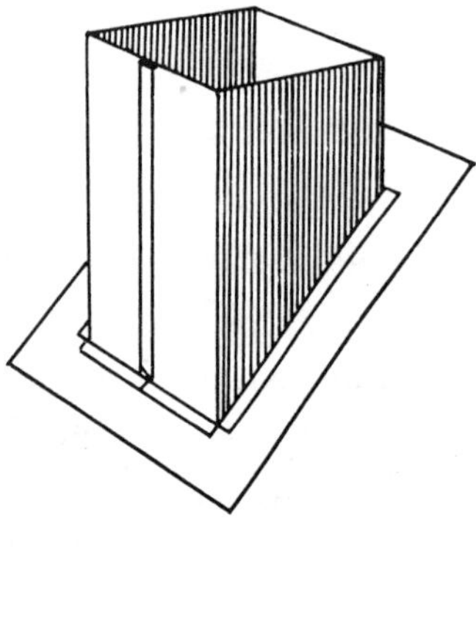

UNIT 38 ARCS OF CIRCLES AND TANGENTS TO CIRCLES

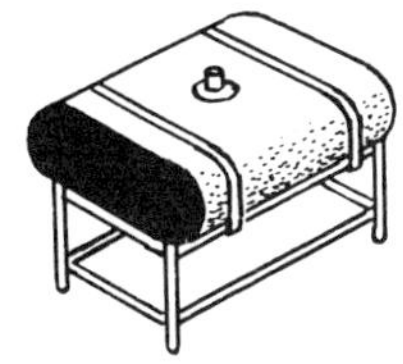

The ability to divide circles and arcs of circles, and to draw tangents to circles, is necessary in the layout of the various jobs containing circles.

ARCS OF CIRCLES

Bisecting an Arc

This is done in a manner similar to bisecting an angle.

PRINCIPLE: A central angle is an angle with its vertex at the center of the circle. If the central angle is bisected, the arc is bisected also.

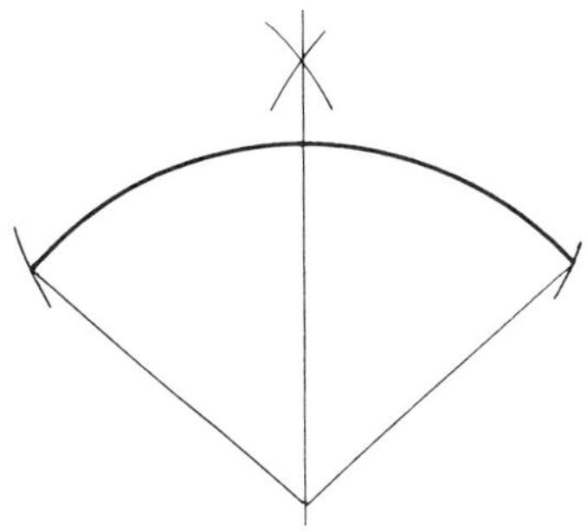

Dividing a Half Circle into Eight Equal Parts

1. Draw two lines at right angles to each other and passing through the center O of the half circle.
2. Divide both 90-degree angles into four parts by bisection.

Note: If lines are extended through the center O, each angle formed will be 90°/4, or 22 1/2°.

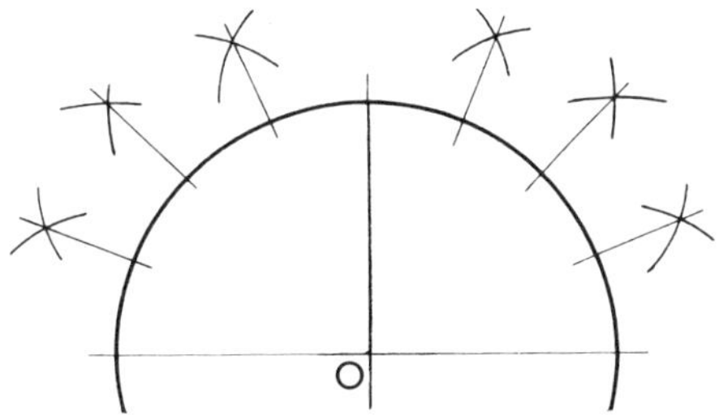

Dividing a Half Circle into Six Equal Parts

STEP 1

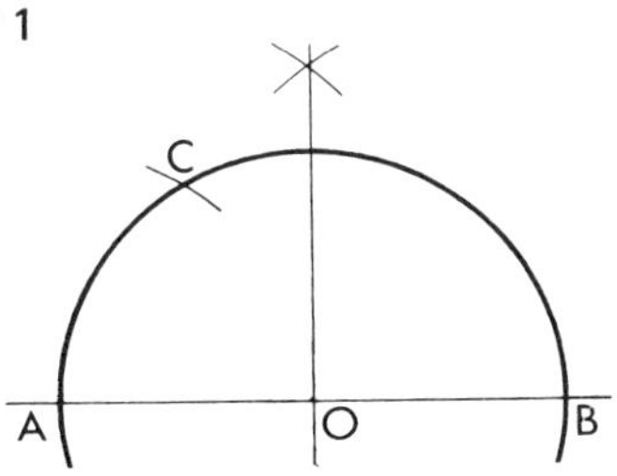

STEP 2

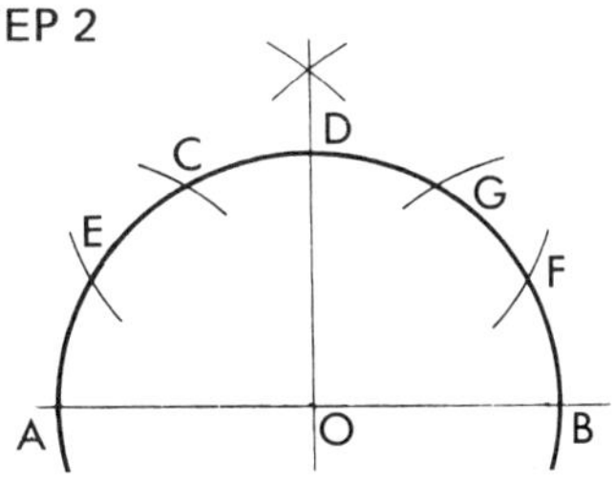

1. Draw two lines at right angles to each other through the center O. Using point A as a center and a radius equal to AO, draw an arc intersecting the half circle at C.
2. Using point D as a center and the same radius, draw arcs intersecting the half circle at E and F. Locate point G, using B as a center and the same radius.

Note: If the lines are extended through the center, each angle formed will be 90°/3, or 30°.

Explanation of Principle

△ EOG and △ COD, are equilateral triangles.

∠ EOG = ∠ COD, or 60°

∠ EOD = 90° - 60°, or 30°.

∴ ∠ COG = ∠ GOD = ∠ EOD, or 30°. (The 90-deg. angle is divided into thirds. Check with a protractor.)

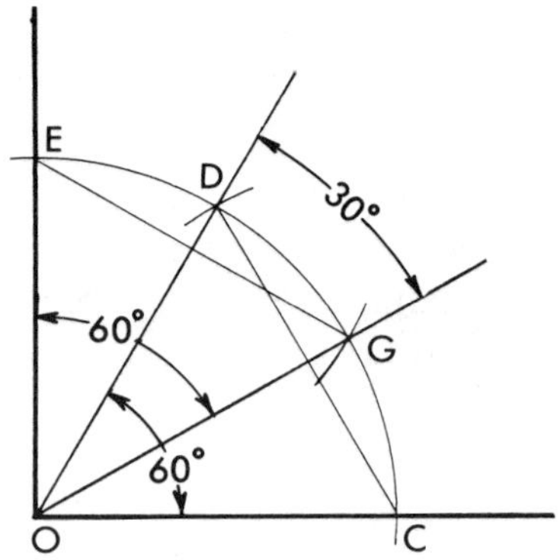

TANGENTS TO CIRCLES

A *tangent* to a circle is a line which touches the circle at one point only and is perpendicular to the radius or the diameter. Tangents to circles are drawn as follows:

Single Circle

1. Draw a line from the center O of the circle through the given point P on the circle.
2. Draw a perpendicular to the line OP at P. P is the point of tangency.

STEP 1

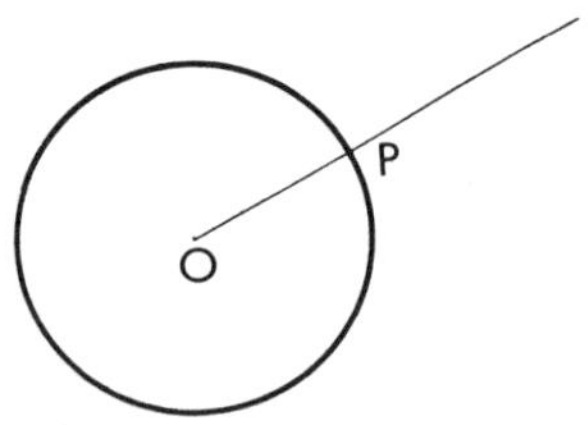

STEP 2

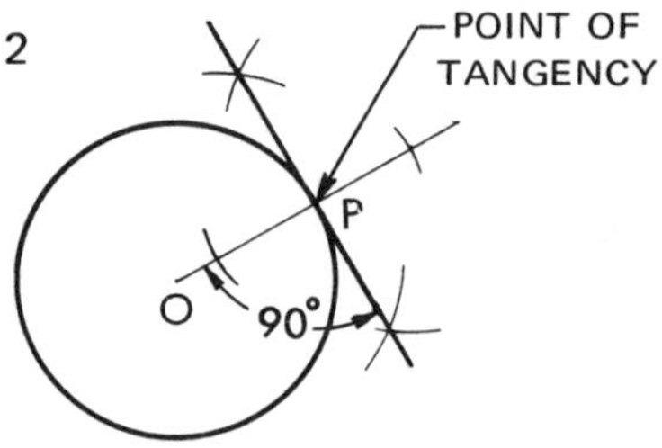

Two Circles

1. Draw perpendiculars to the centerline of both circles.
2. Connect the points of tangency as shown on the drawing.

PRINCIPLE: This forms a semicircular-sided figure which is composed of two semicircles and a rectangle or a square. The total length is two times the radius plus the distance between the centers of the semicircles. The width is equal to twice the radius, or the diameter of the circle.

STEP 1

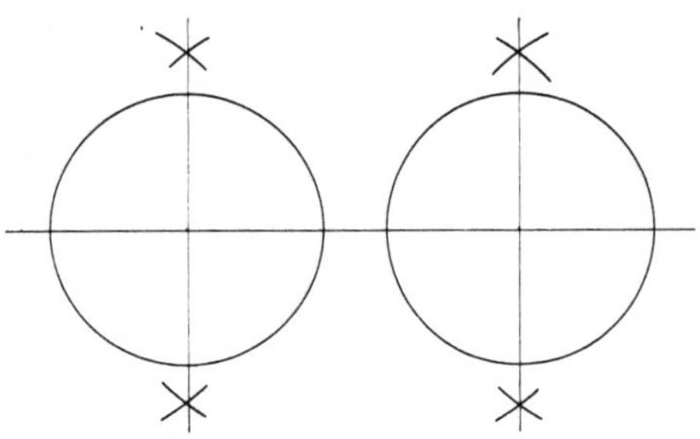

STEP 2

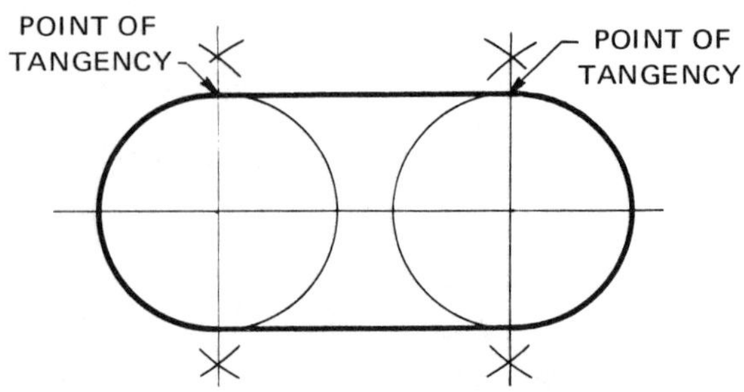

ASSIGNMENT

A. Arcs of Circles

1. Draw an arc, radius 2 in., central angle 30 deg. Bisect the 30-deg. angle. Compute its value and check by measurement.
2. Draw a semicircle, diameter 2 in., and divide it into 8 equal parts. Check the angles formed by computation and measurement.
3. Draw a semicircle, diameter 2 in., and divide it into 6 equal parts. Check the angles formed by computation and measurement.
4. Draw a circle, diameter 4 in., and divide it into 32 equal parts. Check the angles formed by computation and measurement.

B. Tangents to Circles

Solve the following problems:

1. Draw a semicircular-sided figure, diameters of semicircles 1.5 in., distance between centers of semicircles 2 in. Compute and check the overall length and width of the figure.
2. Draw a semicircular-sided figure, overall length 4 in. and overall width 2.5 in.

3.

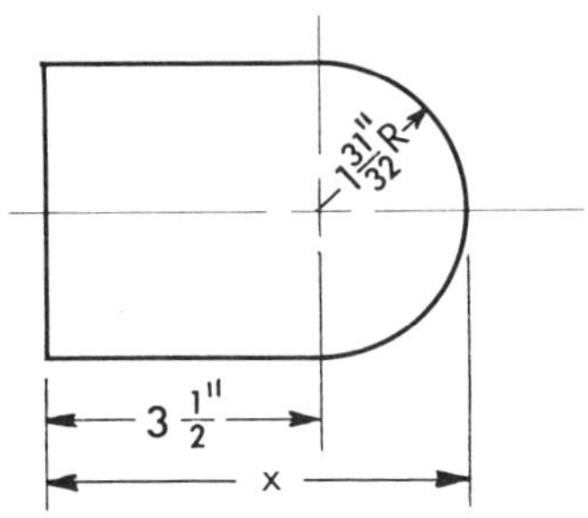

4.

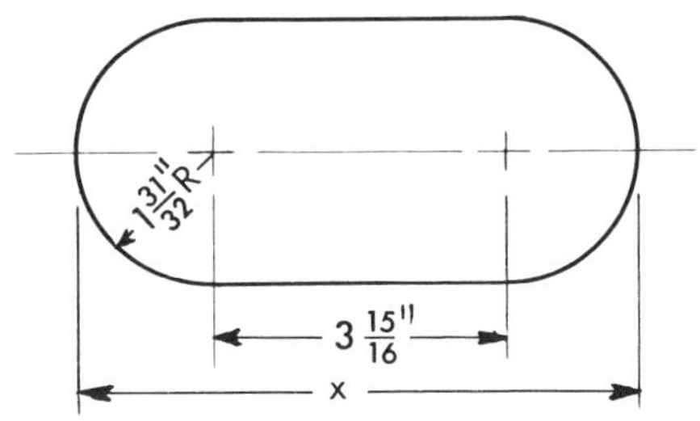

5.

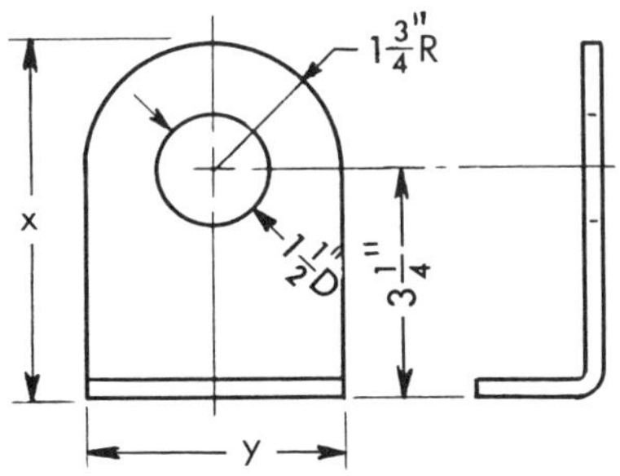

6.

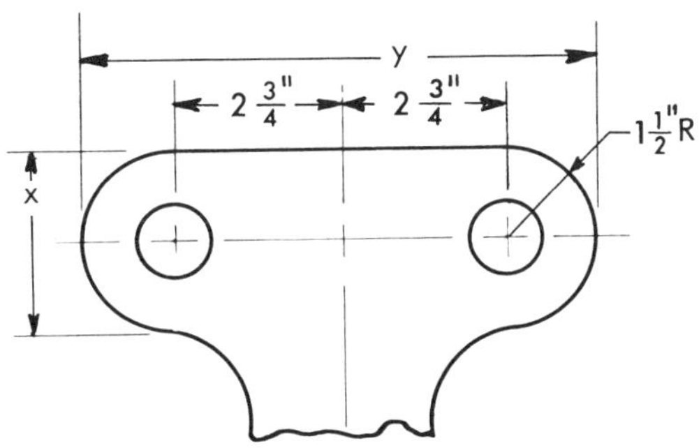

7.

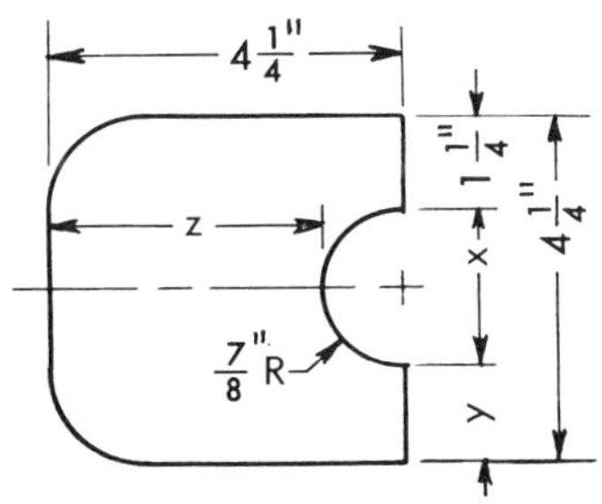

8.

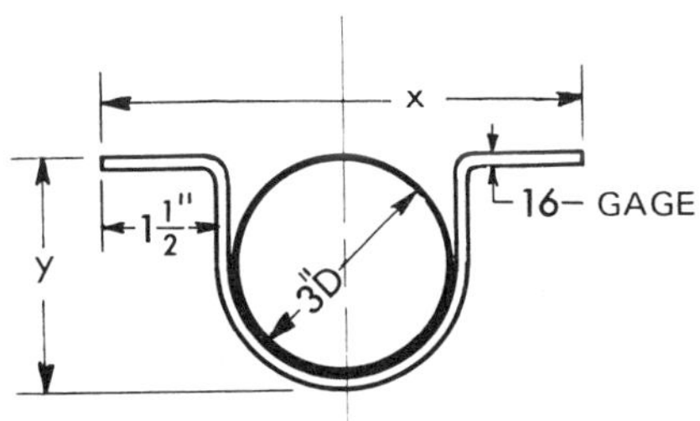

SECTION 5 EQUATIONS

UNIT 39 INTRODUCTION TO SYMBOLS

A knowledge of arithmetic alone is not enough to solve all the problems of the sheet metal trade. The use of letters such as a, b, ℓ, and w to stand for numbers, and symbols such as +, –, x, and / or — to stand for words, enables the mechanic to make his computations more easily and more quickly. It should be noted that a capital letter and a small letter may designate different quantities; for example, the letter *d* may be used for the diameter of a job and the letter *D* for the diameter of a rivet or vice versa.

TRANSLATING WORDS INTO MATHEMATICAL SYMBOLS

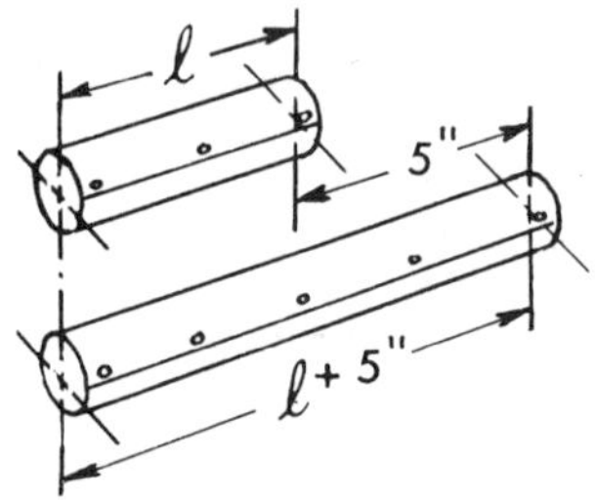

If the length of a pipe is ℓ in. and another pipe is 5 in. longer than the first pipe, the length in inches of the second pipe expressed in terms of the length ℓ of the first pipe is $\ell + 5$.

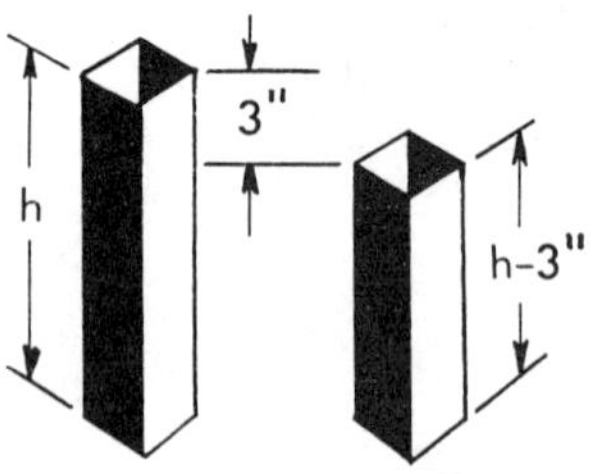

If the height of a square pipe is *h* in. and another square pipe is 3 in. shorter than the first pipe, then the height in inches of the second pipe is h – 3.

One man does a job in 10 hours; another man does the same job in *t* hours less. How long does it take the second man to do the job? The number of hours the second man takes to do the job is 10 – t. Other examples follow:

1. Sum of 2 and a is 2 + a.
2. Difference of 2 and a is 2 –a.
3. Product of a and b is ab.
4. Quotient of x and y is $\frac{x}{y}$.

ADDITION AND SUBTRACTION OF EXPRESSIONS CONTAINING LETTERS

1. If a line is divided into four parts, the length of each part being *a*, the total length of the line is the sum of four a's: a + a + a + a = 4a. (4a means 4 times a, or 4 x a.) **Note:** The factors of 4a are 4 and a. (Factors are numbers or letters which, when multiplied together form a product.) If *a* equals 1 ¼″, the total length of the line is 1 ¼″ + 1 ¼″ + 1 ¼″ + 1 ¼″, or 5″. Check this answer by measuring the drawing.

2. If a job is divided into 3a + 2a parts, the total length is 5a.

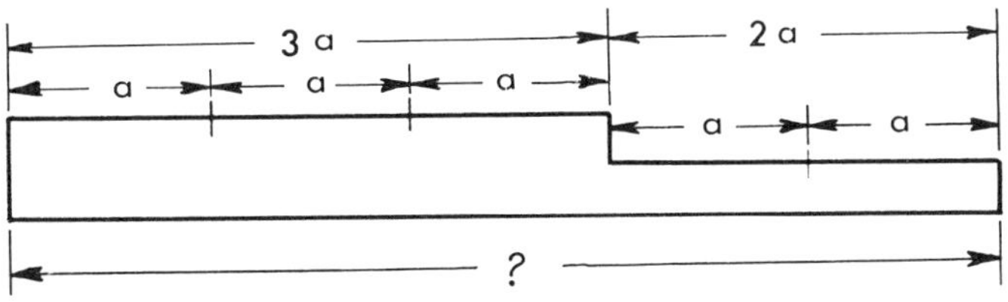

3. Parts may be subtracted in a similar manner: 5a - 3a = 2a.

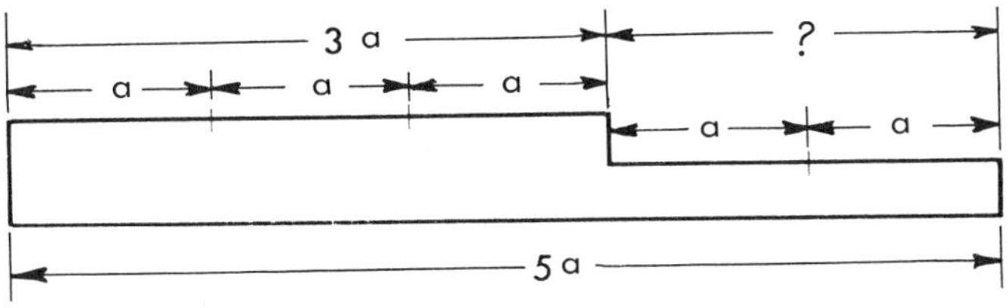

4. If a job is divided into unequal parts, three *a* parts and two *b* parts, then the total length is 3a + 2b.

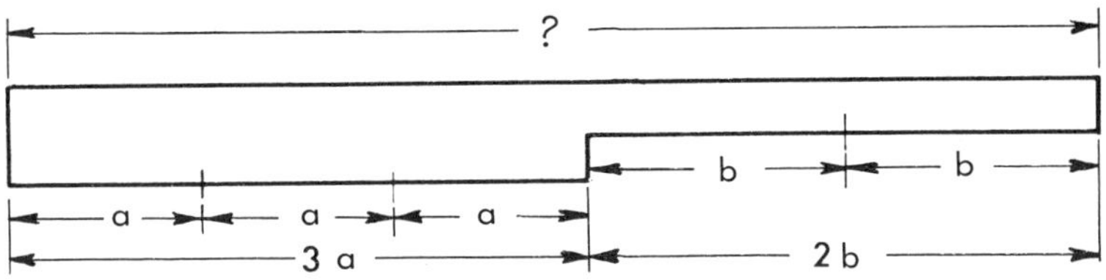

Note: Since *a* and *b* are different lengths, they cannot be added directly. Numerical values must be assigned to 3a + 2b.

If $a = \frac{3''}{4}$ and $b = 1''$, $\therefore 3a + 2b = 3 \times \frac{3}{4} + 2 \times 1$, or $4\frac{1''}{4}$ Ans.

5. If two *b* parts of a job are to be subtracted from three *a* parts of a job, the difference is found in a similar manner. This difference is 3a - 2b.

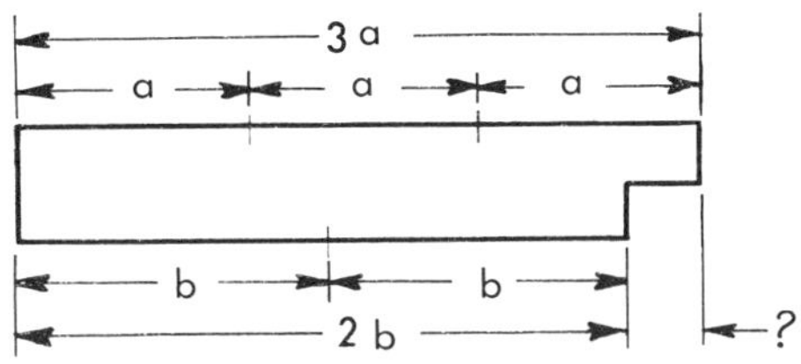

If $a = \frac{3''}{4}$ and $b = 1''$, $\therefore 3a - 2b = 3 \times \frac{3}{4} - 2 \times 1$, or $\frac{1''}{4}$ Ans.

EVALUATION OF EXPRESSIONS CONTAINING LETTERS

1. If $\ell = 4$ and $w = 3$, $\therefore 2\ell + 2w = 2 \times 4 + 2 \times 3$, or 14 Ans.
2. If $a = 2$ and $b = 3$, $\therefore 4ab = 4 \times 2 \times 3$, or 24 Ans.
3. If $a = 3$ and $b = 2$, $\therefore \frac{2a}{5b} = \frac{2 \times 3}{5 \times 2}$, or $\frac{6}{10}$, or $\frac{3}{5}$ Ans.

SUMMARY OF PRINCIPLES

Like quantities such as a, 2a, 3a, etc. can be added or subtracted; unlike quantities such as a, b, etc. must be given numerical values before they can be added, subtracted, multiplied or divided.

ASSIGNMENT

A. Translating Words into Mathematical Symbols

Write the following in symbol form.

1. The sum of 5 and a
2. The difference between 5 and a
3. The product of 5 and a
4. The result of 5 divided by a
5. The weight of a coil of wire is W lb. What is the weight of 1/5 of a coil in terms of W?
6. The length of a piece of band iron is ℓ. What is the length of another piece of iron which is 3 times as long in terms of ℓ ?
7. A line is L inches long. What would be the length of each part, if it is divided into 12 parts in terms of L?
8. Three holes are equally spaced and are t distance apart. What is the distance between the outer holes in terms of t?

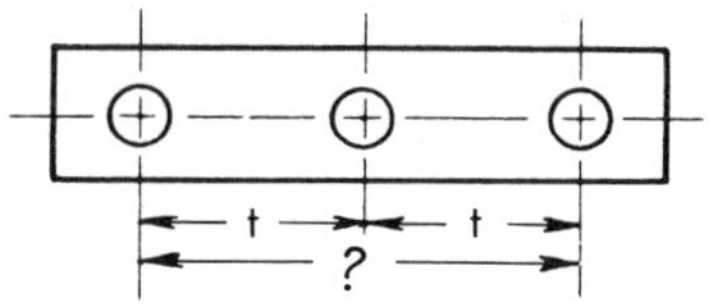

9. Find the spacing of the holes on the drawing in terms of T.

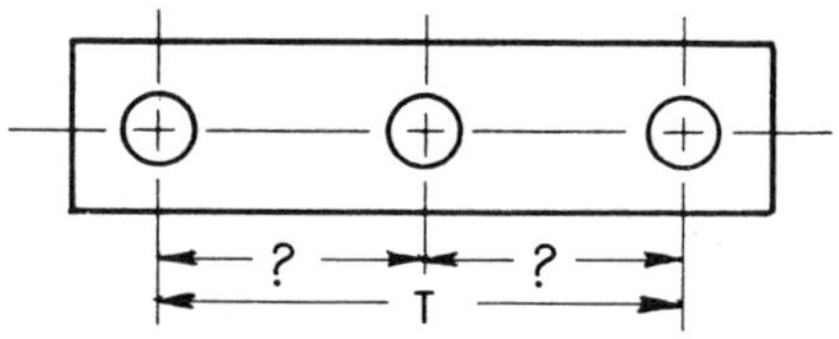

10. If five holes are equally spaced and are each t distance apart, what is the distance apart of the outer holes in terms of t?
11. The width of a rectangle is w inches; the length is 10 inches greater than the width. What is the length of the rectangle in terms of the width?

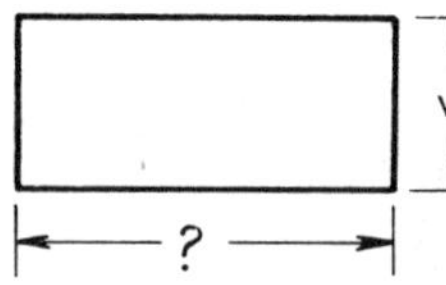

12. If the length of a rectangle is ℓ in. and the width is 5 in. less than the length, what is the width in terms of the length?
13. The width of a rectangle is w and the length is 4 times as great. What is the length in terms of the width?
14. If the length of a rectangle is ℓ and the width w, what is the product of length and width in terms of ℓ and w?
15. The diameter of a circle is d. Find the radius in terms of d.
16. The radius of a circle is r. Find the diameter in terms of r.
17. If the diameter of a gear is D and the diameter of the gear meshing with it is d, express the distance x on the drawing in terms of D and d. **Note:** Put an equal sign between x and the expression you write.

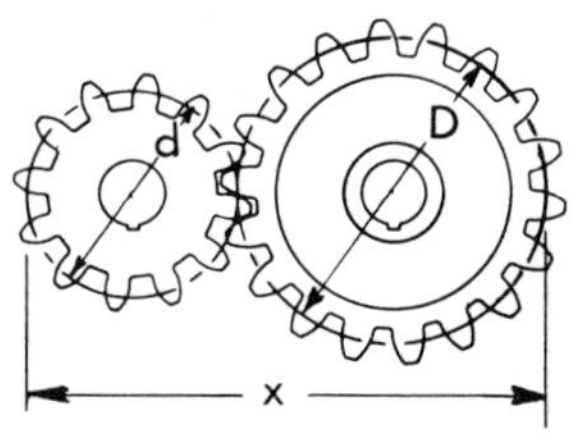

18. It takes t hr. to make a 90° tee and 2 hr. longer to make a 60° tee. How long will it take to make the 60° tee?

B. Addition and Subtraction of Expressions Containing Letters

Collect the like terms in the following expressions.

1. r + r = ?
2. 2r + 2r = ?
3. s + s + s + s = ?
4. $\ell + \ell$ + w + w = ?
5. 4w - 2w = ?
6. 2d - d = ?
7. Express X in terms of D.

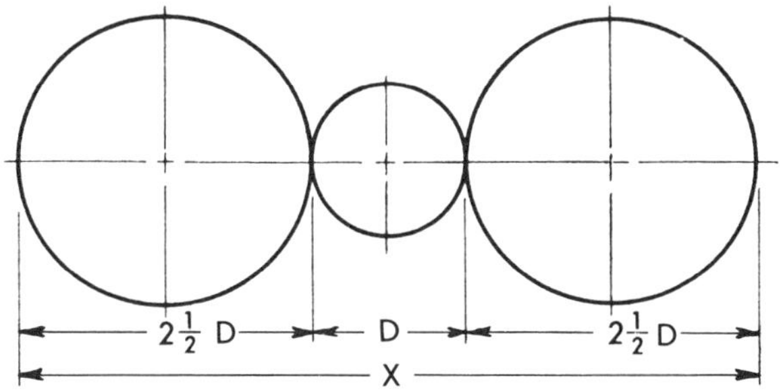

8. Express L in terms of S.

9. Express the length in terms of M.

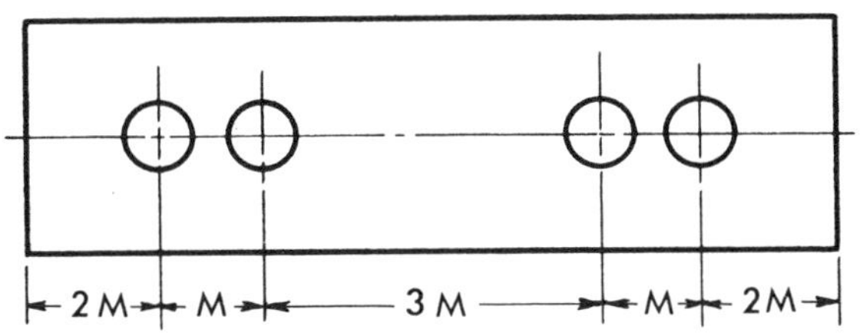

10. Express the length in terms of X.

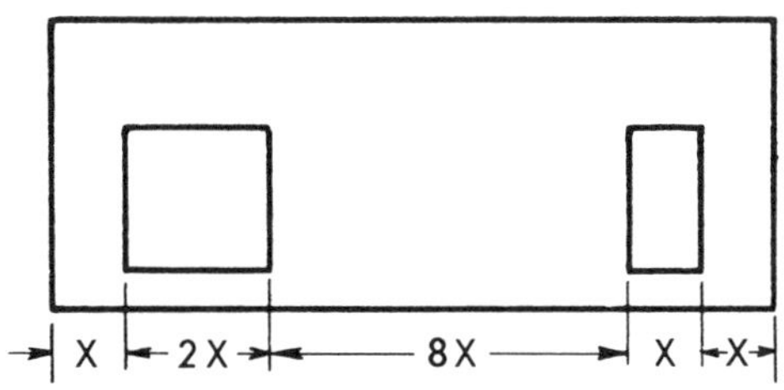

C. Evaluation of Expressions Containing Letters

Using each value given, find values for each of the following expressions:

1. 4s — when s = 2″; 4 1/2″; 5.29″; 5′ 10″; 3 5/16″
2. $\frac{d}{2}$ — when d = 4″; 5″; 8 3/4″; 4.78″; 3′ 10″
3. $2\ell + 2w$ — when ℓ = 3″, w = 2″; ℓ = 5 1/2″, w = 4 3/4″
4. n - 2 — when n = 3; 4; 5; 8; 20

UNIT 40 GROUPING SYMBOLS

Special symbols are used when it is necessary to treat a group of letters or numbers as a whole. Such grouping is indicated by the various symbols below:

1. ( ) parentheses 2. [ ] brackets 3. braces 4. bar

The parenthesis is the most commonly used grouping symbol. The bar is used over numbers; for example, $\overline{2.15}$.

MEANING AND USE OF GROUPING SYMBOLS

1. The average length of the two lines, a and b, may be expressed as $\frac{a + b}{2}$. This really means $\frac{(a + b)}{2}$, with the parenthesis understood. The numerator is to be treated as a whole and the two terms added before being divided by 2.

   If $a = 2''$ and $b = 3''$, $\therefore \frac{a + b}{2} = \frac{2 + 3}{2}$, or $\frac{5''}{2}$, or $2\frac{1}{2}''$

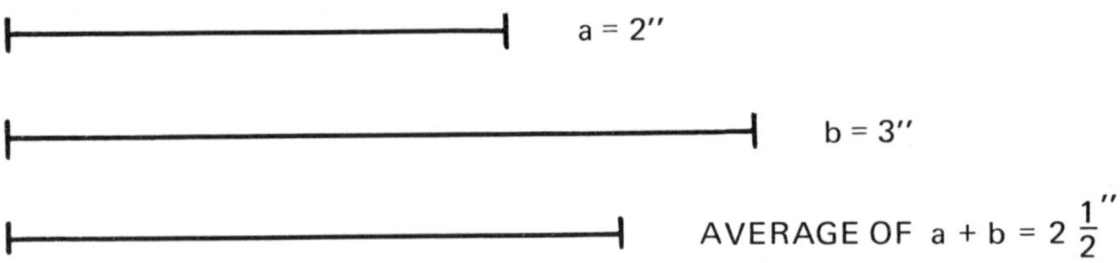

2. In the expression $\frac{3(a + b)}{2}$, the parenthesis is always placed around a and b because a and b must be added before being multiplied by 3, or before being divided by 2.

   If $a = 2''$ and $b = 3''$, $\therefore \frac{3(a + b)}{2} = \frac{3(2 + 3)}{2}$, or $\frac{3 \times 5}{2}$, or $7\frac{1}{2}''$

REMOVAL OF COMMON FACTORS FROM EXPRESSIONS

1. The expression $2a + 2b$ can be written as $2(a + b)$. The number 2 is the common factor: the factor that appears in each term.

   If each term of the expression $(a + b)$ is multiplied by 2, the original expression $2a + 2b$ is obtained.

$$\begin{array}{r} a + b \\ \times\, 2 \\ \hline 2a + 2b \end{array}$$

   If $a = 2$ and $b = 3$, the numerical value of each expression will be the same.

$$2(a + b) = 2 \times (2 + 3), \text{ or } 10$$
$$2a + 2b = 2 \times 2 + 2 \times 3, \text{ or } 10$$

2. Likewise, any common factor may be removed from an expression; for example:

$$3a + 3b = 3(a + b) \qquad ab + ac = a(b + c)$$

ASSIGNMENT

A. Meaning and Use of Grouping Symbols

Use symbols to express each of the following:

1. The average length of lines B and b
2. Four times a minus b
3. The sum of a and b divided by three
4. The difference of a and b divided by three. (The quantity a is the larger.)
5. a times the sum of B and b, all divided by two
6. a times B minus b, all divided by two
7. Two times the sum of a and three times b
8. The average length of lines a, b, and c
9. Twice the sum of two radii, R and r (Assume that this is the distance between the centers of the pulleys in the drawing.)

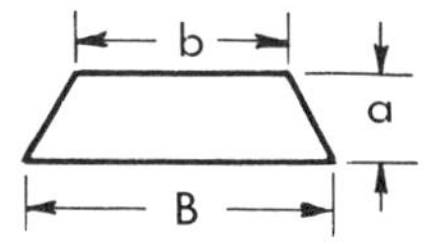

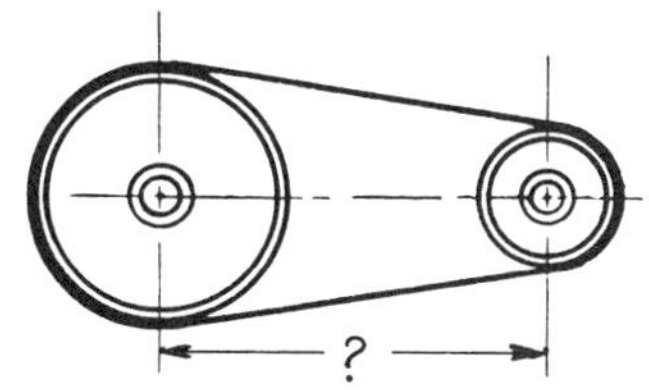

Find the numerical value of each of the following expressions:

10. $2(a + b)$, $a = 1$, $b = 2$
11. $2(R + r)$, $R = 3$, $r = 2$
12. $3(a + 2b)$, $a = 2$, $b = 3$
13. $3(a - 2b)$, $a = 6$, $b = 2$
14. $\frac{(a + b)}{2}$, $a = 3$, $b = 5$
15. $\frac{(B + b)3}{2}$, $B = 4$, $b = 2$
16. $\frac{(2a + b)5}{2}$, $a = 1$, $b = 2$
17. $\frac{(B - b)3}{2}$, $B = 2$, $b = 1$

B. Removal of Common Factors From Expressions

Rewrite each of the following and place the common factor outside the parenthesis:

1. $2\ell + 2w$
2. $3\ell + 3w$
3. $2a - 2b$
4. $ad + ab$
5. $ad + 2ab$
6. $2ad - ab$
7. $2a + 6b$

8. Write an expression for the distance *x* between two gears in terms of the radii R and r.

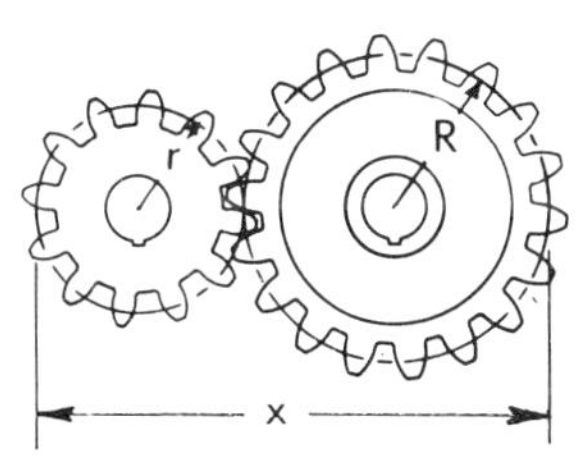

UNIT 41 INTRODUCTION TO EQUATIONS

Equations are useful in solving problems which confront the sheet metal worker. For example, if a workman needs to lay out 7 rivet holes to be equally spaced on the seam of a cone with the side 21 inches long, the first rivet one inch from the bottom, and the last 2 inches from the top, he could use an equation to find the distance between the holes.

x = Distance between the holes

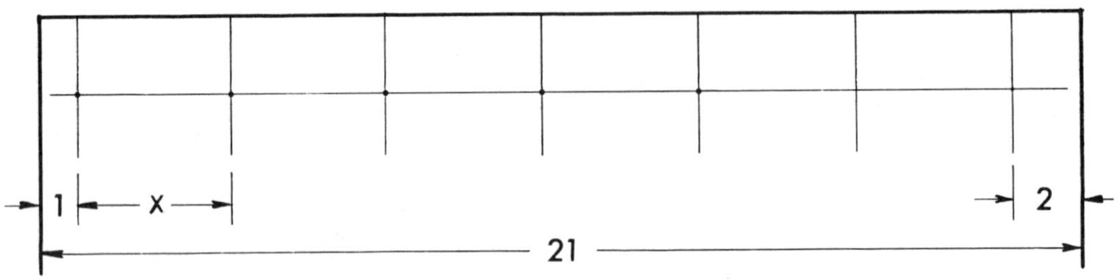

Note: x stands for the unknown number; in this case it is the dimension of the space between the holes.

$$x = \frac{21 - (1 + 2)}{6}$$

$$x = \frac{18}{6}$$

$$x = 3$$

Definition of an Equation

A simple equation is an expression of *equality* between two quantities. The mathematical processes which you have learned can be expressed by equations. Examples of equations are given below.

1. What number added to 3 will give 18?

   What number + 3 = 18?

   Let x = unknown number
   $x + 3 = 18$ (This equality is called an equation.)
   $15 + 3 = 18$
   $\therefore x = 15$ Ans.

2. What number decreased by 4 will give 8?

   What number − 4 = 8?

   Let x = unknown number
   $x - 4 = 8$ (equation)
   $12 - 4 = 8$
   $\therefore x = 12$ Ans.

3. What number multiplied by 7 will give 21?

   What number x 7 = 21 ?

   Let x = unknown number

   $7x = 21$ (equation)

   $\therefore x = 3$ Ans.

   Check

   $7 \times 3 = 21$

   $21 = 21$

4. What number divided by 2 will give 10?

   What number ÷ 2 = 10?

   Let x = unknown number

   $\frac{x}{2} = 10$ (equation)

   $\therefore x = 20$ Ans.

   Check

   $\frac{20}{2} = 10$

   $10 = 10$

PARTS OF AN EQUATION

The *members* of an equation are the expressions on each side of the equality sign.

Simple equation: $4x = 12$

The first member of the above equation is the expression $4x$, to the left of the equality sign; the second member is the number 12, to the right of the equality sign.

The *coefficient* of x (the unknown) is the number or letter which comes immediately before x (the unknown) as a multiplier. In the above equation, the coefficient of x is 4.

EQUALITY OF MEMBERS

In every equation, both members are equal, and the equation balances. To keep the equation in balance, if one member is increased, decreased, multiplied, or divided by a number, the other member must be increased, decreased, multiplied, or divided by the same number. Four methods of solving equations are illustrated below:

Addition

The two members are equal.	6	=	6
Add to each member	+2	=	+2
∴ The members remain equal.	8	=	8

Subtraction

$$\begin{aligned} 6 &= 6 \\ -2 &= -2 \\ \hline 4 &= 4 \end{aligned}$$

Multiplication

$$\begin{aligned} 6 &= 6 \\ \times 2 &= \times 2 \\ \hline 12 &= 12 \end{aligned}$$

Division

$$\begin{aligned} 6 &= 6 \\ \div 2 &= \div 2 \\ \hline 3 &= 3 \end{aligned}$$

SUMMARY OF PRINCIPLES

If equals are increased, decreased, multiplied, or divided by the same number, the results are equal.

ASSIGNMENT

In problems A to D, write an equation for each question, or statement. In problem E, find the missing quantities.

A. Addition Problems

1. What number added to 2 will give 10?
2. What number increased by 5 1/2 will give 8 3/4?
3. If x dollars are increased by \$10 the sum will be \$25.50.
4. What length increased by 10 in. will give 18.5 in.?

B. Subtraction Problems

1. What number decreased by 12 will give 4?
2. What number decreased by 6 1/2 will give 7 1/4?
3. What length decreased by 6 in. will give 6 in.?
4. If x dollars are decreased by \$2.50, the difference will be \$8.00.

C. Multiplication Problems

1. What number multiplied by 2 will give 6?
2. What number multiplied by 15 will be equal to 90?
3. Twelve times a certain number is 60. What is the number ?
4. A certain number multiplied by 4 1/2 is 90. What is the number?

D. Division Problems

1. What number divided by 2 will give 3?
2. One-half of a sheet of metal weighs 20 lb.
3. One-third of a piece of wire is 8 in. long.
4. One-quarter of a man's salary is \$15.25.

E. Equality Problems

1. $\begin{array}{r} 10 \\ +\,? \\ \hline 25 \end{array} = \begin{array}{r} 10 \\ +\,? \\ \hline 25 \end{array}$

2. $\begin{array}{r} 10 \\ \times\,? \\ \hline 20 \end{array} = \begin{array}{r} 10 \\ \times\,? \\ \hline 20 \end{array}$

3. $\begin{array}{r} 10 \\ \div\,? \\ \hline 5 \end{array} = \begin{array}{r} 10 \\ \div\,? \\ \hline 5 \end{array}$

4. $\begin{array}{r} 10 \\ -\,? \\ \hline 8 \end{array} = \begin{array}{r} 10 \\ -\,? \\ \hline 8 \end{array}$

UNIT 42 HOW TO SOLVE EQUATIONS BY DIVISION

Some equations can be solved by dividing both members by the same number. For example, if a scale balances with a 6-lb. weight on each side, and the weight on one side is reduced by dividing by 2, the weight on the other side must be reduced by dividing by 2. Then there will be 3 pounds on each side, and the scale will remain in balance.

This method is used when the equation or formula to be solved contains a product of two quantities, such as $2x = 6$.

Example: Find the value of x in the equation $4x = 12$

1. Write the equation. $4x = 12$

2. Divide each member by the coefficient of x to obtain the value of x. $\frac{4x}{4} = \frac{12}{4}$

3. The equation becomes $x = 3$ Ans.

4. Check by substituting the value of x in the original equation.
$$4x = 12$$
$$4 \times 3 = 12$$
$$12 = 12$$

PRINCIPLE: If both members of an equation are divided by the same quantity, the results are equal.

Example: An area of 40 sq. in. is needed for a rectangular pipe which is 10 in. wide. What must be the other dimension? The area of a rectangle is equal to the product of its length and width, expressed in like units.

1. Select the proper formula. $A = \ell w$
   (A = area; ℓ = length; w = width.)
2. Substitute the known values. $40 = 10w$
3. Reverse the members. $10w = 40$
4. Divide each member by the coefficient of w, which is 10. $\frac{10w}{10} = \frac{40}{10}$
5. The equation becomes $w = 4$ in. Ans.
6. Check by using the original conditions of the problem.
$$A = 4 \times 10$$
$$40 = 40$$

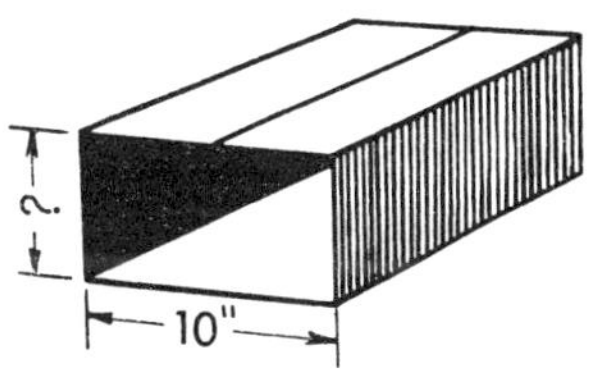

ASSIGNMENT

1. Solve for x in $3x = 6$
2. Solve for a in $3a = 9$
3. Solve for b in $3b = 12$
4. Solve for x in $2x = 5$
5. Solve for y in $8y = 30$
6. Solve for x in $9x = 76\ 1/2$
7. Solve for m in $6\ 1/2m = 19\ 1/2$
8. Solve for n in $1\ 3/4n = 7$
9. Solve for x in $15 = 2.5x$
10. Solve for w in $78.75 = 3.75w$
11. The perimeter of a square pipe is 18 in. Find the length of each side. **Hint:** The perimeter or distance around a square is equal to four times the length of one side.
12. The perimeter of a square pipe must be 4′ 6″. What should be the length of each side?
13. The area of a rectangular shop floor is to be 400 sq. ft., and the width is to be 16 ft. How long must the floor be?
14. The area of a rectangular sheet of metal is 10 sq. ft., the width 2 1/2 ft. How long must it be?
15. The cross-section area of a rectangular pipe is to be 20 sq. in., the width 4 in. How high must the pipe be?
16. The cross-section area of a rectangular pipe is to be 28 sq. in., the height 7 in. How wide should it be?
17. a 30% discount is equal to $120. What is the list price? **Hint:** Let P = list price; d = value of discount, or $120; r = rate of discount, or .30. Then d = r x P. Find P.
18. If 20% of the payroll of an industrial firm is equal to $16,000, what is the total payroll in dollars?
19. If 3% of the profit of a certain roofing contractor amounts to $1500, what is the total profit?
20. If 20% of the annual income of a sheet-metal worker is $2200, what is his total yearly wage?

UNIT 43 HOW TO SOLVE EQUATIONS BY MULTIPLICATION

Some equations can be solved by multiplying both members by the same number. If a 4-lb. weight on one side of a balancing scale is multiplied by 3, then the weight on the other side must be multiplied by 3 to keep the scale in balance. Now there are 12 lb. on each side, and the scale will remain in balance.

This method of solving equations is used when the equation or formula to be solved contains a fraction, such as $x/3 = 8$, or $r/s = p$.

Example: Find the value of x in the equation $x/3 = 8$.

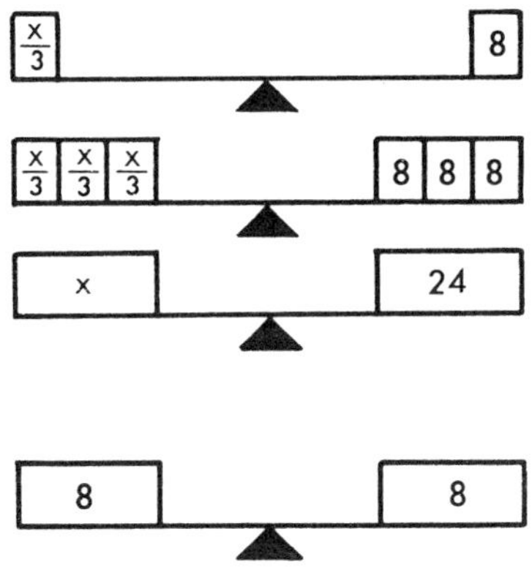

1. Write the equation $\qquad \dfrac{x}{3} = 8$
2. Multiply each member by the denominator of $x/3$ to obtain the value of x. $\qquad \dfrac{3 \times x}{3} = 8 \times 3$
3. The equation becomes $\qquad x = 24$ Ans.
4. Check by substituting the answer in the original equation.

$$\frac{x}{3} = 8$$
$$\frac{24}{3} = 8$$
$$8 = 8$$

PRINCIPLE: If both members of an equation are multiplied by the same quantity, the results are equal.

Example: If a quarter of a line is equal to 5 in., what is the total length of the line?

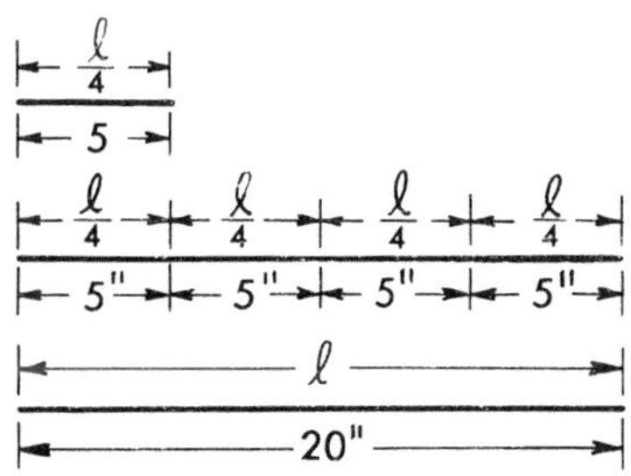

1. Let ℓ = length of line. Then $\ell/4$ = 1/4 of the length of the line, or 5″. $\qquad \dfrac{\ell}{4} = 5$
2. Multiply each member by the denominator of $\ell/4$. $\qquad \dfrac{4 \times \ell}{4} = 5 \times 4$
3. The equation becomes $\qquad \ell = 20''$ Ans.
4. Check by using the original conditions of the problem.

$$\frac{20}{4} = 5$$
$$5 = 5$$

ASSIGNMENT

1. Solve for x in $x/2 = 6$
2. Solve for y in y/3 = 5
3. Solve for a in a/6 = 30
4. Solve for b in b/2 = 6.5
5. Solve for x in $x/11 = .10$
6. Solve for z in z/3 = 5 1/4
7. Solve for m in 2 1/2 = m/21
8. Solve for n in n/7 1/2 = 3
9. Solve for p in p/2.75 = 5.5
10. Solve for x in $.010 = x/.020$
11. One-quarter of a job takes 3 1/2 days. How long will the total job take?
12. A coil of wire 250′ long weighs 27 lb. What will be the weight of a coil containing 1500 ft. of wire?
13. A bundle of 150 sheets of tin plate weighs 317 lb. What will 750 sheets weigh?
14. One-quarter of the yearly income of a worker is $1,740.00. What is his annual wage at this rate?
15. The pitch of a roof is 1/3 and the rise is 10 ft. What is the span?
16. The pitch of a roof is 1/4, the rise 9 ft. What is the span?
17. The pitch of a roof is 5/12, the rise 8 1/2 ft. What is the span? **Hint:** Multiply each member of the equation by 12/5.
18. The pitch of a roof is 7/12, the rise 5 1/4 ft. What is the span?
19. The first month on a new job you receive $684.15. At this average monthly wage, what will be your yearly salary?
20. Your car payments are $87.18 per month. What will be the total of all your auto payments after a period of 24 months?
21. A discount of 1/3 is equal to $120. What is the list price? **Hint:** Let r = rate of discount, or 1/3; d = amount of discount, or $120. Find P in d = r x P.
22. If one-fifth of the payroll of an industrial firm is equal to $20,000, what is the total payroll in dollars?
23. If one-twentieth of the profit of a certain firm amounts to $1500, what is the total profit?

UNIT 44 HOW TO SOLVE EQUATIONS BY SUBTRACTION

If a scale balances with a 5-lb. weight on each side, and if we subtract 2 lb. from one side, we must subtract 2 lb. from the other side to maintain the balance.

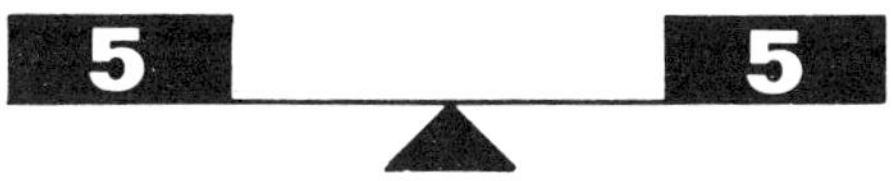

This method of solving equations is used when the equation or formula to be solved contains a sum of two quantities, as $x + 2 = 5$, or $2\ell + 2w = 60$.

Example: Solve for x in the equation $x + 3 = 11$.

1. Write the equation. $x + 3 = 11$

2. From each member, subtract a number large enough to permit x to remain alone on the left-hand side. This number is 3 in this problem.

$$\begin{aligned} x + 3 &= 11 \\ -\ 3 &= -3 \end{aligned}$$

3. The equation becomes $x = 8$ Ans.

4. Check

$$\begin{aligned} x + 3 &= 11 \\ 8 + 3 &= 11 \end{aligned}$$

PRINCIPLE: If the same quantity is subtracted from both members of an equation, the results are equal.

Example: A rectangular plate is 2 ft. longer than it is wide. What is its width if the plate is 9 ft. long?

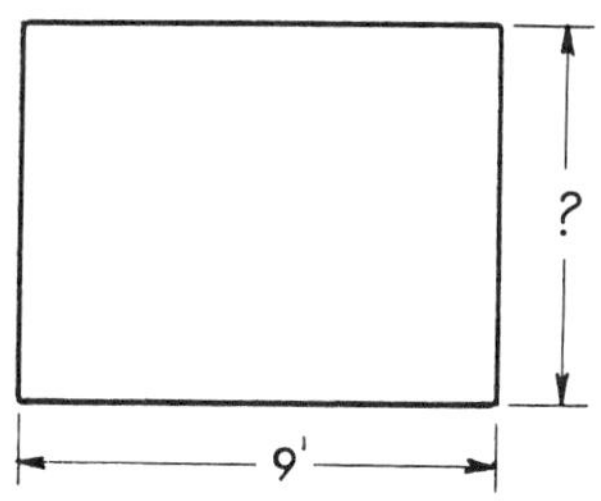

1. Let w = width of plate. Then w + 2 = length of plate, or 9 feet. $w + 2 = 9$

2. Subtract

$$\begin{aligned} w + 2 &= 9 \\ -\ 2 &= -2 \end{aligned}$$

3. The equation becomes $w = 7'$ Ans.

4. Check

$$\begin{aligned} 7 + 2 &= 9 \\ 9 &= 9 \end{aligned}$$

ASSIGNMENT

1. Solve for x in $x + 9 = 13$
2. Solve for y in $y + 4 = 15$
3. Solve for z in $z + 2\frac{1}{2} = 7\frac{1}{2}$
4. Solve for m in $m + 3.75 = 10$
5. Solve for p in $p + 21 = 35\frac{1}{2}$
6. Solve for x in $21 = x + 3$
7. Solve for L in $26 = L + 4$
8. Solve for w in $38\frac{1}{2} = w + 6$
9. Solve for z in $8\frac{1}{2} = z + 2\frac{1}{2}$
10. Solve for m in $13 = 4.5 + m$
11. A rectangular plate is to be 1½ ft. longer than it is wide. What will be its width if the plate is 10 ft. long?
12. A rectangular plate is to be 2¾ ft. longer than it is wide. What will be its width if the plate is 8 ft. long?
13. A piece of wire is 15 in. long. When it is cut, one part is 6½ in. long. How long will the other part be? **Hint:** Let x = length of other part; then $x + 6\frac{1}{2} = 15$.

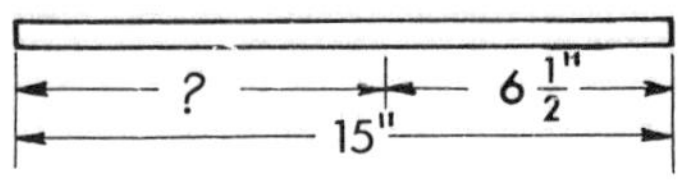

14. A piece of band iron is 19 3/4 in. long. If it is cut so that one part is 5 7/8 in. long, how long will the other part be?
15. Find the distance x in the plate below.

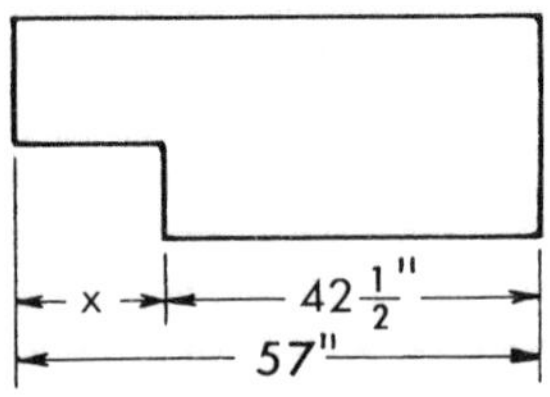

16. A hole is to be drilled 6 3/4 in. from the end of a bar of steel which is 19 in. long. What is the distance from the center of the hole to the opposite end?
17. An increase of 25 cents per hour raises a mechanic's wage to $4.50 per hour. What was his original wage per hour?
18. A firm made a profit of $10,000 in 1971. This amounted to $1250 more than the profit made in 1970. What was the profit in 1970?
19. A man paid $20 interest on a loan. If he paid a total of $140 to the loan company, how much money did he borrow originally?

UNIT 45 HOW TO SOLVE EQUATIONS BY ADDITION

If a scale balances with 5-lb. weights on each side and we add 3 lb. to one side, we must add 3 lb. to the other side to maintain the balance.

This method of solving equations is used when the equation or the formula to be solved contains the difference of two quantities, as $x - 2 = 4$.

Example: Solve for x in the equation $x - 2 = 4$.

1. Write the equation $\quad x - 2 = 4$

2. To each member add a number large enough to permit x to remain alone on the left-hand side. This number is 2.

$$\begin{aligned} x - 2 &= 4 \\ +2 &= +2 \end{aligned}$$

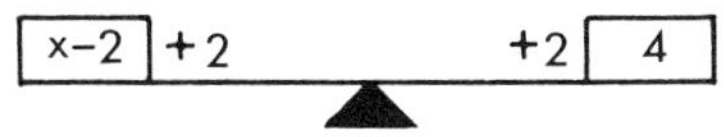

3. The equation becomes $\quad x = 6$ Ans.

4. Check

$$\begin{aligned} x - 2 &= 4 \\ 6 - 2 &= 4 \\ 4 &= 4 \end{aligned}$$

PRINCIPLE: If the same quantity is added to both members of an equation, the results are equal.

Example: A rectangular plate is 10 in. wide. How long is the plate if its width is 6 in. less than its length?

1. Let ℓ = length of plate. Then $\ell - 6$ = the width of plate, or 10 in. $\quad \ell - 6 = 10$

2. Add

$$\begin{aligned} \ell - 6 &= 10 \\ +6 &= +6 \end{aligned}$$

3. The equation becomes (Ans.) $\quad \ell = 16''$

4. Check

$$\begin{aligned} 16 - 6 &= 10 \\ 10 &= 10 \end{aligned}$$

ASSIGNMENT

1. Solve for x in $x - 6 = 2$
2. Solve for m in $m - 2 = 11$
3. Solve for p in $p - 8 = 4\frac{1}{2}$
4. Solve for r in $r - 2\frac{1}{2} = 3\frac{1}{2}$
5. Solve for d in $2 = d - 3$
6. Solve for z in $3 = z - 5$

7. Solve for y in $11\ 1/2 = y - 2$
8. Solve for L in $L - 4 = 6.58$
9. Solve for w in $2 = w - 2$
10. Solve for z in $5\ 1/2 = z - 3\ 1/2$
11. A rectangular plate is 31 1/2 in. wide. How long is the plate if its width is 8 1/4 in. less than its length?
12. A box is 12 in. wide. How long should it be it its width is 8 in. less than its length?

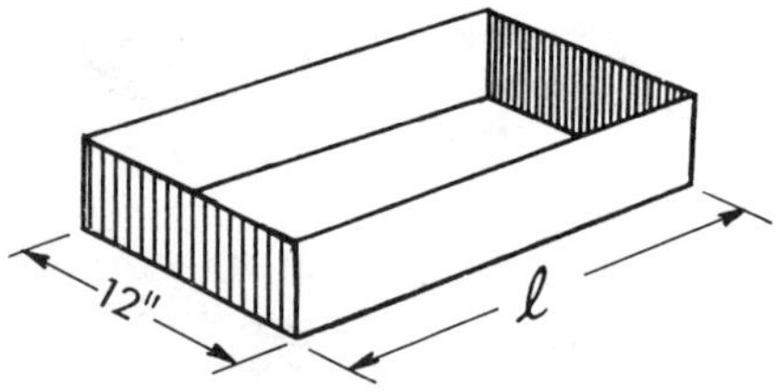

13. If 6 ft. are cut from a piece of wire, and the remaining piece is 5 ft. long, how long was it originally? **Hint:** Let x = original length of the wire.

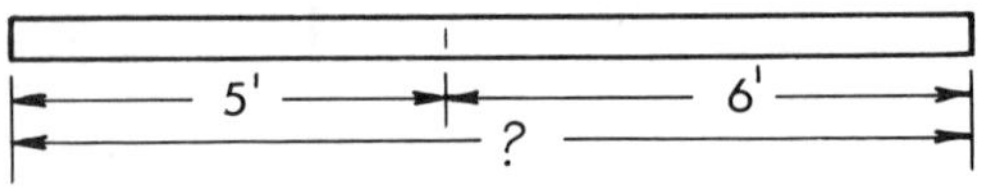

14. If 8 1/2 in. are cut from a piece of wire, and the remaining piece is 4 1/4 in. long, how long was it originally?
15. If 2 3/4 in. are cut from a piece of wire, and the remaining piece is 5 1/8 in. long, how long was it originally?
16. Deductions for taxes decrease a man's weekly pay by $31.66. He takes home $138.34. What is his earned pay? **Hint:** Let "x" = earned pay; then x - 31.66 = $138.34.
17. After drawing $60 from a bank account, a man had $550 left. How much did he have originally?
18. The net price of a micrometer is $16.50. The discount is $3.75. What is the list price?

UNIT 46 EXPONENTS

Equations often contain *exponents.* An exponent is a small symbol (or a number) written above and to the right of another symbol (or a number). It denotes how many times the latter is repeated as a factor. In the expression 4^2, the small number ^{2} is the exponent and indicates that 4 is to be taken as a factor twice; therefore, 4^2 means 4 x 4 and equals 16.

MEANING OF EXPONENTS

If 3 is the exponent of 2, then 2 is to be taken as a factor three times: 2^3 (read as "two, cubed") means 2 x 2 x 2 and equals 8.

2^4 (two to the fourth power) means 2 x 2 x 2 x 2 and equals 16

2^5 (two to the fifth power) means 2 x 2 x 2 x 2 x 2 and equals 32

$\left(\frac{2}{3}\right)^3$ (two-thirds cubed) means 2/3 x 2/3 x 2/3 and equals 8/27

$\frac{2^2}{3}$ (two squared, divided by 3) means $\frac{2 \times 2}{3}$ and equals 4/3

a^2 (a squared) means a x a

a^3 (a cubed) means a x a x a

a (a to the first power – means a, the 1 being understood)

Note: The exponent of a is 1, or a^1 (a to the first power).

EVALUATION OF EXPONENTIAL EXPRESSIONS

Exponential expressions may contain both numbers and letters: $2a^2b = 2 \times a \times a \times b$. If a = 4 and b = 5, $\therefore 2a^2b = 2 \times 4^2 \times 5$, or 2 x 4 x 4 x 5, or 160

ADDING, SUBTRACTING, MULTIPLYING, DIVIDING EXPONENTIAL EXPRESSIONS

Addition. The numerical factors are added; the exponents remain the same: $2a^2 + a^2 = 3a^2$.

Subtraction. The numerical factors are subtracted; the exponents remain the same: $4x^2 - 2x^2 = 2x^2$.

Multiplication. The numerical factors are multiplied; the exponents are added: $5a^2 \times 2a = 5 \times a \times a \times 2 \times a$, or $10a^3$.

Division. The numerical factors are divided; the exponents are subtracted:

$\frac{2a}{2} = \frac{\overset{1}{\cancel{2}}a}{\underset{1}{\cancel{2}}}$, or a $\qquad \frac{4a}{a} = \frac{4\overset{1}{\cancel{a}}}{\underset{1}{\cancel{a}}}$, or 4 $\qquad \frac{5a^2}{2a} = \frac{5 \times a \times \overset{1}{\cancel{a}}}{2 \times \underset{1}{\cancel{a}}}$, or $\frac{5a}{2}$

ASSIGNMENT

A. Meaning of Exponents

1. Find the value of each of the following expressions:

a. 4^2	d. 8^4	g. $\left(12\frac{1}{2}\right)^2$	j. $\overline{.20}^3$
b. 5^2	e. $\left(\frac{3}{8}\right)^2$	h. $\left(\frac{1}{2}\right)^4$	k. $\overline{.03}^2$
c. 5^3	f. $\left(\frac{5}{6}\right)^2$	i. $\overline{13.6}^2$	l. $\overline{.25}^3$

2. Write each of the following as an exponential expression:

a. s x s	c. d x d	e. 4w x w	g. s x s x s
b. r x r	d. 2r x 2r	f. $\frac{\ell}{2} \times \ell$	h. a x b x b

3. If the length of the side of a square is s, write an exponential expression for the product of the lengths of the two sides.
4. If the radius of a circle is r, write an exponential expression for the product of the radius times the radius.
5. The width of a rectangle is w and the length is 4 times as great. Write an exponential expression for the product of the length and width.
6. The length of a rectangle is ℓ and the width is 1/4 as much. Write an exponential expression for the product of the length and the width.

B. Evaluation of Exponential Expressions

Find the value of each of the following expressions.

1. s^2, $s = 2\frac{1}{2}$	4. d^2, $d = 4$	7. s^2h, $s = 4$, $h = 5$
2. r^2, $r = 4\frac{1}{4}$	5. $\frac{d^2}{2}$, $d = 4$	8. $R^2 - r^2$, $R = 5$, $r = 3$
3. $4w^2$, $w = 3$	6. $\left(\frac{d}{2}\right)^2$, $d = 4$	9. $(R - r)^2$, $R = 5$, $r = 3$

C. Adding, Subtracting, Multiplying, Dividing Exponential Expressions

Perform the operations indicated in the problems below.

1. $r^2 + r^2$	4. $d^2 - \frac{d^2}{2}$	7. $\frac{8a}{4}$
2. $\frac{r^2}{2} + \frac{r^2}{2}$	5. $s^2 + s$	8. $\frac{6a}{a}$
3. $2s^2 - s^2$	6. $2a^2 \times 2a$	9. $\frac{r^3}{r}$

UNIT 47 HOW TO SOLVE EQUATIONS BY SQUARE ROOT

Another method of solving equations must be used when the exponent *two* appears in the unknown as in $x^2 = 4$. When square root is used in solving equations, the square root of each side must be taken to keep the equation in balance.

The *square root* of a number is that result which, when taken twice as a factor, gives the original number. The determination of the square root of a number can be considered as an operation opposite to that of squaring a number. Square root is indicated by the symbol $\sqrt{\ }$. For example, the square root of 4 is 2 because $2 \times 2 = 4, \sqrt{4} = 2$ Ans. Table viii, pages 267 - 270 gives the square roots of numbers. Refer to table viii when working the problems which follow.

Example: Find the value of x in the equation $x^2 = 49$.

1. Write the equation. $x^2 = 49$
2. Take the square root of both members. $\sqrt{x^2} = \sqrt{49}$ $(\sqrt{x^2} = \sqrt{x \times x}$, or $x)$
3. The equation becomes $x = 7$ Ans.
4. Check
$$x^2 = 49$$
$$7^2 = 49$$
$$49 = 49$$

PRINCIPLE: If the square root is taken of both members of an equation, the results are equal.

Example: If the cross-section area of a square ventilating pipe must be 60 sq. in., what must be the length (to the closest 16th of an in.) of each side?

1. Select the proper formula. (A = area; s = length of side.) $A = s^2$
2. Substitute the known values in the equation. $60 = s^2$
3. Reverse the members. $s^2 = 60$
4. Take the square root of both members. $\sqrt{s^2} = \sqrt{60}$
5. The equation becomes $s = 7.745''$, or 7 3/4'' Ans.
6. Check. (Square 7.745; add remainder.)
$$(7.745)^2 = 60$$
$$60 = 60$$

ASSIGNMENT

1. Solve for x in $x^2 = 64$
2. Solve for y in $y^2 = 121$

3. Solve for z in $z^2 = 169$
4. Solve for a in $a^2 = 50$
5. Solve for b in $b^2 = 75$
6. Solve for s in $s^2 = 1021$
7. Solve for x in $21 = x^2$
8. Find the length of the side (to the nearest 16th of an in.) of a square pipe if the cross-section area is to be 50 sq. in. (Use the equation $50 = x^2$.)

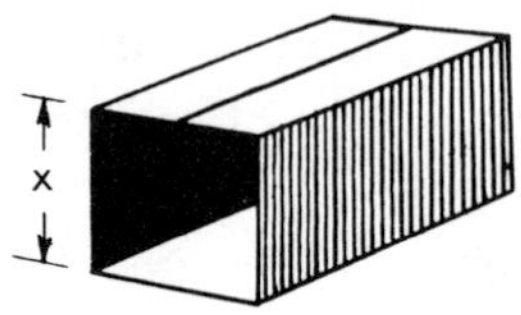

9. Find the length of the side (to the nearest 16th of an in.) of a square pipe if the cross-section area is to be 80 sq. in.
10. Find the length of the side (to the nearest 16th of an in.) of a square pipe if the cross-section area is to be 75.32 sq. in.
11. A square plate is to have an area of 230 sq. in. What will be the length of each side to the closest 16th of an inch?

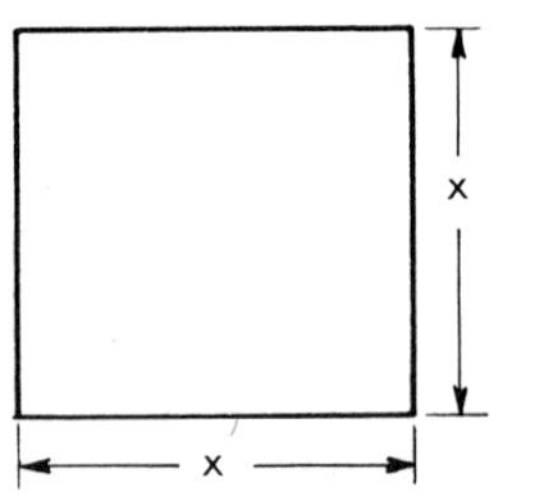

12. A square plate is to have an area of 20 sq. ft. What will be the length of each side in ft. and in. (closest 16th)?

UNIT 48 SUMMARY OF EQUATIONS

Sometimes the solution of a problem requires the use of two or more methods of computation within one equation. Therefore, it is helpful to follow a definite procedure in analyzing a problem.

METHOD OF ANALYZING THE PROBLEM

1. Read the problem carefully. Represent the unknown quantity by a letter.
2. If several unknown quantities are to be found, try to express them in terms of the same letter. For instance, if one side of a rectangle is w, and the other side is 4 times as long, the length of the longer side is 4w.
3. Try to discover two expressions or an equality which expresses the relationship; such as P = 4s (perimeter of square).
4. Solve the equation, using the result obtained to find the value of each unknown quantity.
5. Check results by using the original statement of the problem.

Problems Involving the Combining of Terms

If several terms, in numerical and in literal form, occur in an equation, combine the numbers, and combine the letters before proceeding to solve the equation.

$$(7 \times 3)n = 100 \qquad x + 2x = 9 \qquad 6x - 2x = 8$$
$$\therefore 21n = 100 \qquad \therefore 3x = 9 \qquad \therefore 4x = 8$$

Example: The width of a piece of sheet metal is 3/4 of its length, and the distance around it is 140 in. Find the length and the width.

$$x = \text{length}$$
$$\frac{3}{4}x = \text{width}$$

Substitution $$x + x + \frac{3}{4}x + \frac{3}{4}x = 140$$

Evaluation $$3\frac{1}{2}x = 140$$

$$\frac{7}{2}x = 140$$

$$x = \overset{20}{\cancel{140}} \times \frac{2}{\cancel{7}}$$

$$x = 40$$

$$\frac{3}{4}x = \frac{3 \times \overset{10}{\cancel{40}}}{\cancel{4}}$$

$$x = 30$$ Ans. Length 40 in., width 30 in.

Check: $$3\,1/2 \times 40 = 140$$
$$140 = 140$$

Example: If a 60° elbow is to be made in four pieces, find the angle of each piece. **Note:** The angle of the inner spaces is made double the angle of the outer spaces.

1. Let a = number of degrees in each of the two outer angles. Then 2a = number of degrees in each of the two inner angles.

   Equation is $a + 2a + 2a + a = 60$

2. Combine the terms. $6a = 60$
3. Divide each member by the coefficient of a. $\frac{6a}{6} = \frac{60}{6}$
4. The equation becomes $a = 10°$ Ans.
   $\therefore 2a = 20°$ Ans.
5. Check $10° + 20° + 20° + 10° = 60°$
   $60° = 60°$

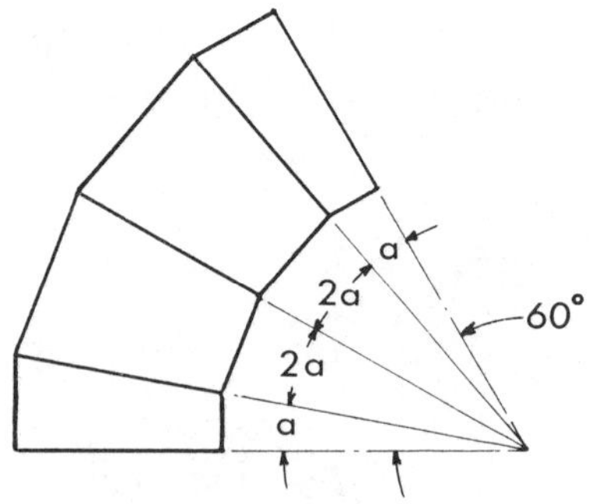

Problems Involving Square Root

If a problem involves an equation such as $4x^2 = 100$, divide each member by the coefficient of x^2 before taking the square root of each member.

Example: One side of a rectangular pipe is to be 4 times the other, and the cross-section area is to be 100 sq. in. Find the dimensions of the opening of the pipe.

1. Let w = length of one side of pipe; then 4w = length of other side of pipe, and the cross-section area = 4w x w, or $4w^2$.

   Equation is $4w^2 = 100$

2. Divide each member by the coefficient of w. $\frac{4w^2}{4} = \frac{100}{4}$
3. The equation becomes $w^2 = 25$
4. Take the square root of each member. $\sqrt{w^2} = \sqrt{25}$
5. The equation becomes $w = 5''$ Ans.
   $\therefore 4w = 20''$ Ans.
6. Check Area $= 5 \times 20$
   $100 = 100$

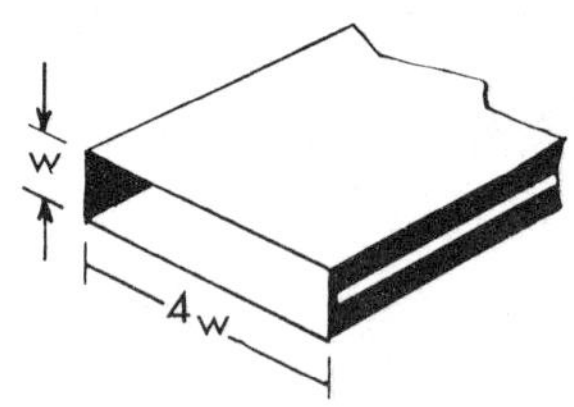

ASSIGNMENT

Find the value of the unknown in problems 1 to 8.

1. $(5 \times 6)x = 60$
2. $\left(\frac{10 \times 14}{7}\right)n = 45$
3. $\frac{3.14d}{2} = 6.28$
4. $a + 2a + a = 80$
5. $4w + w + 4w + w = 20$
6. $2.75y + 3.5y = 12.5$
7. $5z - 3z = 20$
8. $6a - 2a = 50.6$

9. What number divided by 8 is equal to 62?
10. One-half of a number plus 11 equals 37. Find the number.
11. The sum of two numbers is 43. One number is 7 greater than the other. Find the two numbers.
12. Two boys bought an electric drill for $45. One furnished 4 times as much money as the other. How much money did each boy furnish?
13. Find the angle of each piece of a three-piece elbow of 30°.
14. Find the angle of each piece of a four-piece elbow of 90°.
15. Find the angle of each piece of a five-piece elbow of 90°.
16. If 4 parts tin and 5 parts lead are to be used to make 36 lb. of solder, how much tin and lead must be used? **Hint:** Let x = weight of one part; then the tin = $4x$, and the lead = $5x$.
17. If 1 part silver, 2 parts tin and 3 parts lead are to be used to make 36 lb. of solder, how many pounds of each must be used?
18. A 24-in. piece of wire is to be cut into 2 parts, one part being 5 times the length of the other. Find the length of each.
19. A truck loaded with sheet iron weighed 7200 lb. The sheet iron weighed 1800 lb. more than the truck. How much did the truck weigh? The sheet iron?

B. Problems Involving Square Root

Find the value of the unknown in problems 1 to 6.

1. $2x^2 = 50$
2. $5y^2 = 500$
3. $8z^2 = 1600$
4. $80 = 4w^2$
5. $\frac{\ell^2}{4} = 16$
6. $\frac{m^2}{3} = 15$

C. Additional Problems

1. Find the wall thickness (t) of the tubing in each of the following problems:

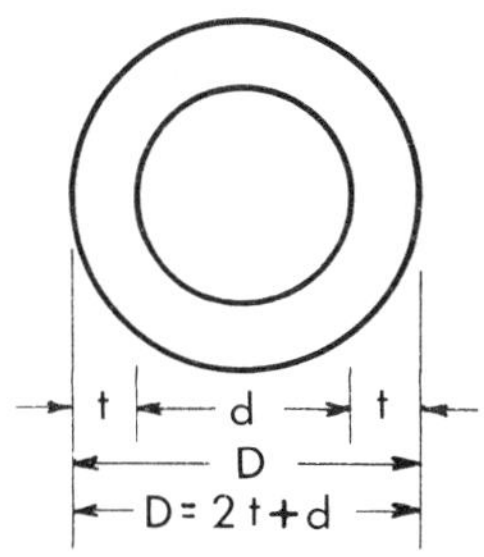

	Outside Diameter - D	Inside Diameter - d
a.	1.315″	1.000″
b.	1.600″	1 1/4″
c.	1.900″	1 1/2″

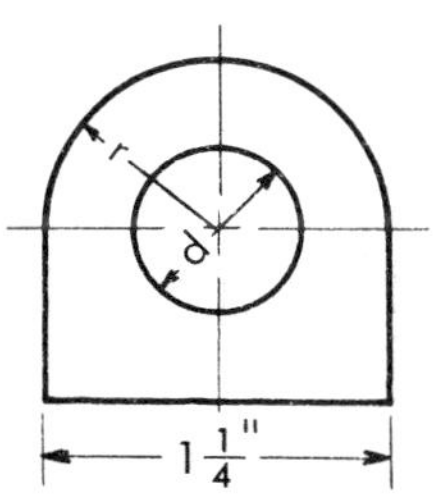

2. If a mechanic's rate is double that of his helper, and altogether they receive $90 for a job, how much should each receive?
3. If r = d, find the value of r and of d in the drawing at the left.
4. 132 = 3 1/7D   D = ?
5. 84 = 1/2h x 8   h = ?
6. 2L + 30″ = 7′8″   L = ?

7. $78 = h/5 \quad h = ?$

8. Using the equation $\frac{D}{d} = \frac{r}{R}$, find R or r as indicated below. Give your answers in r.p.m.

   a. R when D = 12   d = 6 1/2   r = 1400 r.p.m.

   b. r when D = 5   d = 2   R = 1725 r.p.m.

   c. r when D = 4   d = 3   R = 1725 r.p.m.

   d. r when D = 3   d = 4   R = 1725 r.p.m.

   e. r when D = 2   d = 5   R = 1725 r.p.m.

9. The equation for electric power is P = I E.

   P = power in watts   I = current in amps   E = voltage

   a. How many 250-watt soldering irons can be plugged into a circuit which is fused for 15 amps at 110 volts?

   b. How much current will each iron draw when voltage is 110?

   c. Using the above formula, how many watts would you be using in an electric welder consuming 75 amps at 18 volts?

   d. How many amps would be used for lighting if the shop uses five 200-watt, nineteen 150-watt, twelve 100-watt bulbs? Assume 110 volts.

   e. How many amps will a 1/2-horsepower motor consume under full load - 115 volts? (1 hp. = 746 watts.)

SECTION 6 RATIO AND PROPORTION

UNIT 49 INTRODUCTION TO RATIO

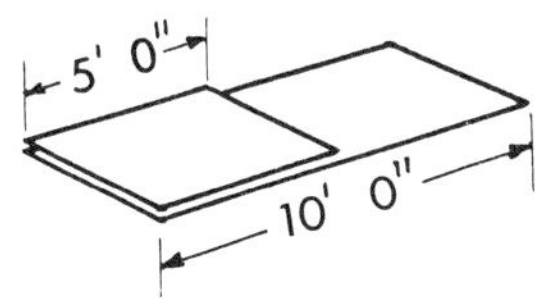

When it becomes necessary to compare two things which are expressed in like units of measurement, but which differ in respect to size, number, weight, etc., the comparison is often expressed as a *ratio.* In the drawing, the ratio of the length of the longer sheet to that of the shorter sheet is 2 to 1 (10 ft./5 ft. = 2).

Ratio is one of several ways of stating the relationship of one number to another. The numbers 10 and 5 may be compared by saying that the first number is twice as large as the second number. The same relationship may be shown by stating that the first number is to the second number as 2 is to 1.

Both numbers must be expressed in the same units of measurement, as feet, inches, yards, etc. There can be no ratio between two numbers that are not of the same unit of measurement; for example, there can be no ratio of 20 *dollars* to 6 *inches.*

FORMS OF RATIO

1. Ratio may be indicated by the use of a colon, for example, 5 : 2, which is read, "the ratio of 5 to 2."

   Ratio may also be indicated either in the fractional form or with division sign, thus: 5/2 or 5 ÷ 2.

2. The two terms being compared are the *terms of the ratio.*

First Term	Second Term
5	2

3. The ratio 10 : 4 can be reduced to its lowest terms.

   $\frac{10}{4} = \frac{10 \div 2}{4 \div 2}$, or $\frac{5}{2}$, or 5 : 2

RATIOS OF LENGTHS OF LINES

If line a = 2 in. and line b = 6 in., then

1. Ratio of a to b = 2/6, or 1/3, or 1 : 3, or 1 to 3.
2. Ratio of b to a = 6/2, or 3/1. or 3 : 1; or 3 to 1.

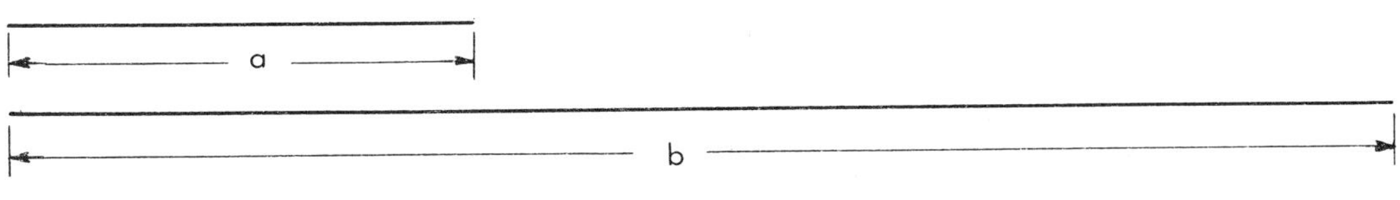

ASSIGNMENT

Forms of Ratio

1. Change the following ratios to fractional form:

   a. 3 : 6

   b. 12 : 4

   c. 25 to 40

   d. 5 : 4

   e. 3/8 : 3/4

   f. 1.125 to 2.25

   g. 5 1/4 to 15 3/4

   h. 1 1/4 to 5

   i. 3.375 to 1.125

2. Find the ratios of the following dimensions in fractional form:

   a. 2 ft. to 8 in.

   b. 8 yd. to 3 ft.

   c. 6 in. to 3 ft.

   d. 1 yd. to 1 in.

   e. 1 in. to 1 yd.

   f. 4 yd. to 6 in.

3. Find the ratios (fractional form) in the following:

   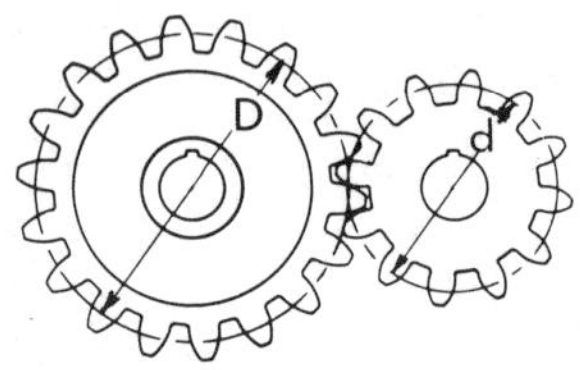

   a. A gear 5 in. in diameter to a gear 10 in. in diameter

   b. A gear 18 in. in diameter to a gear 4 1/2 in. in diameter

   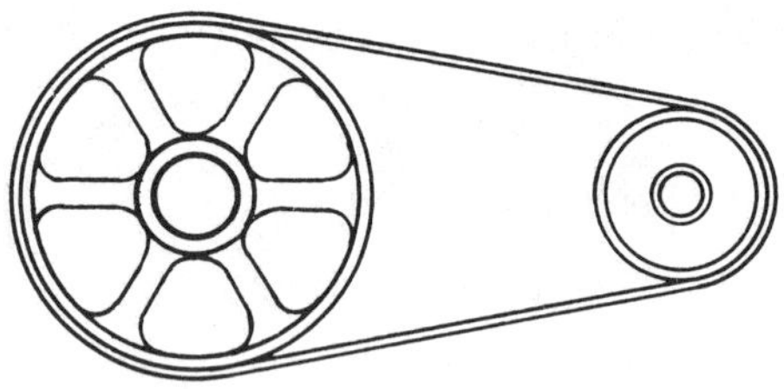

   c. A pulley with a 15-in. diameter to a pulley with a 6-in. diameter

   d. A pulley with a 12-in. radius to a pulley 48 in. in diameter

B. Ratios of Lengths of Lines

1. Measure the following lines:

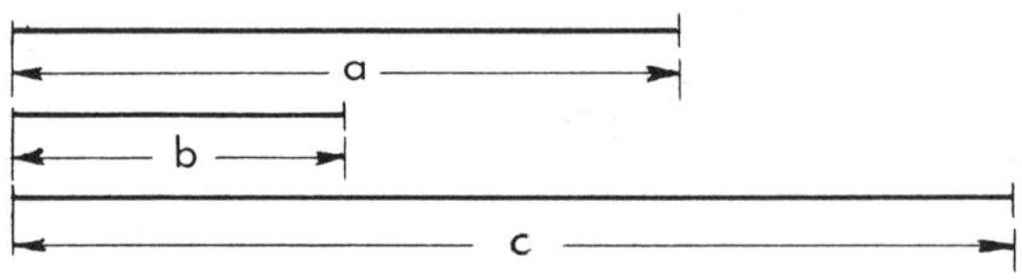

2. Using the values for the lengths of the lines above, indicate in fractional form, the ratios of the lines in the following problems:

   a. a to b =

   b. a to c =

   c. b to c =

   d. c to a =

   e. b to a =

   f. c to b =

3. One hot-air duct is 20 ft. long and another is 4 ft. long. What is the ratio?

4. A cold-air register is 36 in. long and a hot-air register is 9 in. long. What is the ratio?

UNIT 50 RATIO APPLIED TO SCALE DRAWINGS

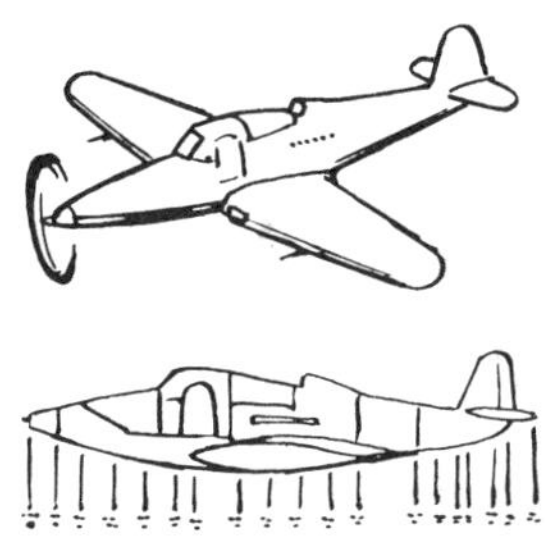

It is not always practical to make a full-scale drawing of a job. Sometimes the job is so large that a full-scale drawing would be unwieldy; on the other hand, the details of a very small job are made clearer and easier to interpret by means of a larger drawing. Drawings are, for example, made half-size, quarter-size, and double-size, etc.

SCALES

The terms *half-size, quarter-size,* etc., refer to the scale of the drawing and indicate the ratio of the length of the lines on the drawing to those on the job. The mechanical engineers' method and the architects' method of giving the scale of the drawing, together with examples of each, are listed below:

	TERM	SCALE	RATIO
Mechanical Engineers	Full-size	1″ = 1″	1 : 1, or 1/1
	Half-size	1/2″ = 1″	1 : 2, or 1/2
	Double-size	2″ = 1″	2 : 1, or 2/1
Architects	Full-size	12″ = 1′ 0″	1 : 1, or 1/1
	Half-size	6″ = 1′ 0″	1 : 2, or 1/2
	One-forty- eighth size	1/4″ = 1′ 0″	1 : 48, or 1/48

LENGTHS OF LINES

If the lines on the drawing and the job are measured, we find the following results:

a divided by A = 1/2

b divided by B = 1/2

c divided by C = 1/2

d divided by D = 1/2

∴ the scale ratio is 1 : 2.

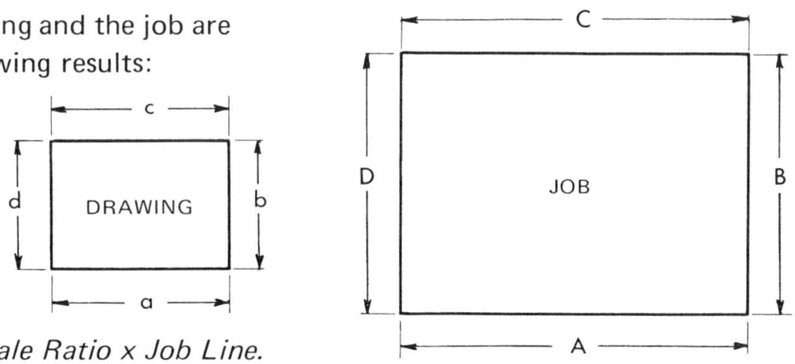

RULE 1: Drawing Line = Scale Ratio x Job Line.

Example 1: If a line is 2 in. on a job, the same line on the drawing, scale 1/2″ = 1″, is 1/2 x 2 in., or 1 in.

RULE 2: Job line = Drawing Line ÷ Scale Ratio.

Example 2: If the actual length of a line on a drawing is 3/4 in., scale 1/2″ = 1″, the length of the same line on the job is 3/4 in. ÷ 1/2, or 1 1/2 in.

ASSIGNMENT

A. Scales

Express the following scales as ratios:

1. 1/2″ = 1″	3. 1/8″ = 1″	5. 3″ = 1″	7. 3″ = 1′ 0″
2. 1/4″ = 1″	4. 2″ = 1″	6. 1/8″ = 1′ 0″	8. 4″ = 1′ 0″

B. Lengths of Lines

1. Find the scale for each drawing and its corresponding job:

a.

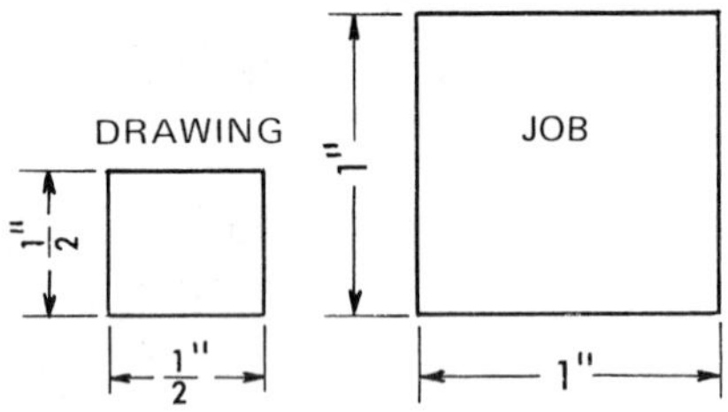

b.

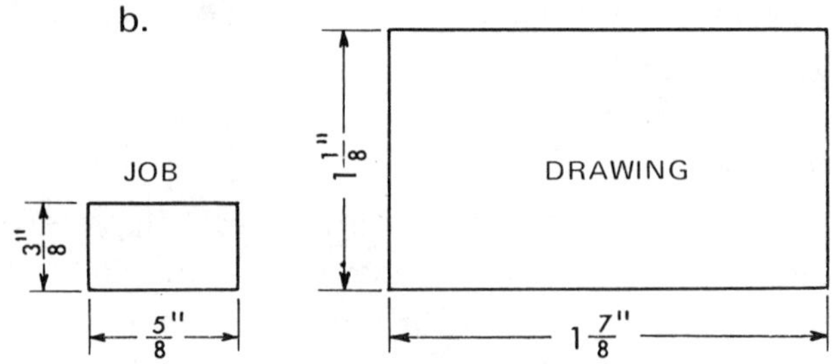

2. A dimension on a job is 24 in. Using a scale of 3″ = 1′0″, compute the length of the same line for a drawing.
3. The length of a line on the drawing is 4 1/2 in. and the scale is 1/4″ = 1″. Compute the length of the same line on the job.
4. Compute the actual lengths of the dimensions that must be used to make a drawing of a belt guard with the scale given below.

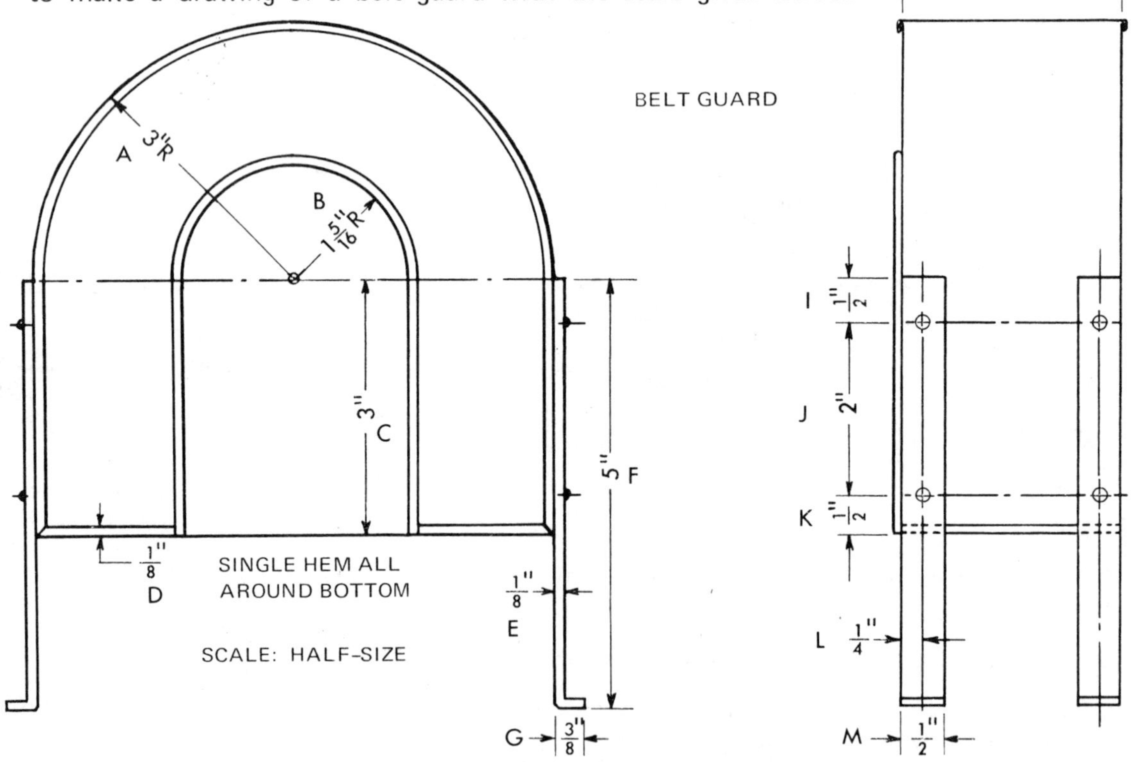

5. On a drawing, the length of a line is 3 3/8 in. If the scale used is 1/4′ = 1′, what is the length of this on the completed job?
6. On a plan drawing using a scale of 1/8″ = 1′, a pipe length is represented by a line 4 7/16 in. long. How long will the pipe be on the actual job?
7. A dimension on a job is 24 inches. Using a scale of 1″ = 3/4″, compute the length of the same line on the drawing.
8. The length of a line on the blueprint is 3 1/4 in., and the scale is 1/8″ = 1′. Compute the length of the line on the job.
9. Compute the dimensions that must be used to make a drawing of two concentric circles with diameters 9 in. and 18 in. Scale: half-size.
10. Repeat question no. 9 using scale: double-size.

UNIT 51 DIRECT PROPORTION

Proportion may be used when certain problems involving pulleys and gears, weights, and costs and time are to be solved. When two ratios, equal in value, are placed on opposite sides of an equal sign, the statement becomes an equation and may be called a proportion. For example, the equation 1/2 = 4/8 is a proportion.

RULE 1: In any proportion, if the members are cross-multiplied, the equality will be maintained.

$$\frac{1}{2} = \frac{4}{8}$$

$$1 \times 8 = 2 \times 4$$

$$8 = 8$$

RULE 2: Each ratio can be in a different unit of measure.

$$\frac{1 \text{ ft.}}{2 \text{ ft.}} = \frac{4 \text{ in.}}{8 \text{ in.}}$$

RULE 3: The proportion is called direct when the values of both ratios increase either from the numerator to the denominator or from the denominator to the numerator.

$$\frac{1}{2} = \frac{4}{8} \text{ or } \frac{2}{1} = \frac{8}{4}$$

Example: If 100 ft. of wire weighs 18 lb., what will 150 ft. weigh?

Analysis: Since 150 ft. of wire will weigh more than 100 feet of the same kind of wire, both ratios should have the larger numerator, and both ratios the smaller denominator.

	Weights		**Lengths**
1. The proportion is	$\frac{\text{heavier weight}}{\text{lighter weight}}$	=	$\frac{\text{longer length}}{\text{shorter length}}$
2. Let x = wt. of 150′ of wire. Substitute the given values in the proportion above.	$\frac{x}{18}$	=	$\frac{150}{100}$
3. Cross-multiply the members and then divide each member by the coefficient of x.	$x \times 100$	=	18×150
	$\frac{x \times 100}{100}$	=	$\frac{18 \times 150}{100}$
	x	=	27 lb. Ans.

ASSIGNMENT

Find the unknowns in problems 1 to 12.

1. $\frac{x}{4} = \frac{6}{12}$
2. $\frac{x}{6} = \frac{5}{15}$
3. $\frac{x}{5} = \frac{3}{7}$
4. $\frac{y}{5} = \frac{8}{30}$
5. $\frac{3}{2} = \frac{z}{9}$
6. $\frac{2}{w} = \frac{5}{9}$
7. $\frac{2}{3} = \frac{17}{\ell}$
8. $\frac{p}{6} = \frac{2.875}{5.75}$
9. $\frac{x}{9} = \frac{2.5}{7.5}$
10. $\frac{x}{5.75} = \frac{20}{1}$
11. $\frac{2\ 1/2}{x} = \frac{5}{10}$
12. $\frac{3.5}{1\ 3/4} = \frac{20}{x}$
13. If 125 ft. of wire weighs 13 lb., how much will 375 ft. weigh?
14. If 50 ft. of band iron weighs 11.50 lb., what will 125 ft. weigh?
15. If 75 ft. of wire weighs 8.70 lb., how much will 25 ft. weigh?
16. If 100 ft. of band iron weighs 31.8 lb., what will 79 ft. weigh?
17. The actual length of a line on a job is 60 in. If the scale is 1 : 4, what will be the length of the line on the drawing?
18. The actual measurement of a line on a drawing is 3/4 in. The scale is 1/4″ = 1′ 0″. What will be the length of the line on the job?
19. It takes a sheet metal worker 2 hours to make 3 elbows. How many will he make in an 8-hour day?
20. If 12 drills cost $4.26, how much will 15 drills cost?
21. A sheet metal apprentice is paid $19.11 for 7 hours of work. What will be his pay for 40 hours of work at the same rate?
22. If 50 gallons of oil flow through a feeder pipe in 20 minutes, how long will it take to fill a tank of 950-gallon capacity?
23. A cylindrical tank holds 500 gallons when it is filled to its full depth of 7 ft. When the gage shows that it contains oil at a height of 2′ 6″, how many gallons are in the tank?

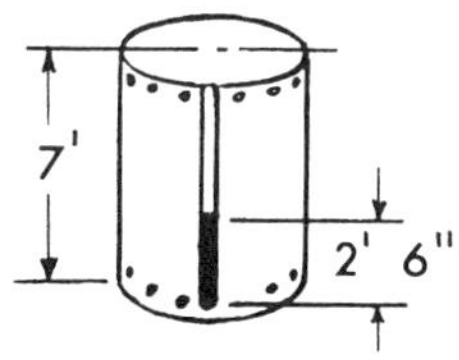

24. A bronze alloy contains 9 parts of copper by weight to 2 parts of tin. Find the amount of each metal in 220 lb. of this bronze.

    Hint: x = lb. of tin; y = lb. of copper

    $\frac{x}{220} = \frac{2}{2 + 9}$ and $\frac{y}{220} = \frac{9}{2 + 9}$
25. A victory solder contains 13 parts tin, 2 parts silver, 85 parts lead. Find the amount of each metal in 50 lb. of this solder.

UNIT 52 INVERSE PROPORTION

Proportion can be direct or *inverse* (indirect). In a direct proportion, the ratios vary directly (an increase in one ratio produces an increase in the other ratio). A proportion is called inverse when an increase in one ratio produces a decrease in the other ratio. In the illustration of the change pulleys on a drill press, the smallest driver pulley will rotate the largest driven pulley at a lesser r.p.m.

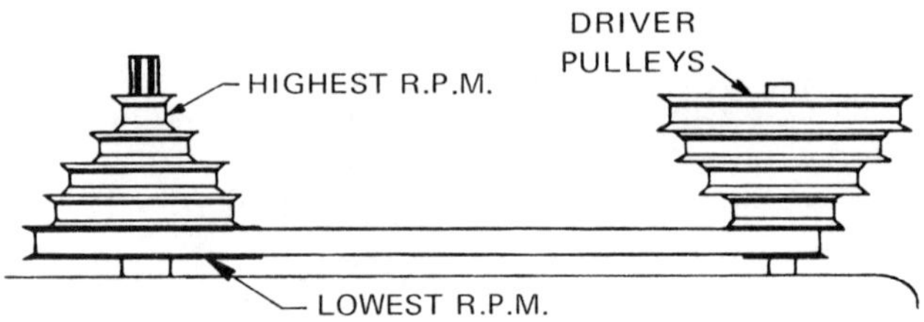

Inverse proportion is calculated in the same manner as direct proportion, except that in writing the inverse proportion, the first and second, or the third and fourth terms are reversed, or inverted.

RULE: The diameters of the driven pulley and the driver pulley vary inversely as the r.p.m. of the driven pulley and the driver pulley

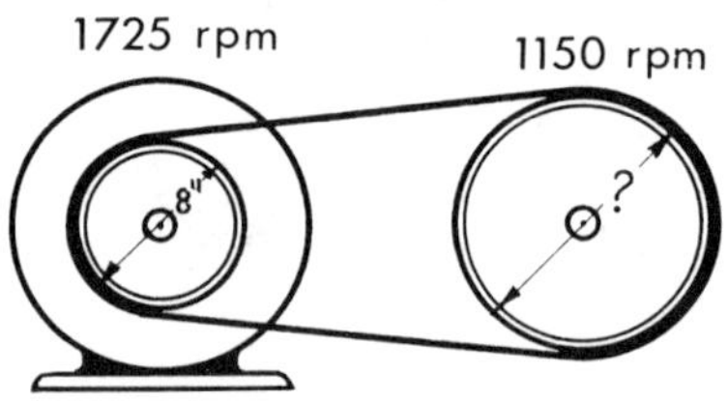

Example: The 8-in. pulley (driver) on a motor turns at 1725 r.p.m. If the driven pulley turns at 1150 r.p.m., find its diameter.

	Diameters		Turns
1. The proportion is written with the third and fourth terms inverted.	dia. driven / dia. driver	=	r.p.m. driver / r.p.m. driven
2. Let x = dia. of driven pulley. Substitute known values in the proportion above.	$\frac{x}{8}$	=	$\frac{1725}{1150}$
3. Cross-multiply the members.	$x \times 1150$	=	8×1725
Divide each member by the coefficient of x.	$\frac{x \times 1150}{1150}$	=	$\frac{8 \times 1725}{1150}$
	x	=	12″ Ans.

Note: Since the surface speed of both pulleys must be the same, then $\pi dn/12 = \pi DN/12$, or $dn = DN$ where d, n, and D, N equal the diameters and turns of the driver and driven pulleys respectively.

ASSIGNMENT

A. Inverse Proportion

1. Compute the missing value in each problem.

	DRIVEN PULLEY		DRIVER PULLEY	
	Dia.	**r.p.m.**	**Dia.**	**r.p.m.**
a.	?	300	6″	400
b.	?	250	9″	750
c.	12″	150	?	450
d.	8″	600	?	200
e.	4″	?	3″	1600
f.	12″	?	8″	800
g.	6″	900	9″	?
h.	12″	450	9″	?

2. Compute the missing value in each problem. **Hint:** Use "No. of teeth" instead of "dia."

	DRIVER GEAR		DRIVEN GEAR	
	No. of Teeth	**r.p.m.**	**No. of Teeth**	**r.p.m.**
a.	16	100	?	50
b.	48	200	?	100
c.	?	300	96	200
d.	?	420	28	360
e.	56	250	70	?
f.	28	390	32	?
g.	24	?	20	135
h.	14	?	21	1200

3. Solve each problem. **Hint:** Fewer men will take a longer time to do the job.
   a. If 10 men can do a job in 4 hr., how long will it take 5 men?
   b. If 3 men can do a job in 9 hr., how long will it take 6 men?
   c. If 8 men can do a job in 3 hr., how long will it take 5 men?

B. Summary of Proportion

1. If 4 men can assemble 200 elbows in one day, how many elbows can 35 men, working at the same rate, assemble in one day?
2. A pile of 105 sheets of metal is 16 in. high. How many sheets would there be in a 40-in. pile?
3. A man can drill 105 holes in 3 hours. How many holes will he be able to drill in an 8-hour day?
4. Find the diameter of the small pulley shown below.

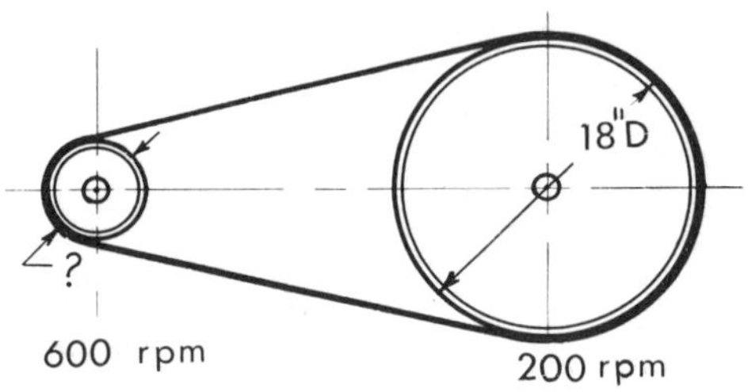

5. A 7-in. pulley revolving at 820 r.p.m. must be belted to a 28-in. pulley. How many r.p.m. will the 28-in. pulley make?
6. A pulley must be driven at the rate of 850 r.p.m. If the motor r.p.m. is 2125 and the driven pulley is 17.5 in., what must be the size of the pulley on the motor?

7. If the operating handle of a slip roll forming machine with a gear reduction drive is turned 10 revolutions, how many revolutions will the rolls make? The gear attached to the operating handle has 12 teeth and the gear attached to the rolls has 36 teeth.

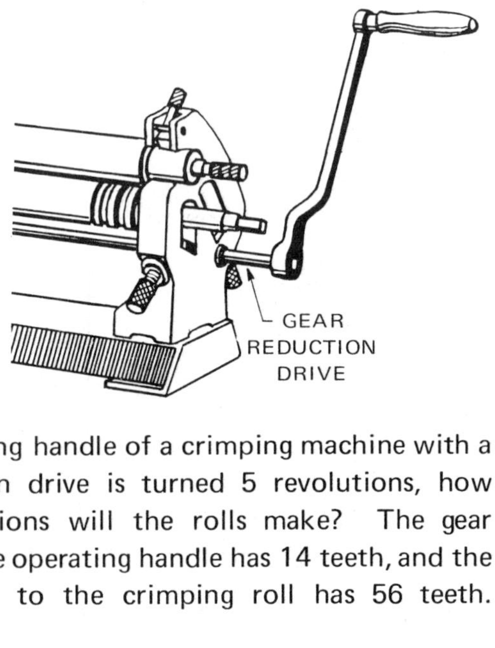

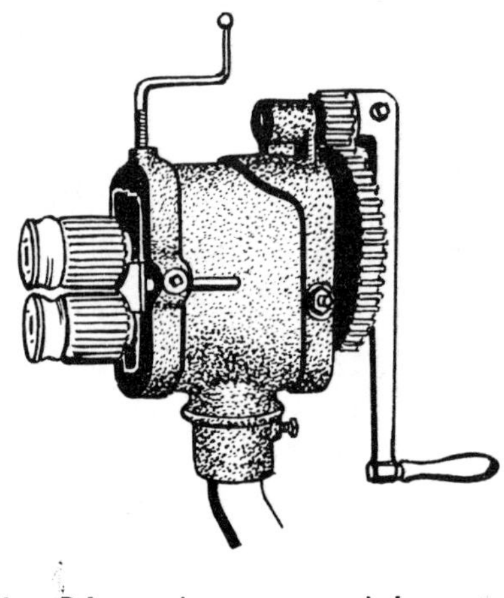

8. If the operating handle of a crimping machine with a gear reduction drive is turned 5 revolutions, how many revolutions will the rolls make? The gear attached to the operating handle has 14 teeth, and the gear attached to the crimping roll has 56 teeth.

9. An 84-tooth gear revolving at 450 r.p.m. drives a gear which revolves at 350 r.p.m. How many teeth has the second gear?

UNIT 53 INTRODUCTION TO FORMULAS

A formula is an equation which represents a rule or principle expressed in symbols. The perimeter of a rectangle, for example, is found by adding twice the length and twice the width. This rule may be written in the form of an equation as follows: $P = 2\ell + 2w$

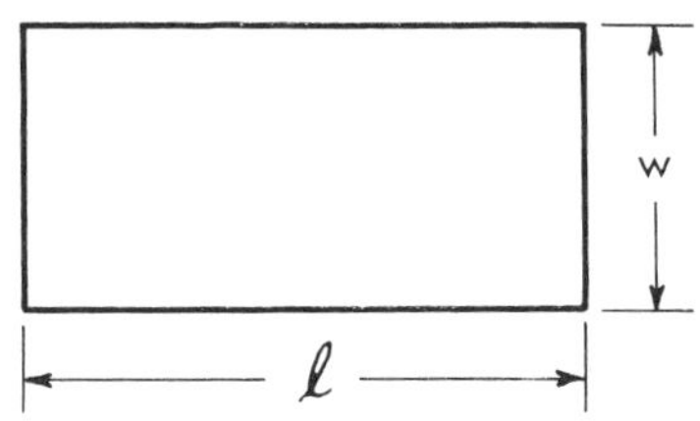

When formulas are written in simple forms, numerical values can be substituted for them in order to solve the problem. The formula is used extensively in shop work and other technical work. Formulas make problems more easily understood and more easily applied than rules.

In order to explain how to solve a problem by the use of a formula, a problem which has been solved previously by arithmetic will be solved by using a formula.

Example: How many pieces, each 8 3/4 in. long, can be cut from a piece 60 in. long? How much waste will be left?

PROCEDURE - USING A FORMULA

1. Analyze the problem.

   What is to be found in this problem? The example calls for the number of pieces to be cut from a piece of known length. Think of the problem as an equation:

$$\text{Number of Pieces} = \frac{\text{Length of stock}}{\text{Length of one piece}}$$

2. Write the formula.

   Assign one letter to represent each of the quantities in the equation. Let *N* signify the number of pieces because it is the first letter of the word "number." Use a capital *L* to represent the total length of stock, and a small ℓ to represent the length of one piece. These letters are symbols which express the problem in the form of an equation. (See Unit 31, Introduction to Symbols.)

$$N = \frac{L}{\ell}$$

3. Substitute the known values of the letters in the formula.

$$N = \frac{60}{8\ 3/4} \qquad 8\frac{3}{4}\text{ in.} = 8.75\text{ in.}$$

4. Find the result by performing the computations as indicated on the right side of the equal sign.

$$N = \frac{60}{8.75} \qquad 8.75\overline{)\,60.00_\wedge} \text{ in.} = 6 \text{ pieces}$$
$$\begin{array}{r} 52\ 50 \\ \hline 7.50 \end{array}$$

5. Label the answer correctly. N = 6 pieces, 7.5 in. waste Ans.

ARRANGEMENT OF WORK – USING A FORMULA

1. Formula $N = \frac{L}{\ell}$

2. Substitution $N = \frac{60}{8\ 3/4}$ $\qquad 8\frac{3}{4} \text{ in.} = 8.75 \text{ in.}$

3. Evaluation

   Note: The term evaluation means doing the computations indicated.

$$N = \frac{60}{8.75} \qquad 8.75\overline{)\,60.00_\wedge} \text{ in.} = 6 \text{ pieces}$$
$$\begin{array}{r} 52\ 50 \\ \hline 7.50 \text{ in.} \end{array}$$

N = 6 pieces, 7.5-in. waste Ans.

UNIT 54 PERIMETER OF A SQUARE

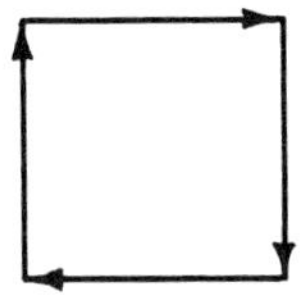

The distance around a polygon is called the *perimeter.* It is expressed in units of linear measure. Sheet metal workers must know how to find the perimeters of plane figures so that the length of material, such as sheet metal, wire, etc., for a job can be determined. The length of the stretchout of a pipe is equal to the perimeter of the end of the pipe without the allowance for the seam.

A square is a four-sided plane figure with the sides equal and the adjacent sides meeting at angles of 90 degrees. The ends of pipes and elbows are frequently square-shaped. (To construct a square, see page 96.)

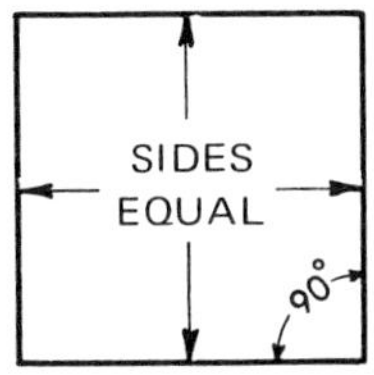

SQUARE

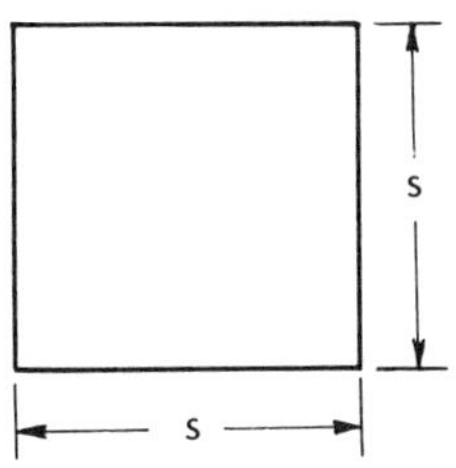

RULE: The perimeter of a square is equal to four times the length of one side:

Formula: $P = 4s$

Symbols: P = perimeter

s = length of one side of square

EXPLANATION OF RULE AND FORMULA

Since the perimeter is the distance around the square, then

1. $P = s + s + s + s$ 2. $\therefore P = 4s$

PERIMETER OF

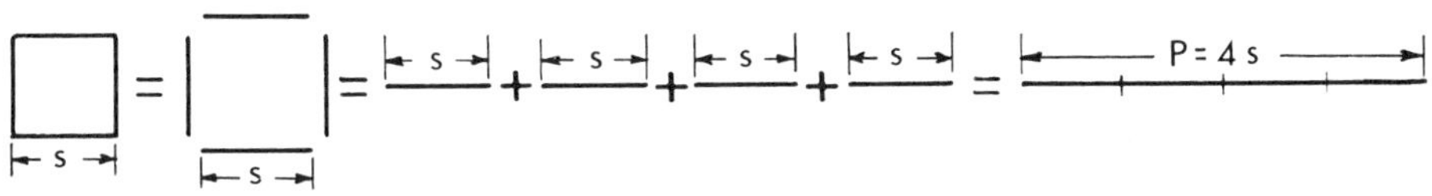

Example: If the side of a square is 2 3/4 in., find the perimeter.

Note: The perimeter of a square can be computed using either fractions or decimals.

1. Formula	$P = 4s$	
2. Substitution	$P = 4 \times 2\,3/4$	2 3/4 in. = 2.75 in.
3. Evaluation	$P = 4 \times 2.75$	2.75 in.
		x4
	$P = 11$ in. Ans.	11.00 in.

ASSIGNMENT

1. Draw the squares listed below and find the perimeter of each:

   a. side = 1 1/4″ c. side = 1 11/16″

   b. side = 2 1/8″ d. side = 2.58″

2. Find the perimeter of each square listed below:

a. s = 3 1/2″

b. s = 5 3/8″

c. s = 9 5/16″

d. s = 18 31/32″

e. s = 9 63/64″

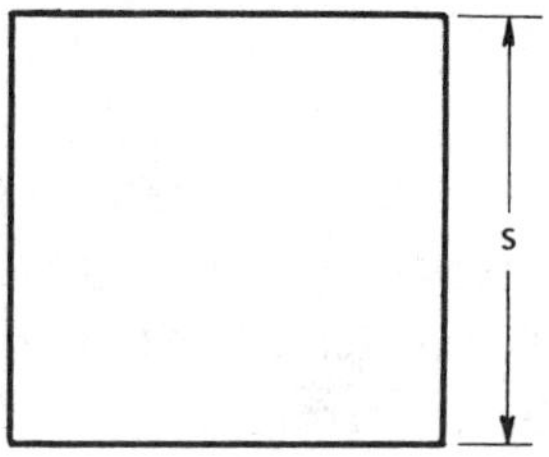

f. s = 9.30″

g. s = 2′ 9″

h. s = 3′ 4 1/2″

i. s = 5′ 10 3/4″

j. s = 2′ 9 1/8″

3. Find the length of the stretchout (no allowance for seam) of each square pipe listed below:

a. s = 1 3/8″

b. s = 1 1/16″

c. s = 5/8″

d. s = 13/16″

e. s = 3 1/2″

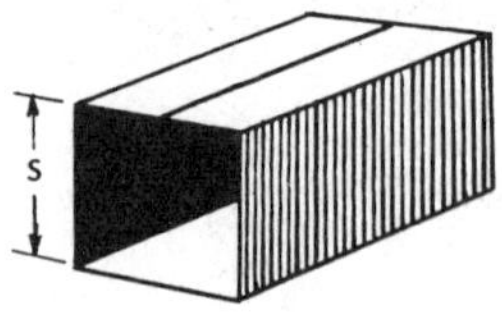

f. s = 7 1/4″

g. s = 10 1/8″

h. s = 5.8″

i. s = 2′ 1 1/2″

j. s = 4′ 4 3/8″

4. Compute the length of metal molding required for a room, 12′ 6″ x 12′ 6″.

5. Find the approximate length of metal edging needed for a square table; the length of one side is 4′ 8 1/2″.

UNIT 55 PERIMETERS OF RECTANGLES AND TRIANGLES

PERIMETER OF A RECTANGLE

A rectangle is a four-sided plane figure whose opposite sides are equal and whose adjacent sides meet at angles of 90 deg. Jobs, such as pipes and elbows, may be rectangular in shape. The perimeter of a rectangle is found in a manner similar to that of the square. (To draw a rectangle, see page 97.)

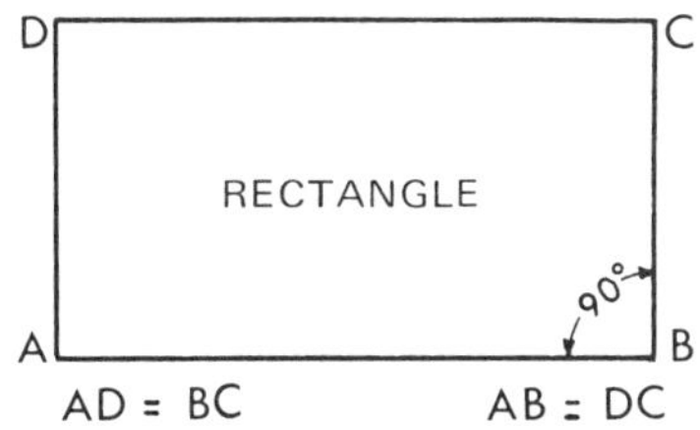

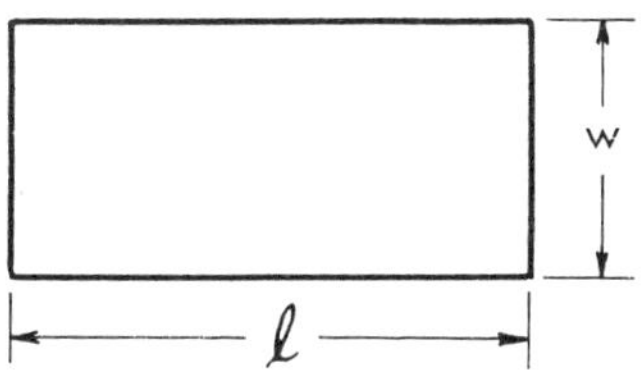

RULE: The perimeter of a rectangle is equal to two times the length plus two times the width.

Formula: $P = 2\ell + 2w$

Symbols: P = perimeter

ℓ = length of longer side, or length

w = length of shorter side, or width

EXPLANATION OF RULE AND FORMULA

1. $P = \ell + w + \ell + w$ (See drawing below.)
2. $P = \ell + \ell + w + w$ (rearranging terms)
3. $P = 2\ell + 2w$ (adding like terms), or $P = 2(\ell + w)$

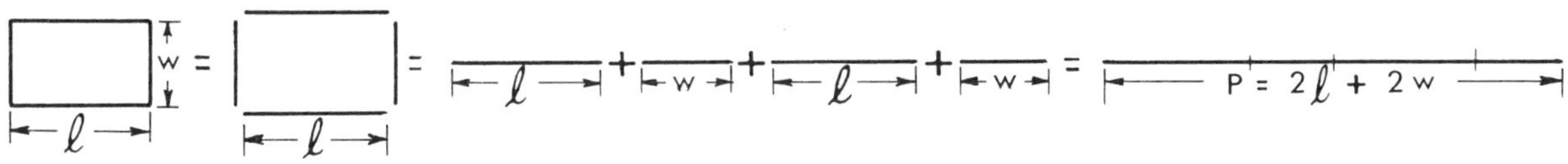

Example: Find the perimeter of a rectangle, 2 1/2″ x 10″

Note: The perimeter of a rectangle can be computed using either fractions or decimals.

1. Formula $\quad P = 2\ell + 2w$
2. Substitution $\quad P = 2 \times 10 + 2 \times 2\ 1/2 \qquad 2\ 1/2$ in. = 2.5 in.
3. Evaluation $\quad P = 2 \times 10 + 2 \times 2.5$

$P = 20 + 5$

$P = 25$ in. Ans.

2.5 in.
x 2
———
5.0 in.

PERIMETER OF A TRIANGLE

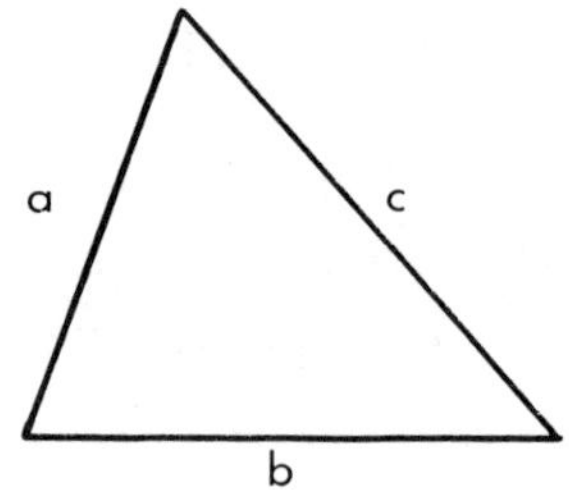

A triangle is any three-sided figure. The length of the sides may be equal or unequal. The perimeter of a triangle, like that of all polygons, is the distance around the figure.

RULE: The perimeter of a triangle is equal to the sum of the lengths of its sides.

The perimeter in this figure is a + b + c. If a = 12″, b = 14″, c = 15″, perimeter is 12 + 14 + 15 = 41″.

ASSIGNMENT

A. Perimeter of a Rectangle

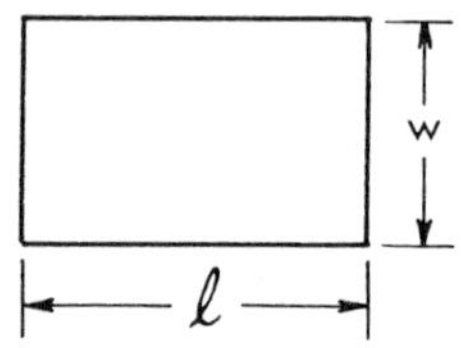

1. Draw the rectangles listed below and find the perimeter of each:

   a. ℓ = 3″, w = 2″ c. ℓ = 2 3/8″, w = 1 9/16″
   b. ℓ = 2 1/4″, w = 1 7/8″ d. ℓ = 2.5″, w = 1.9″

2. Find the perimeter of each rectangle listed below:

   a. ℓ = 5″, w = 4″
   b. ℓ = 8″, w = 6″
   c. ℓ = 10 1/2″ w = 9 1/4″
   d. ℓ = 15 3/4″ w = 11 7/8″
   e. ℓ = 18 5/8,″ w = 15 11/16″
   f. ℓ = 18 9/32″ w = 12 63/64″
   g. ℓ = 6.58″, w = 4.20″
   h. ℓ = 8.36″, w = 6.94″
   i. ℓ = 12′9″, w = 5′ 8″
   j. ℓ = 6′ 5″, w = 3′ 4″
   k. ℓ = 9′ 10″, w = 7′ 11″
   l. ℓ = 15′ 6 1/2″ w = 9′ 8 1/2″

3. Find the length of the stretchout of each rectangular pipe (no allowance for seam) listed below:

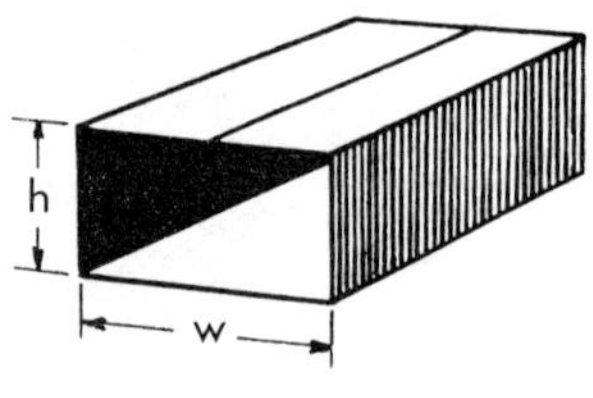

Note: Construct the formula by using the symbols on the drawing.

a. w = 10″, h = 3 1/4″
b. w = 12″, h = 5 1/2″
c. w = 14 1/2″, h = 8 1/4″
d. w = 5 3/8″, h = 3 7/32″
e. w = 5.32″, h = 2.36″
f. w = 2′ 1 1/2″, h = 1′ 9 3/16″
g. w = 3′ 5 1/4″, h = 2′ 9 7/8″
h. w = 2′ 10 1/2″, h = 1′ 7 3/4″

4. Compute the linear feet of metal molding required for a room size 8′ 4″ x 13′ 9″.

5. Find the approximate length of metal edge for a table of rectangular shape, 2′ 6 1/4″ x 5′ 9 7/8″.

B. Perimeter of a Triangle

1. Find the perimeters of triangles with sides of the following lengths:

a. a = 7″	b = 10″	c = 15″
b. a = 3′ 6″	b = 2′ 4″	c = 4′ 9″
c. a = 4′ 2″	b = 7.25′	c = 6.67′
d. a = 8.6′	b = 9.3′	c = 14.85′

2. If the perimeter of a triangle is 9′ 8 1/2″ and side a = 3.75′ and b = 2.33′, how many inches is side c?

3. The perimeter of a triangle is 44″. Side b is 1′ 4″; side c is .92′. What is side a to the nearest 1/4″?

UNIT 56 CIRCUMFERENCE OF A CIRCLE

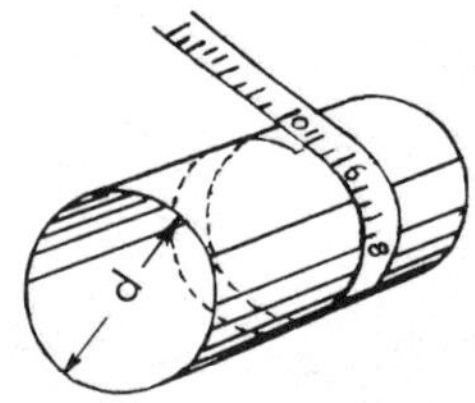

The circumference of a circle is the length of the curved line, expressed in units of linear measure. Since the length of the stretchout of a round pipe is equal to the circumference of the end, metal workers must be able to compute circumferences of circles.

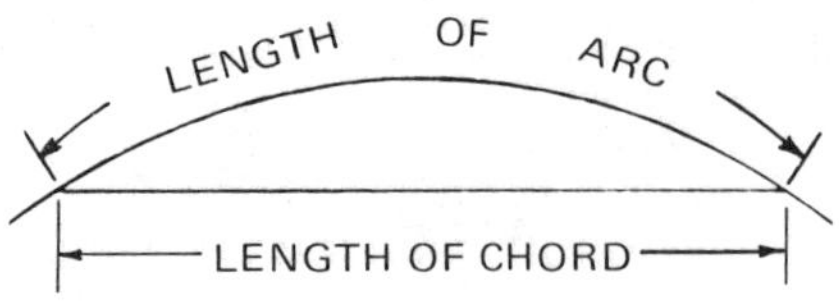

In the case of circles of large diameters, the approximate length of stretchout can be found with dividers, but since the length of the arc of a circle is greater than the chord, this method is very inaccurate with circles of small diameters. This can be proved by measuring an arc and a chord of a circle with a flexible rule.

RULE: The circumference of a circle is found by multiplying the diameter by a number called pi. (The value of pi, symbol π, is usually 3.1416 for accurate work, 3.14 or 3 1/7 for approximate work.)

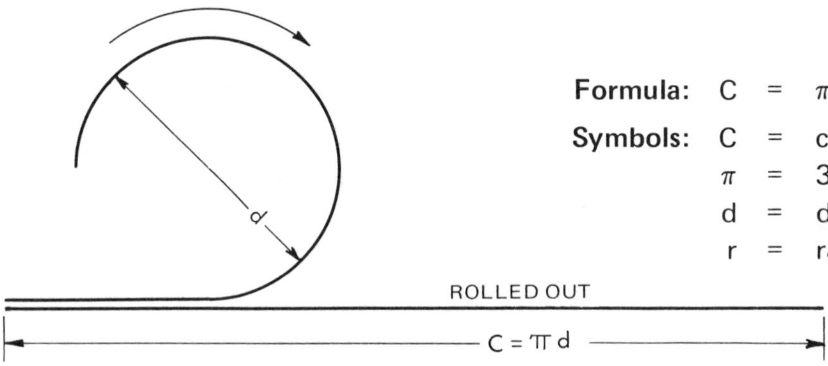

Formula: $C = \pi d$, or $2\pi r$

Symbols:

C = circumference

π = 3.14

d = diameter, or $2r$

r = radius

Example: If the diameter of a circle is 5 3/4 in., compute the circumference to the closest 16th of an inch.

Note: Since π is in decimal form, change the diameter to decimal form. All problems will be calculated using 3.14 for π.

1. Formula $C = \pi d$
2. Substitution $C = 3.14 \times 5\,3/4$
3. Evaluation $C = 3.14 \times 5.75$

$C = 18.0550$ in.

$C = 18\,1/16$ in. Ans.

5 3/4 in. = 5.75 in.

```
     3.14
   x 5.75
     1570
    2198
   1570
  18.0550
```

ASSIGNMENT

1. Find the circumference of each of the following circles to the nearest 1/32nd of an inch.

a. diameter	=	3″	g. radius	=	6 3/4″
b. diameter	=	4″	h. diameter	=	5 7/8″
c. diameter	=	6.12″	i. diameter	=	10 1/8″
d. diameter	=	12.24″	j. radius	=	2 5/16″
e. diameter	=	3 1/2″	k. radius	=	3 9/16″
f. diameter	=	3 3/4″	l. diameter	=	5 31/32″

2. Find the circumference of each of the following circles to the nearest 16th of an inch. Change the feet to inches before working the problems.

a. diameter	=	2′ 6″	d. diameter	=	9′ 11 1/2″
b. diameter	=	1′ 3″	e. radius	=	5′ 3 1/2″
c. diameter	=	3′ 9″	f. radius	=	1′ 2 1/2″

3. Find the length of stretchout (no allowance for seam) for each of the following circular pipes to the closest 64th of an inch. Work all problems in inches.

a. diameter	=	9 1/2″	e. radius	=	4 5/32″
b. diameter	=	10 1/2″	f. radius	=	9.88″
c. diameter	=	4 3/8″	g. diameter	=	4′ 9 1/2″
d. diameter	=	6 15/16″	h. diameter	=	2′ 10 3/4″

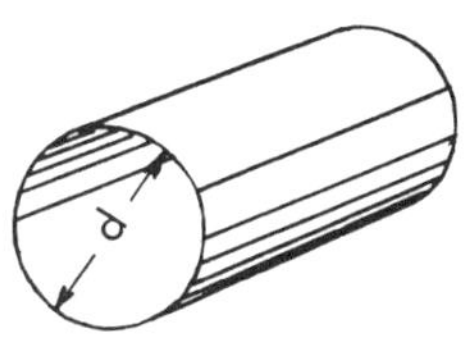

4. Find the length of wire in inches (closest 16th) in a single loop of a coil, if the average diameter is 25 1/2 in.

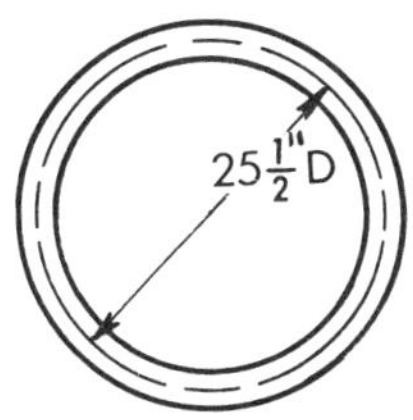

5. Find the approximate length of wire in feet and the closest inch in a coil of 20 turns, if the average diameter of the coil is 39 3/4 in.

6. Find the length of a band of iron in feet and inches (closest 16th) to go around a circular tank, if the average diameter of the tank is 10 ft. 6 1/2 in.

UNIT 57 PERIMETER OF A SEMICIRCULAR-SIDED FIGURE

A semicircular-sided figure consists of two half circles and a rectangle or a square. Twice the radius, or the diameter of one of the half circles equals the height of the figure. (For the construction of a semicircular-sided figure, see page 100.) Since this shape is used for tanks, pipes, etc., it is important to know how to find the perimeter.

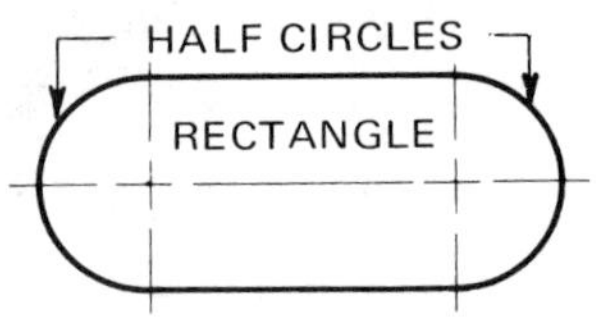

RULE: The perimeter is equal to the length of the arcs of the two half circles plus twice the distance between the half circles.

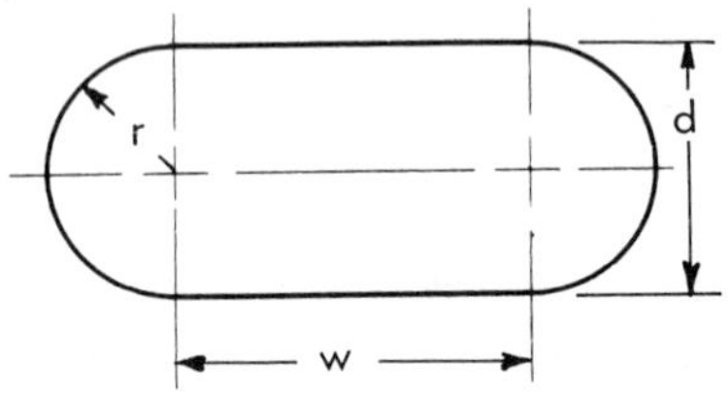

Formula: $P = \pi d + 2w$

Symbols:

P = perimeter

π = 3.14

r = radius of half circles

d = diameter of half circles, or 2r

w = distance between the half circles

EXPLANATION

1. $P = \dfrac{\pi d}{2} + w + \dfrac{\pi d}{2} + w$ (See drawing.)

2. $P = \dfrac{\pi d}{2} + \dfrac{\pi d}{2} + w + w$ (rearranging)

3. $\therefore\ P = \pi d + 2w$ (adding the like terms)

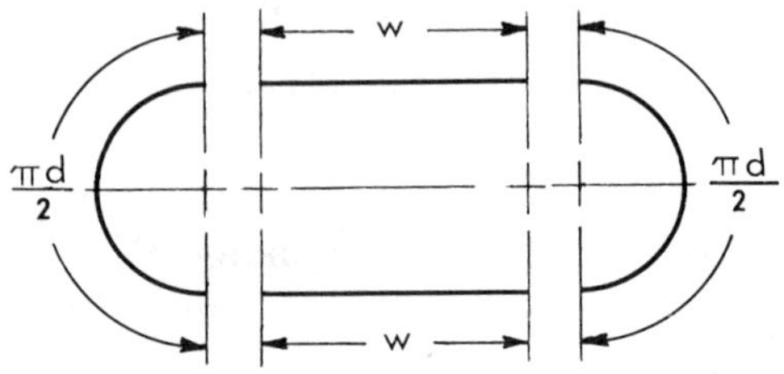

Example: If r = 6 in. and w = 5 1/4 in. in the figure shown above, find the perimeter to the closest 16th of an inch.

Note: Work problem decimally.

1. Formula $P = \pi d + 2w$ d = 2 x 6, or 12 in.

2. Substitution P = 3.14 x 12 + 2 x 5 1/4 5 1/4 in. = 5.25 in.

3. Evaluation P = 3.14 x 12 + 2 x 5.25

P = 37.68 + 10.5

P = 48.18 in., or 48 3/16 in. Ans.

```
  3.14
 x 12
  628
 314
 37.68 in.
```

ASSIGNMENT

1. Find the perimeter of each of the following semicircular-sided figures to nearest 32nd of an inch. Make full-size drawings of the figures for problems *a* to *e.*

a. r = 2", w = 1"
b. r = 1", w = 2"
c. r = 1 1/2", w = 2"
d. r = 1 1/4", w = 1 1/2"
e. r = 1 3/4", w = 2 1/2"
f. r = 6" w = 7 1/4"
g. r = 5", w = 9 3/4"
h. r = 7.6", w = 10.9"
i. d = 10 1/2", w = 12 3/4"
j. r = 5 3/8", w = 6 7/8"

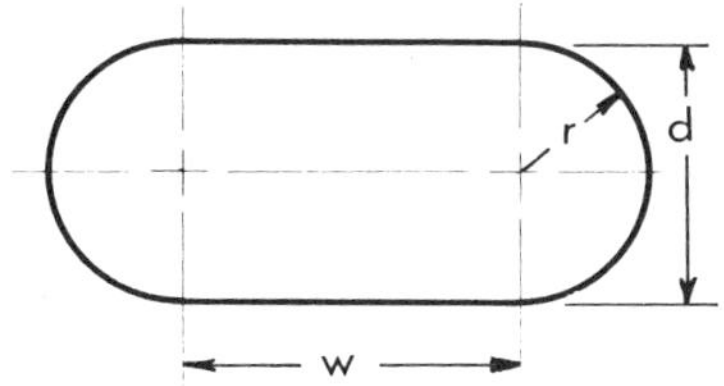

2. Find the perimeter of each of the following semicircular-sided figures to the nearest 16th of an inch. Change to inches before working.

a. d = 3', w = 2'
b. d = 5', w = 3'
c. d = 6', w = 4'
d. d = 6.50', w = 3.75'
e. d = 3' 6", w = 2' 6"
f. r = 4' 3", w = 3' 9"
g. d = 2' 6", w = 4' 4"
h. r = 1' 9", w = 2' 10"

3. Find the length of the stretchout (no allowance for seam) for each of the following semicircular-sided pipes to the nearest 64th of an inch.

a. w = 3", r = 2"
b. w = 4", r = 3"
c. w = 5", r = 4"
d. w = 6", r = 2 1/2"

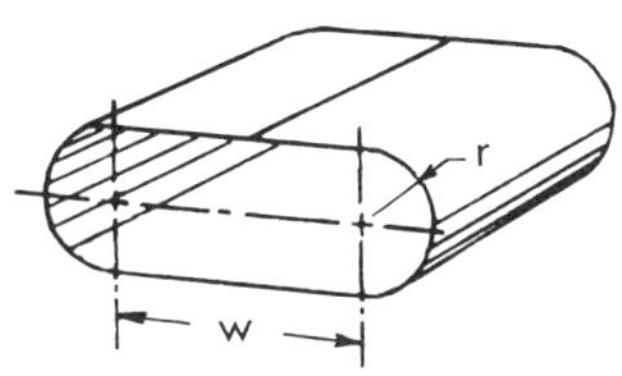

e. w = 6.25", d = 5.75"
f. w = 2' 1", d = 1' 6"
g. w = 4' 2", d = 2' 9"
h. w = 12.39", d = 8.60"

4. Find the approximate length of belt in feet and inches (closest 16th) required for the job shown below.

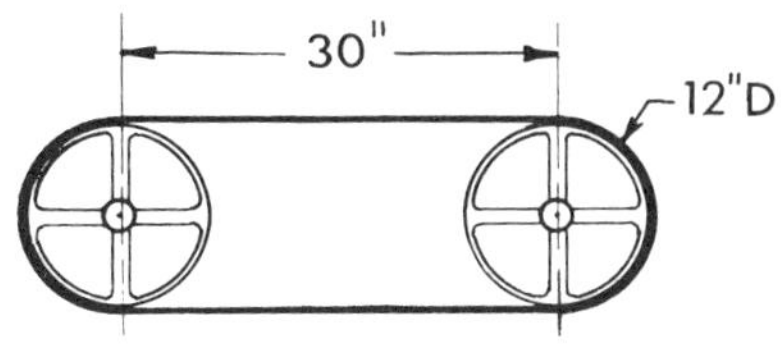

SECTION 8 — AREA MEASURE

UNIT 58 THE UNITS OF AREA MEASURE

The area of a plane figure is the number of square units which that figure contains. The ability to compute areas is of great importance to the sheet metal worker who may be called upon to determine the size, weight, cost, and amount of waste of the sheet metal he uses. Finding the air-carrying capacity of the pipes required for a ventilating system is a good illustration of the practical application of this work. For additional illustrations one needs only to consider the use of lighter materials and lightening holes in the aircraft, shipbuilding, and railway industries.

TABLE 58-1 AREA OF SQUARE MEASURE

144 square inches	=	1 square foot (sq. ft.)
9 square feet	=	1 square yard (sq. yd.)
100 square feet	=	1 square (for roofing, etc.)

The left drawing below shows the number of square inches in a square foot. If the 1-in. squares are counted, it will be found that there are 144 of them in a square foot. This result can be obtained more easily by multiplying 12 rows by 12 in a row (12 x 12 = 144).

Note: in. x in. = sq. in., or □″; ft. x ft. = sq. ft., or □′; yd. x yd. = sq. yd.

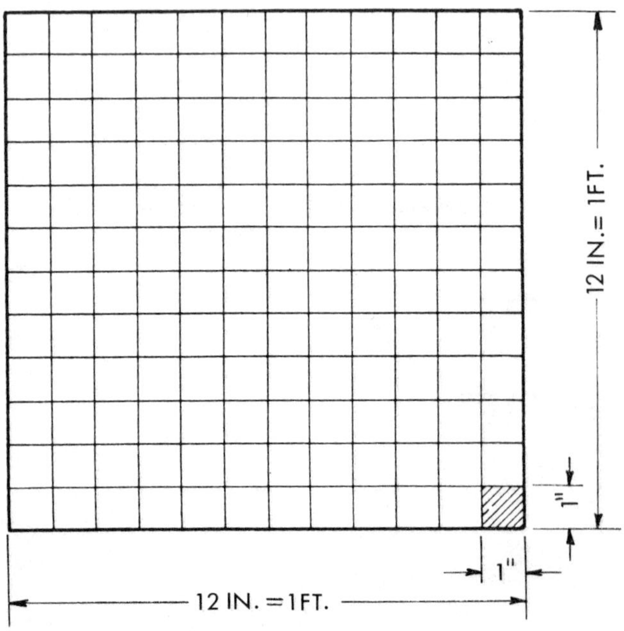

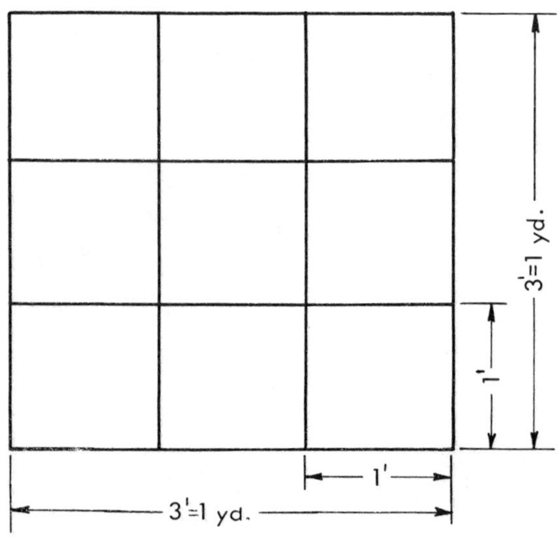

In a similar manner, it will be found that there are 9 square feet in one square yard, since 3 x 3 = 9.

Note: To verify this answer, count the squares in the right drawing on the preceding page.

Conversion Rules and Problems

RULE 1: To change sq. in. to sq. ft., divide the number of sq. in. by 144.

Example: Change 1420 sq. in. to sq. ft.

Since there are 144 sq. in. in one sq. ft., divide the sq. in. (1420) by 144 to obtain square feet. The answer is 9.86 sq. ft.

9.861 sq. ft.
144) 1420.000

RULE 2: To change sq. ft. to sq. in., multiply the number of sq. ft. by 144.

Example: Change 2.5 sq. ft. to sq. in.

Since there are 144 sq. in. in one sq. ft., multiply the sq. ft. (2.5) by 144 to obtain the number of square inches. The answer is 360 sq. in.

144
x2.5
720
288
360.0 sq. in.

RULE 3: To change sq. in. to sq. yd., change sq. in. to sq. ft., and then to sq. yd.

Example: Change 21,402 in. to sq. yd.

1. Change the sq. in. (21,402) to sq. ft.
2. Change the sq. ft. (148.625) to sq. yd. The answer is 16.51 sq. yd.

148.625 sq. ft.
144) 21,402.000

16.513 sq. yd.
9) 148.625

ALTERNATE METHOD

1. Find the number of sq. in. in a sq. yd.
2. Divide the given sq. in. (21,402) by this result, to find the sq. yd. The answer is 16.51 sq. yd.

144 x 9 = 1296 sq. in.

16.513 sq. yd.
1296) 21,402.000

TABLE 58-2 BLACK AND GALVANIZED IRON

	BLACK IRON		GALV. IRON		BLACK IRON		GALV. IRON
Gage U.S.	Thick. In.	Lb. per Sq. Ft.	Lb. per Sq. Ft.	Gage U.S.	Thick In.	Lb. per Sq. Ft.	Lb. per Sq. Ft.
32	.0097	.40625	.56250	19	.0418	1.75	1.90625
31	.0105	.4375	.59375	18	.0478	2.	2.15625
30	.0120	.5	.65625	17	.0538	2.25	2.40625
29	.0135	.5625	.71875	16	.0598	2.5	2.65625
28	.0149	.625	.78125	15	.0673	2.8125	2.96875
27	.0164	.6875	.84375	14	.0747	3.125	3.28125
26	.0179	.75	.90625	13	.0897	3.75	3.90625
25	.0209	.875	1.03125	12	.1046	4.375	4.53125
24	.0239	1.	1.15625	11	.1196	5.	5.15625
23	.0269	1.125	1.28125	10	.1345	5.625	5.78125
22	.0299	1.25	1.40625	9	.1495	6.25	6.40625
21	.0329	1.375	1.53125	8	.1644	6.875	7.03125
20	.0359	1.5	1.65625				

TABLE 58-3 GAGES AND THICKNESSES OF TIN PLATE

Base Wt. Lb.	Trade Term	Approx. U.S. Gage	Approx. Thick. In.	Lb. Per Sq. Ft.	Base Wt. Lb.	Trade Term	Approx. U.S. Gage	Approx. Thick. In.	Lb. Per Sq. Ft.
55		38	.0061	.253	155	2X	27	.0175	.712
60		37	.0068	.276	168	3XL	26	.0189	.771
65		36	.0073	.298	175	3X	26	.0197	.804
70		35	.0079	.321	180	DX	25	.0203	.827
75		34	.0084	.344	188	4XL	25	.0211	.863
80		33	.0090	.367	195	4X	25	.0220	.895
85		32	.0095	.390	208	5XL	25	.0238	.955
90		31	.0101	.413	210	D2X	25	.0241	.964
95		31	.0107	.436	215	5X	25	.0246	.987
100	1CL	30½	.0113	.459	228	6XL	24	.0261	1.047
107	1C	30	.0120	.491	235	6X	24	.0269	1.079
118		29	.0133	.542	240	D3X	24	.0275	1.102
123		29	.0138	.564	248	7XL	23	.0284	1.139
128	1XL	28	.0144	.588	255	7X	23	.0292	1.171
135	1X	28	.0152	.620	268	8XL	23	.0307	1.231
139	DC	28	.0156	.638	270	D4X	23	.0310	1.240
148	2XL	27	.0167	.680	275	8X	22	.0315	1.263
150		27	.0169	.688					

Note: To find the weight of 4 square feet of 26-gage black iron, multiply .75 by 4. The weight is 3 lb. Ans.

Note: A sheet labeled 26 x 24 x 120 means that the sheet is 26 gage, 24 in. wide, and 120 in. long.

ASSIGNMENT

A. Conversion Rules and Problems

1. Change 576 sq. in. to sq. ft.
2. Change 1000 sq. in. to sq. ft.
3. The area of the opening of a pipe is 535 1/2 sq. in. Find this area in sq. ft.

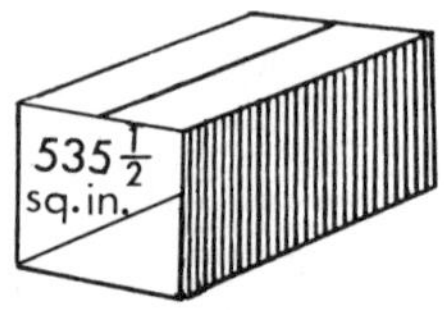

4. The area of the opening of a pipe is 200.6 sq. in. Find the area in sq. ft.
5. Change 2 sq. ft. to sq. in.
6. Change 5.75 sq. ft. to sq. in.
7. The area of the opening of a heating pipe is 7.5 sq. ft. Find this area in sq. in.
8. If the cross-section area (area of the opening) of a heating pipe is 12.125 sq. ft., find this area in sq. in.
9. Change 4000 sq. in. to sq. yd.
10. The area of a window is 2100 sq. in. Find its area in sq. yd.
11. Change 2 sq. yd. to sq. in.
12. Change 3.25 sq. yd. to sq. in.

B. Tables of Weight of Sheet Metal

Note: Refer to tables 58–2 (page 149) and 58–3 (page 150).

1. Find the weight of 5 sq. ft. of 26-gage black iron.
2. Find the weight of 8 sq. ft. of 24-gage galvanized iron.
3. Find the weight of 3 1/2 sq. ft. of 107-lb. tin plate.
4. If 10 sheets of 20-gage black iron, each 4 sq. ft. in area, are to be ordered, how many pounds of black iron must be ordered?
5. Find the difference in weight between 40 sq. ft. of 26-gage black iron and 40 sq. ft. of 26-gage galvanized iron.
6. Find the difference in weight between 36 sq. ft. of 28-gage black iron and 36 sq. ft. of 24-gage black iron.

UNIT 59 AREA OF A SQUARE

In order to find the weight and cost of square pieces of metal, it is necessary for the sheet metal worker to know how to compute the area of a square. Such computations are also necessary when he must find the area of the walls and floors of rooms, and the air-carrying capacity of pipes in heating and ventilating work.

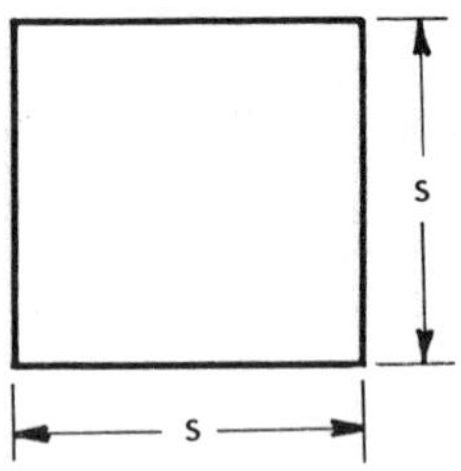

RULE: The area of a square is equal to the length of one side squared.

Formula: $A = s^2$

Symbols: A = area of square

s = length of one side of square

EXPLANATION OF RULE AND FORMULA

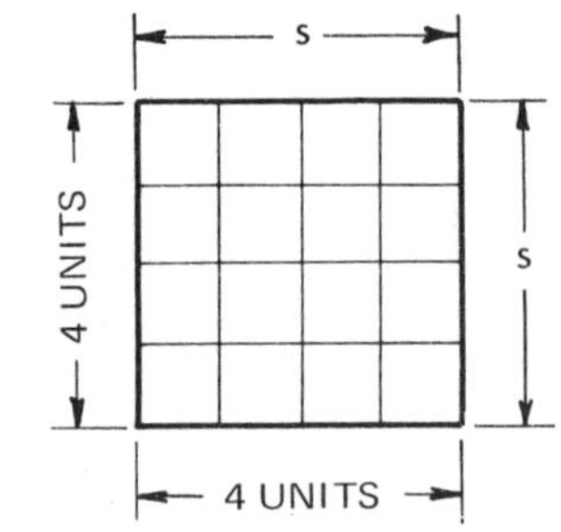

If a square is 4 units long on each side, then there will be 4 rows of 4 small (unit) squares in a row. The total unit squares will be 4 x 4, or 4^2, or 16. The number of unit squares will be equal to the area of the large square.

Since *s* equals number of units in one side, then *A* will equal s x s, or s^2. If the lengths of the sides are given in inches, feet, or yards, then the area will be expressed respectively in sq. in., sq. ft., or sq. yd.

Note: To check this answer, count the squares in the drawing above.

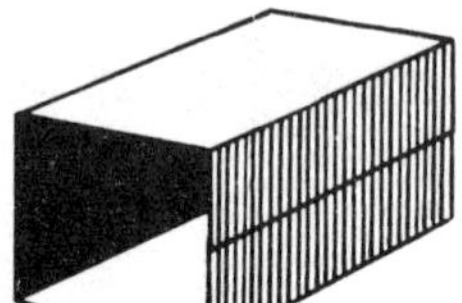

Example: Compute the cross-section area (area of opening) of a square pipe, 5 1/2 in. on a side.

1. Formula $A = s^2$
2. Substitution $A = \left(5\frac{1}{2}\right)^2$
3. Evaluation $A = 5.5 \times 5.5$

$A = 30.25$ sq. in. Ans.

$5\frac{1}{2}$ in. = 5.5 in.

```
    5.5 in.
  x 5.5 in.
  ---------
    2 7 5
  2 7 5
  ---------
  3 0.2 5  sq. in.
```

ASSIGNMENT

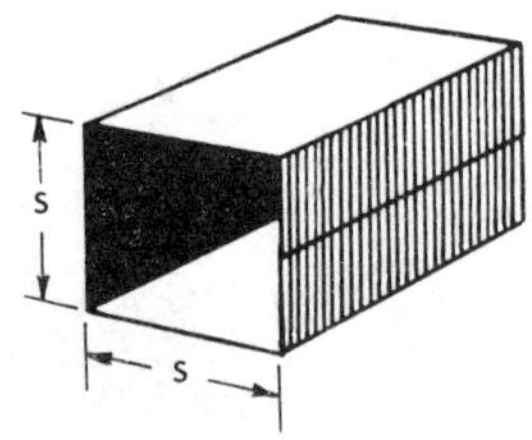

1. Compute the cross-section area in sq. in. of each pipe:

a. s = 9″

b. s = 4 1/2″

c. s = 5 1/2″

d. s = 8 1/2″

e. s = 9.25″

f. s = 10 3/4″

g. s = 14 5/8″

h. s = 50 1/4″

i. s = 5 11/16″

j. s = 1.503″

k. s = 1′ 9″

l. s = 3′ 4″

2. Compute the weight of each square piece of sheet metal. Refer to tables 58-2 (page 149) and 58-3 (page 150).

   a. s = 5′, 24-gage black iron
   b. s = 4 1/2′, 28-gage galvanized iron
   c. s = 32″, 10-gage black iron
   d. s = 36″, 14-gage galvanized iron
   e. s = 20″, 100-lb. tin plate
   f. s = 28″, 107-lb. tin plate

3. Compute the cost of sheet metal for each problem below. Refer to tables 58-2 (page 149) and 58-3 (page 150).

   a. s = 4′, 1 piece of 18-gage galvanized iron at $12.30 cwt.
   b. s = 3 1/2′, 10 pieces of 22-gage black iron at $7.15 cwt.
   c. s = 25″, 50 pieces of 28-gage galvanized iron at $8.65 cwt.
   d. s = 40″, 3 pieces of 14-gage black iron at $6.45 cwt.

4. Draw a square 5 7/8 in. on a side and compute its area.

5. Draw a square 4 15/16 in. on a side and compute its area.

6. When a square hole, 10 in. on a side, is cut in 10-gage black iron, what is the amount of weight reduction? Refer to table 58-2, page 149.

7. When a square hole, 20 in. on a side, is cut in 12-gage galvanized iron, what is the weight of the material that is eliminated? Refer to table 58-2, page 149.

8. Compute the area in sq. ft. of a shop floor 40′ 6″ by 40′ 6″.

9. Compute the area in sq. ft. of a wall 10′ 3″ by 10′ 3″.

UNIT 60 AREA OF A RECTANGLE

As in the case of the area of the square, there are similar practical applications for the mathematics of finding the area of a rectangle. In all sheet metal jobs of rectangular shape, there are the problems of areas, weights, costs, waste, etc.

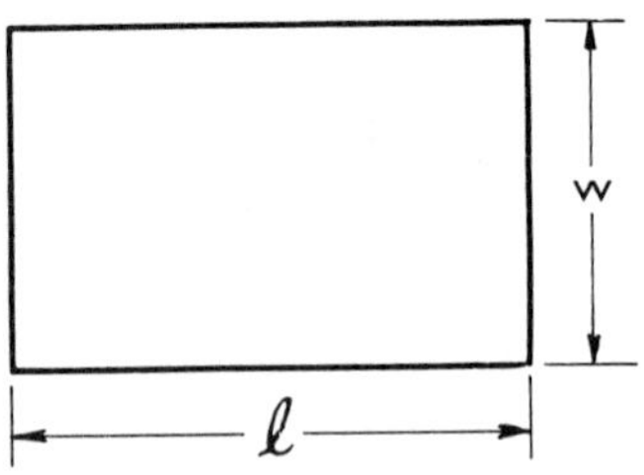

RULE: The area of a rectangle is equal to the product of its length and its width.

Formula: $A = \ell w$

Symbols: A = area
ℓ = length of longer side, or length
w = length of shorter side, or width

EXPLANATION OF RULE AND FORMULA

If a rectangle is 2 units wide and 6 units long, there will be 2 rows with 6 unit squares in each row, or 2 x 6 or 12 unit squares. The area of the rectangle is equal to the number of unit squares. Since ℓ = number of unit squares in the length and w = number of unit squares in the width, then $A = \ell w$.

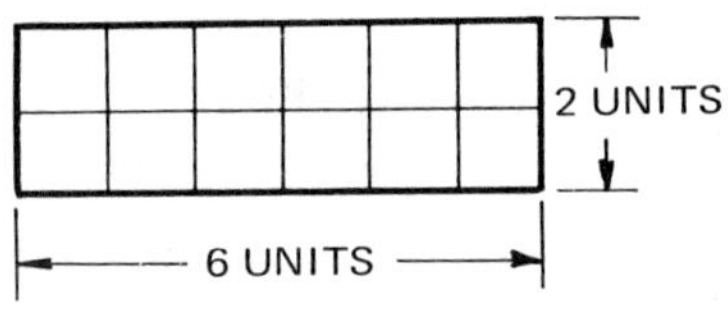

Example: If the length of a rectangle equals 12 in. and its width equals 3 1/4 in., compute the area in sq. in.

1. Formula $\quad A = \ell w$
2. Substitution $\quad A = 12 \times 3\frac{1}{4}$
3. Evaluation $\quad A = 12 \times 3.25$

$A = 39$ sq. in. Ans.

$3\frac{1}{4}$ in. = 3.25 in.

$$\begin{array}{r} 3.25 \text{ in.} \\ \times\ 12 \text{ in.} \\ \hline 650 \\ 325 \\ \hline 39.00 \text{ sq. in.} \end{array}$$

AREA OF A PARALLELOGRAM

The illustration below shows how a parallelogram can be converted into an equivalent rectangle. The area of the parallelogram is equal to the area of that rectangle, or base x height, or bh.

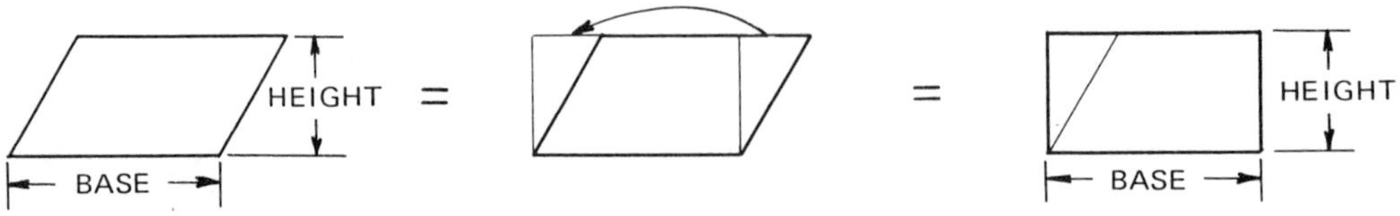

ASSIGNMENT

1. Compute the area of each rectangular piece of sheet metal:

   a. ℓ = 120″, w = 30″ (sq. in.)
   b. ℓ = 96″, w = 24″ (sq. in.)
   c. ℓ = 96″, w = 30″ (sq. in.)
   d. ℓ = 96″, w = 32″ (sq. ft.)
   e. ℓ = 120″, w = 26″ (sq. ft.)
   f. ℓ = 28″, w = 20″ (sq. ft.)

2. Find the cross-section area in sq. in. of each pipe listed below. Construct the formula by using the symbols on the drawing.

   a. w = 16″, h = 8″
   b. w = 10″, h = 3 1/4″
   c. w = 12″, h = 5 1/4″
   d. w = 10 5/8″ h = 3 7/8″

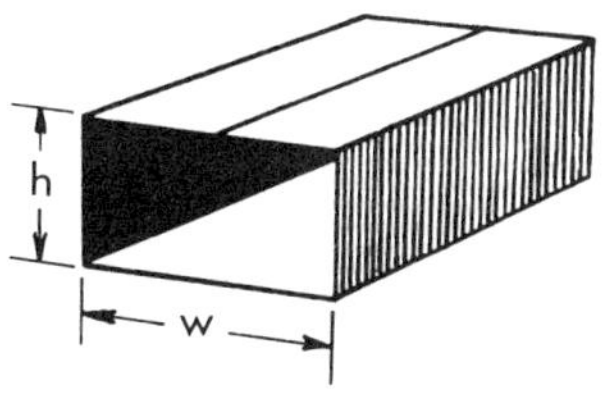

   e. w = 6.87″, h = 4.59″
   f. w = 3′ 6″, h = 1′ 6″
   g. w = 3′ 4″, h = 1′ 10″
   h. w = 6′ 3″, h = 2′ 4″

3. Compute the weight in pounds and the cost of sheet metal for each problem listed below. Refer to tables on pages 149 and 150.

   a. One sheet of 24″ x 96″, 28-gage black iron at $15.65 cwt.
   b. One sheet of 36″ x 120″, 26-gage black iron at $11.75 cwt.
   c. Six sheets of 36″ x 120″, 26-gage galvanized iron at $14.05 cwt.
   d. Ten sheets of 30″ x 96″, 20-gage galvanized iron at $14.65 cwt., 30% discount.

4. Draw a rectangle 2 1/8″ x 3 1/2″ and compute its area in sq. in.

5. Find the area in sq. ft. of a shop floor 40′ 6″ by 80′ 9″.

6. What is the cross-section area in sq. in. of band iron 3/4″ x 1/8″?

7. Find the cross-section area in sq. in. of band iron 1/2″ x 12 gage.

8. How much weight would be saved by cutting a rectangular hole in the 10-gage metal plate shown?

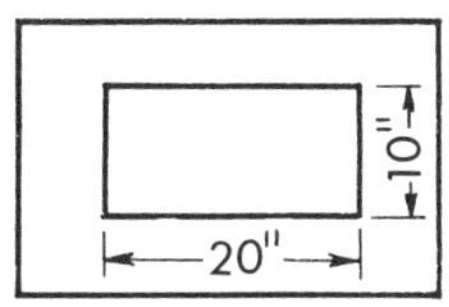

9. What is the area in sq. ft. of the side of the house? Disregard the windows.

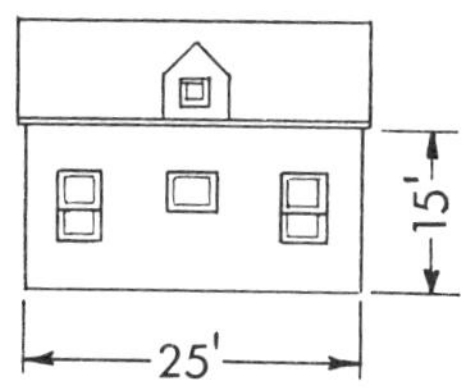

10. Find the area in sq. in. of a piece of sheet metal of parallelogram shape, base 10″ and height 5 3/4″.

UNIT 61 AREA OF A TRIANGLE

The computation of the area of triangles is needed for much the same reasons as that of the square and rectangle except there are no pipes of triangular shape. In some cases, as in roof ends, walls are triangular shaped.

RULE: The area of a triangle is equal to one-half the product of the base and the altitude. Some of the triangles to which this rule applies are shown below:

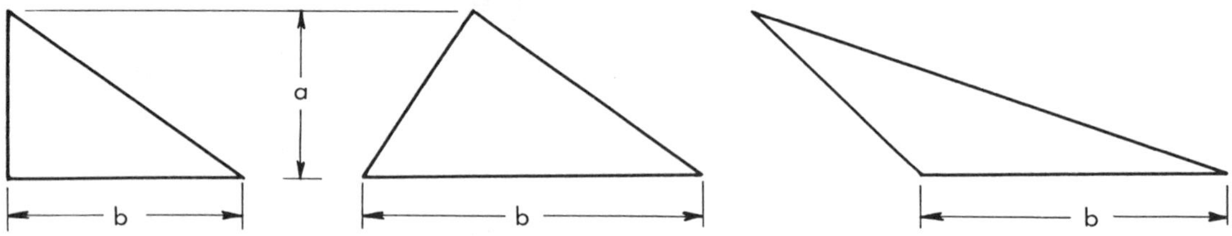

Formula: $A = \frac{ab}{2}$ **Symbols:** A = area; a = altitude; b = base

EXPLANATION OF RULE AND FORMULA

A right triangle is equal to one-half the rectangle formed by drawing lines parallel to the sides of the triangle as shown below:

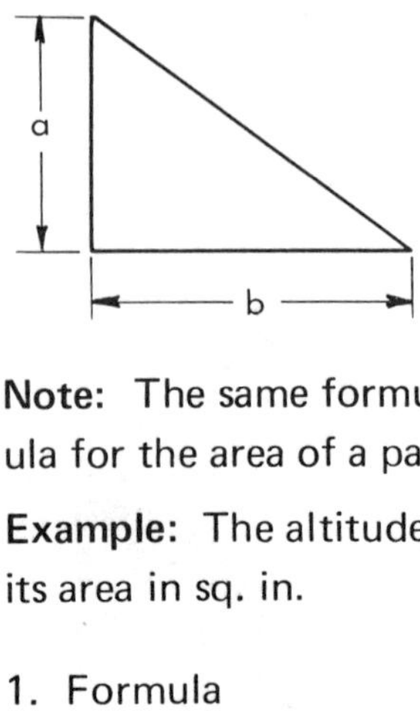

is $\frac{1}{2}$ of

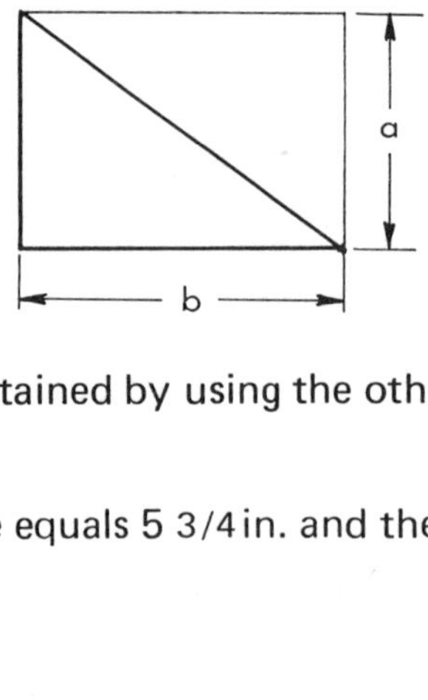

A (rectangle) = ab

A (triangle) = $\frac{ab}{2}$

Note: The same formula can be obtained by using the other types of triangles and the formula for the area of a parallelogram.

Example: The altitude of a triangle equals 5 3/4 in. and the base equals 10 1/2 in. Compute its area in sq. in.

1. Formula $A = \frac{ab}{2}$

2. Substitution $A = \frac{5\ 3/4 \times 10\ 1/2}{2}$

3. Evaluation $A = \frac{5.75 \times 10.5}{2}$

$A = 30.19$ sq. in. Ans.

```
    5.7 5 in.
  x 1 0.5 in.
    2 8 7 5
  5 7 5 0
2) 6 0.3 7 5 sq. in.
   3 0.1 8 7 5 sq. in.
```

ASSIGNMENT

1. Compute the area in sq. in. of each triangular piece of sheet metal below. Construct right triangles for problems a to d.

a. altitude = 3″,	base = 4″	c. altitude = 3″	base = 2 1/2″
b. altitude = 3″,	base = 5″	d. altitude = 4 1/2″	base = 2″

e.

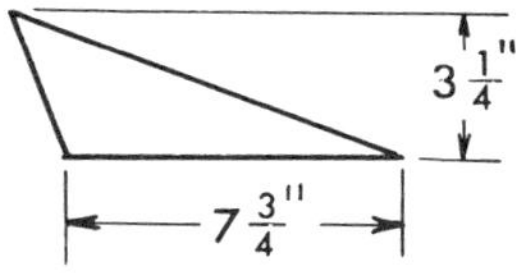

f.

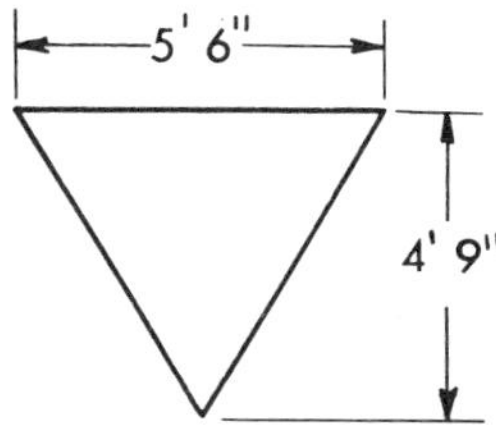

g.

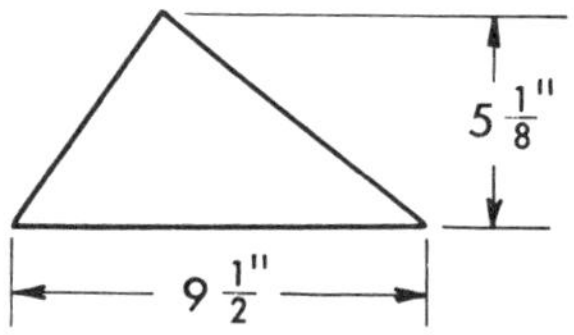

h.

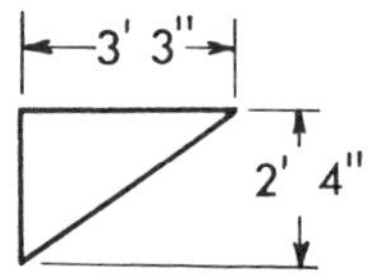

2. Compute the weight in pounds of each of the following triangular pieces of sheet metal:

a. a = 60″,	b = 24″,	28-gage black iron
b. a = 36″,	b = 12″,	26-gage black iron
c. a = 24″,	b = 96″,	20-gage black iron
d. a = 51″,	b = 39″,	10-gage galvanized iron

3. How much weight is saved by cutting a triangular hole in the bracket of 8-gage metal shown?

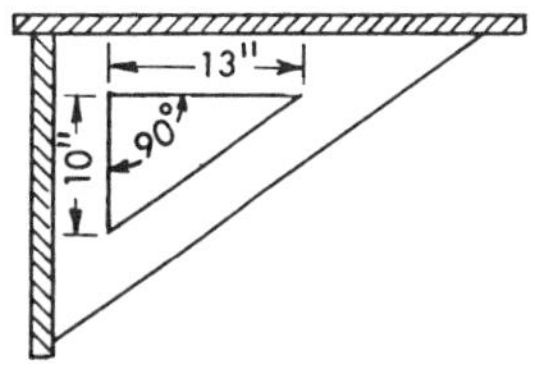

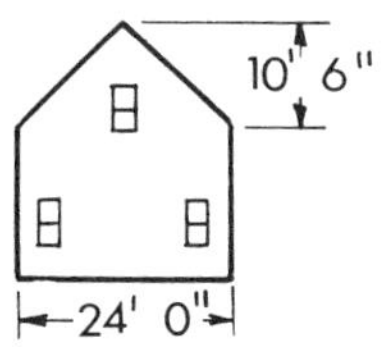

4. Compute the area in sq. ft. of the triangular-shaped part of the house shown on the drawing.

5. Find the area in sq. in. of triangle A and triangle B on the drawing to the right.

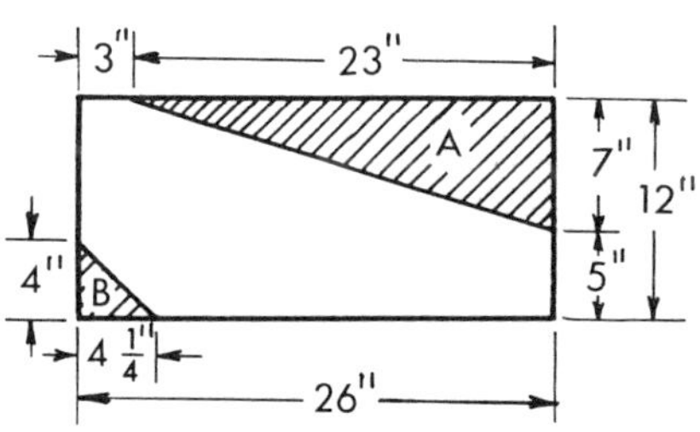

UNIT 62 AREA OF A TRAPEZOID

Finding the area of a trapezoid has the same applications to the sheet metal trade as finding the area of the triangle has, namely, to compute the weight and cost of pieces of sheet metal of that shape.

RULE: The area of a trapezoid is equal to the sum of the lengths of the bases divided by two and then multiplied by the altitude.

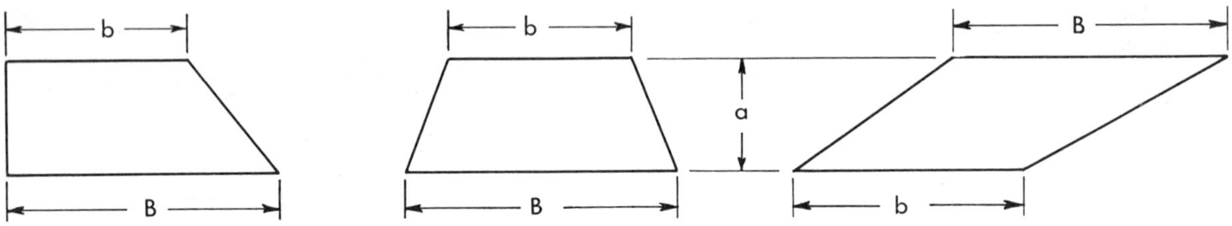

Formula: $A = \frac{(B + b)\,a}{2}$

Symbols: A = area, a = altitude, B = length of large base, b = length of small base

EXPLANATION OF RULE AND FORMULA

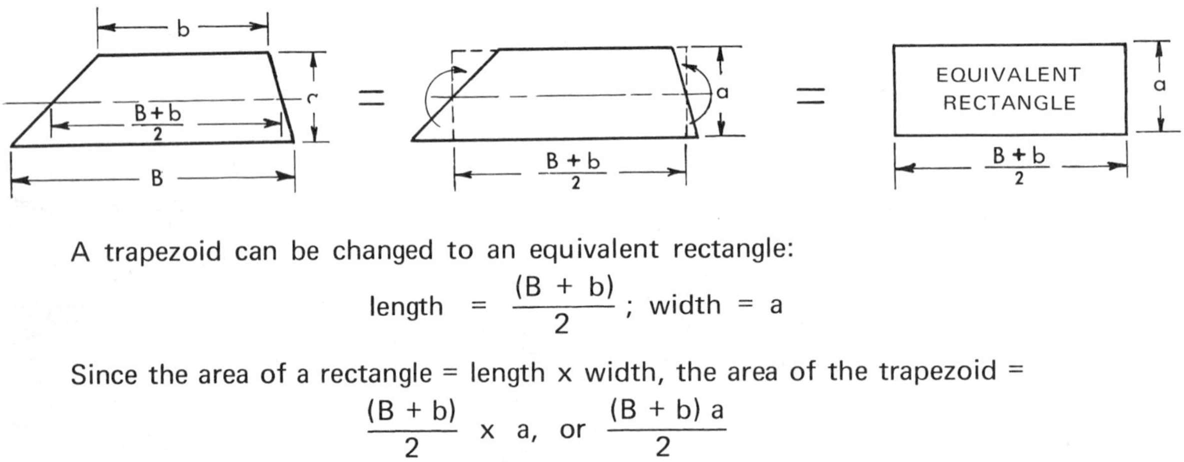

A trapezoid can be changed to an equivalent rectangle:

$$\text{length} = \frac{(B + b)}{2}\,;\ \text{width} = a$$

Since the area of a rectangle = length x width, the area of the trapezoid =

$$\frac{(B + b)}{2} \times a, \text{ or } \frac{(B + b)\,a}{2}$$

Example: The large base of a trapezoid is 8 1/2″, the small base 7 1/2″, and the altitude 3 1/4″. Find its area in sq. in.

1. Formula $A = \frac{(B + b)\,a}{2}$

2. Substitution $A = \frac{(8\frac{1}{2} + 7\frac{1}{2})\,3\frac{1}{4}}{2}$

3. Evaluation $A = \frac{(8.5 + 7.5)\,3.25}{2}$

$A = 8 \times 3.25$

$A = 26$ sq. in. Ans.

```
   8.5 in.
 + 7.5 in.
2)16.0 in.
   8.0 in.

   3.25 in.
  x 8   in.
  26.00 sq. in.
```

ASSIGNMENT

1. Compute the area of each trapezoid in the unit of measure given:

	B	b	a		B	b	a
a.	20″	10″	5″	e.	20 1/2″	10 1/4″	8 1/2″
b.	30″	20″	15″	f.	25 1/4″	22 1/2″	12 1/8″
c.	50″	28″	9″	g.	9.32″	7.89″	5.21″
d.	19″	8″	9 1/2″	h.	3′ 9″	2′ 6″	4′ 6″

2. Compute the weight in pounds of each piece of sheet metal:

a.

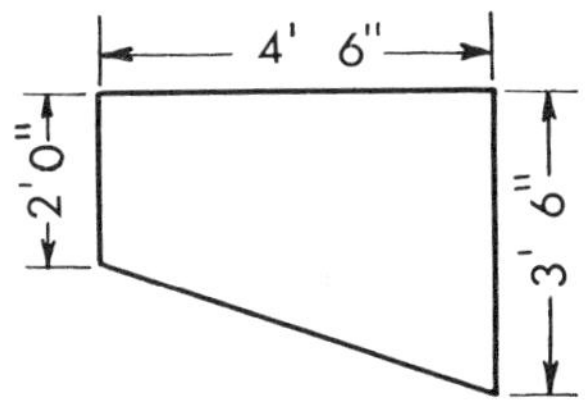

26-GAGE GALV. IRON

b.

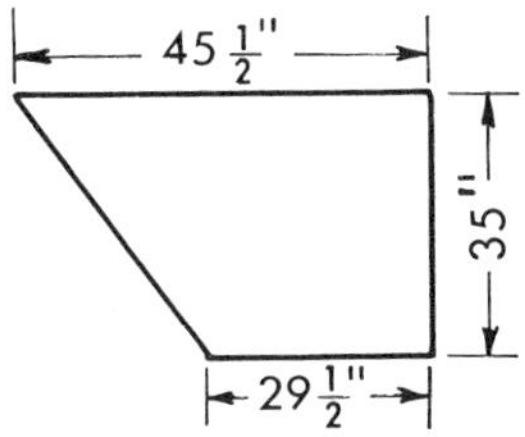

20-GAGE GALV. IRON

c.

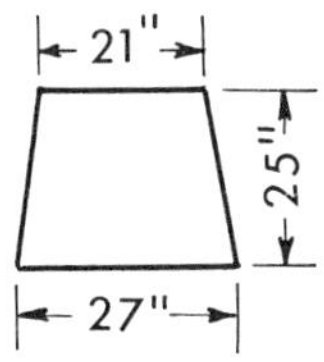

26-GAGE BLACK IRON

3. Draw an isosceles trapezoid: large base 5 in., small base 3 in., altitude 2 in. Find the area in sq. in.

4. Construct a right trapezoid: large base 3 1/2 in., small base 2 1/2 in., altitude 1 1/2 in. Find the area in sq. in.

5. A hole of trapezoidal shape (large base 35 in., small base 25 in., altitude 21 in.) is cut in 10-gage black iron plate. What is the weight reduction in lbs.? Refer to the table on page 149.

6. Compute the area in sq. ft. of the roof end shown.

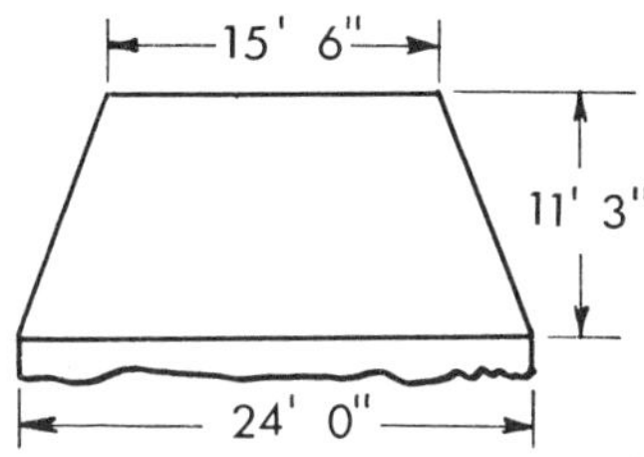

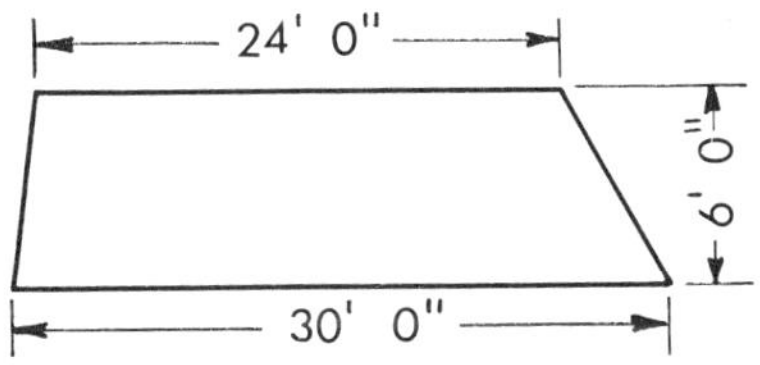

7. Find the area in sq. ft. of the section of main deck shown at the left.

8. Compute the area in sq. ft. of the airplane wing (approximate shape) shown at the right.

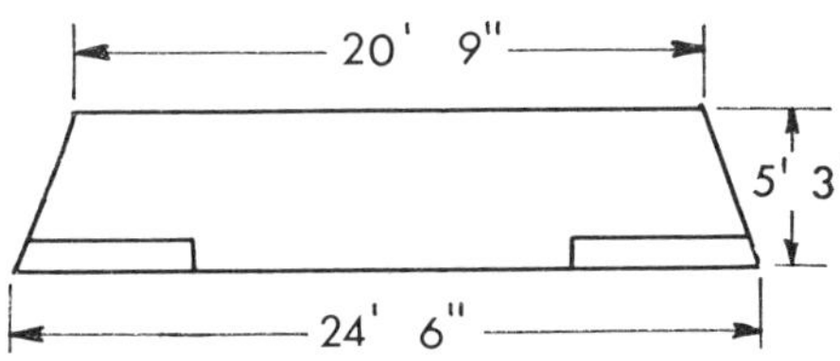

UNIT 63 AREA OF A CIRCLE

The computation of the area of a circle is needed for weight and cost problems, and for the air-carrying capacity of circular heating and ventilating pipes.

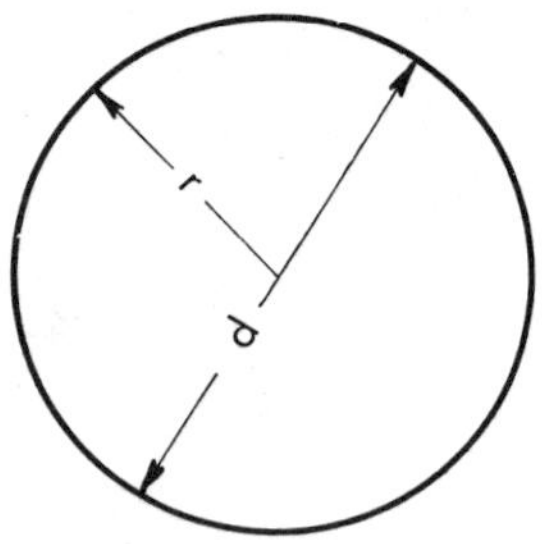

RULE: The area of a circle is equal to the product of pi (π) and the radius squared, or to the product of π/4 and the diameter squared.

Formula: $A = \pi r^2$ or $A = \dfrac{\pi d^2}{4}$

Symbols: A = area π = 3.14
r = radius d = diameter

EXPLANATION OF RULE AND FORMULA

If a circle is divided into an even number of equal parts, and the parts are fitted together as shown below at right, a figure that is approximately a parallelogram in shape will be formed.

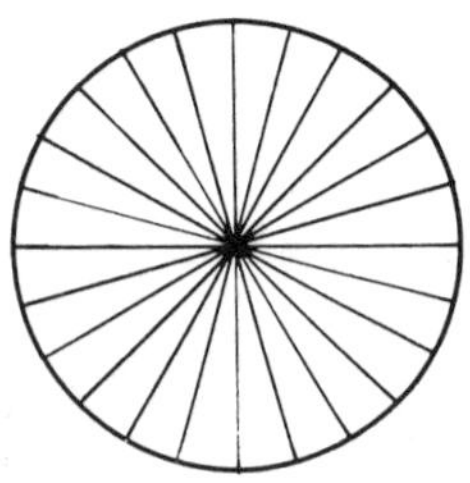

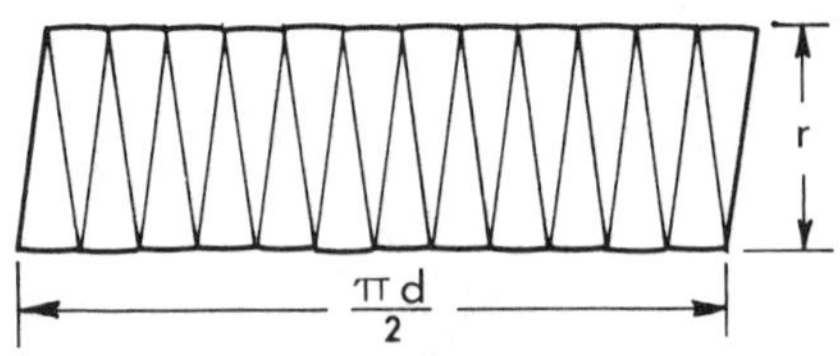

If the number of parts are made sufficiently large, the figure formed will approach a parallelogram in shape whose base is equal to one-half the circumference of the circle and whose height is equal to the radius. One-half the circumference = $\pi d/2$ or πr. Thus the area of the circle equals the area of the parallelogram, or $\pi r \times r$, or πr^2.

Example: The diameter of a circle equals 10 in. Compute its area in sq. in.

1. Formula $A = \pi r^2$
2. Substitution $A = 3.14 \times 5^2$ $r = \frac{10}{2}$ in., or 5 in.
3. Evaluation $A = 3.14 \times 5 \times 5$

$A = 3.14 \times 25$

$A = 78.5$ sq. in. Ans.

```
  5
x 5
---
 25
```

```
   3.14
  x  25
  -----
   1570
   628
  -----
  78.50  sq. in.
```

ASSIGNMENT

1. What is the cross-section area in sq. in. of each pipe?

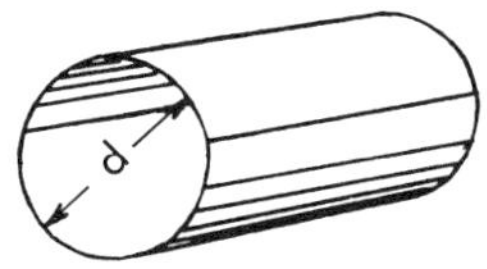

a. d = 2″
b. d = 3″
c. d = 4″
d. d = 5″
e. d = 6″
f. d = 12″
g. d = 14″
h. d = 16″
i. d = 18″
j. d = 20″
k. d = 4 1/2″
l. d = 11″
m. d = 3 1/2″
n. d = 10 1/2″
o. d = 4 1/4″
p. d = 8 3/4″
q. d = 20 1/4″
r. d = 15.13″
s. d = 10.52″
t. d = 3′ 6″
u. d = 2′ 9″

2. Compute the weight in pounds of each piece of metal.

a.

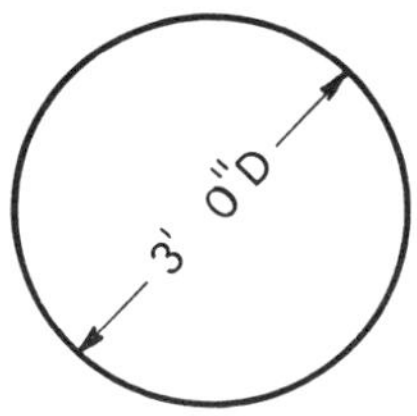

26–GAGE GALV.

b.

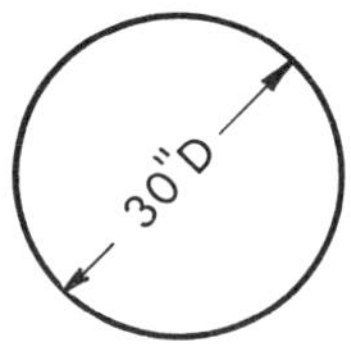

10–GAGE BLK. IRON

c.

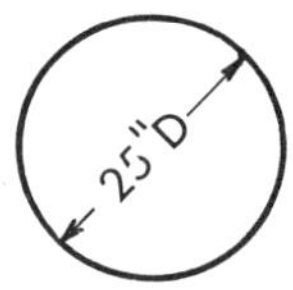

107–LB. TIN PLATE

d.

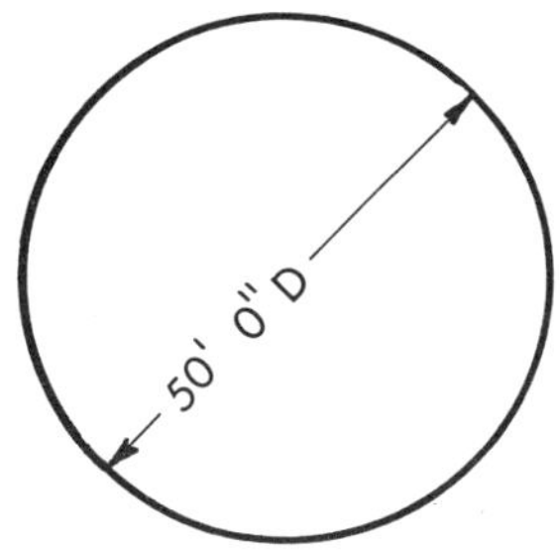

26–GAGE BLK. IRON

3. Find the cross-section area in sq. in. of wire 1/4 in. in diameter.

4. Find the cross-section area in sq. in. of wire 1/8 in. in diameter.

5. Compute the cross-section area in square inches of a rivet, diameter .120 in.

6. How much weight in pounds is saved by cutting the circular hole in the 12-gage channel made of black iron? Refer to the table on page 149.

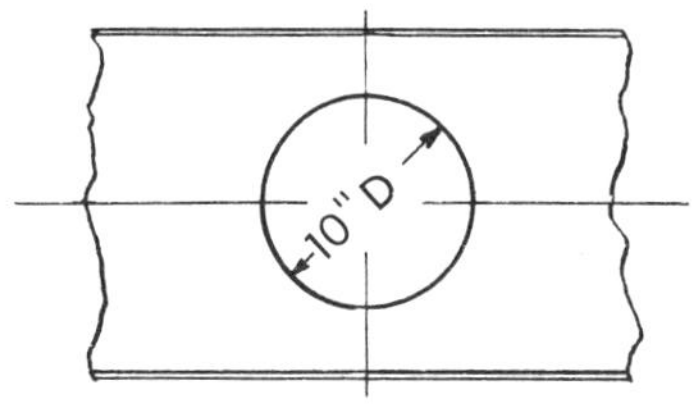

7. Compute the weight in pounds saved by cutting two 10-in. holes in a plate of 16-gage black iron. Refer to the table on page 149.

UNIT 64 ADDITION AND SUBTRACTION OF AREAS

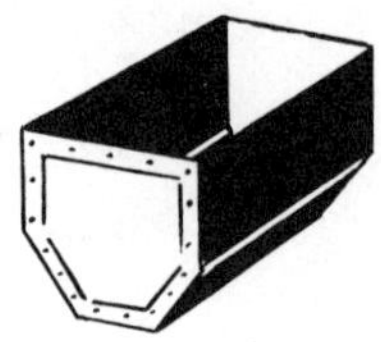

Many jobs or parts of jobs are composed of two or more plane figures. For instance, the end of the tank shown consists of a rectangle and a trapezoid.

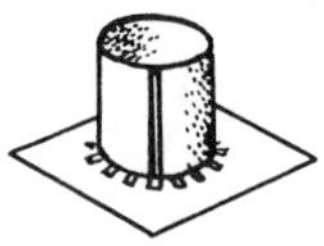

Likewise, jobs or parts of jobs having one or more holes, may be composed of plane figures. The flange part shown consists of a square piece with a circular hole in the center.

To find the area of such jobs, the area of each plane figure is found, and then these areas are added or subtracted, depending on shape of job.

ADDITION OF AREAS

Example: Find the total area of a rectangle 20 in. long by 10 in. wide and a trapezoid with a large base of 20 in., a small base of 12 in. and an altitude of 8 in.

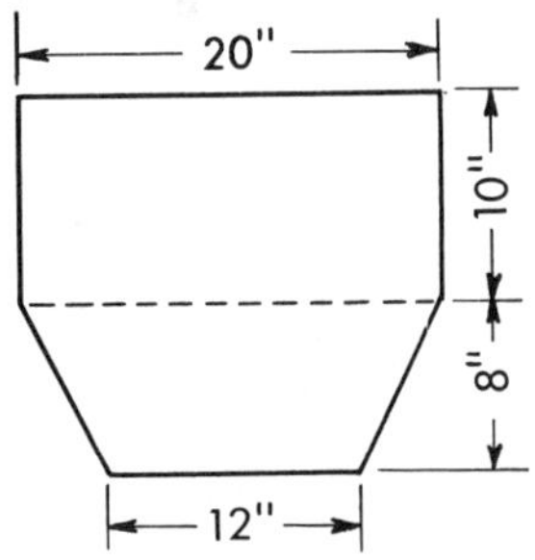

Formula: $A \text{ (sum)} = \ell w + \dfrac{(B + b)\,a}{2}$

1. a. $A = \ell w$

   b. $A = 20 \times 10$

   c. $A = 200$ sq. in.

2. Area of a trapezoid

   a. $A = \dfrac{(B + b)\,a}{2}$

   b. $A = \dfrac{(20 + 12)\,8}{2}$

   c. $A = 16 \times 8$

   d. $A = 128$ sq. in.

3. Total area = area of rectangle + area of trapezoid
   = 200 + 128 = 328 sq. in. Ans.

Subtraction of Areas

Example: Find the net area of a circular hole 4 in. in diameter to be cut from a piece of metal 8in. square.

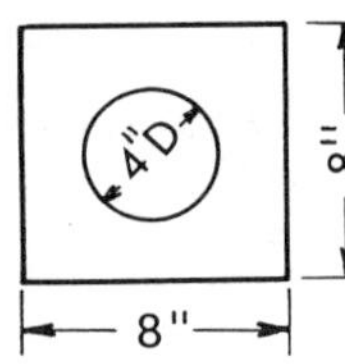

Formula: $A \text{ (difference)} = \ell w - \pi r^2$

1. Area of square

   a. $A = s^2$

   b. $A = 8^2$

   c. $A = 64$ sq. in.

2. Area of circle

   a. $A = \pi r^2$

   b. $A = 3.14 \times 2^2$

   c. $A = 12.56$ sq. in.

3. Net area = area of square - area of circle
   = 64.00 - 12.56 = 51.44 sq. in. Ans.

ASSIGNMENT

A. Addition of Areas: Find the total area of each job in sq. ft.

1. Marine Tank

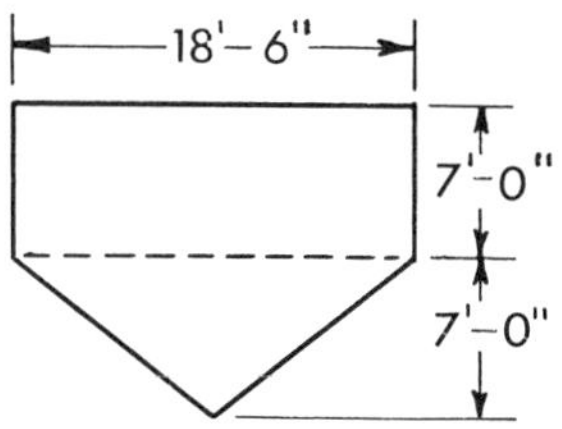

2. House

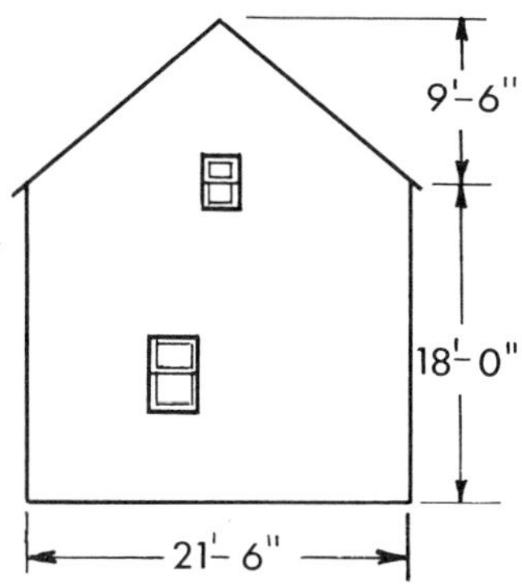

3. Marine Tank

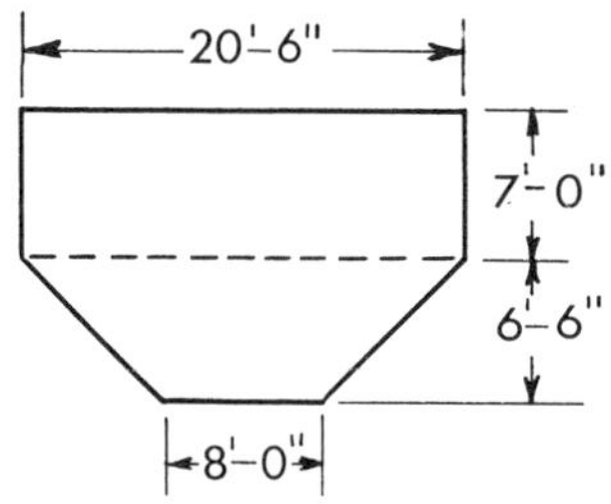

B. Subtraction of Areas: Find the net area of each job in sq. in.

1. Flange

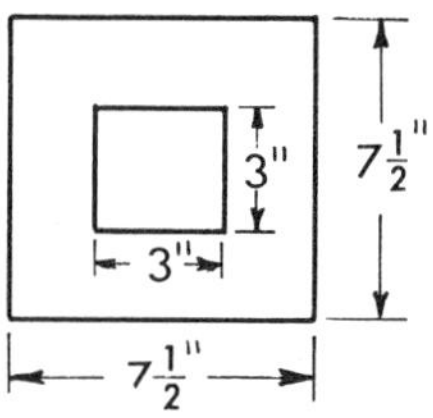

2. Flange

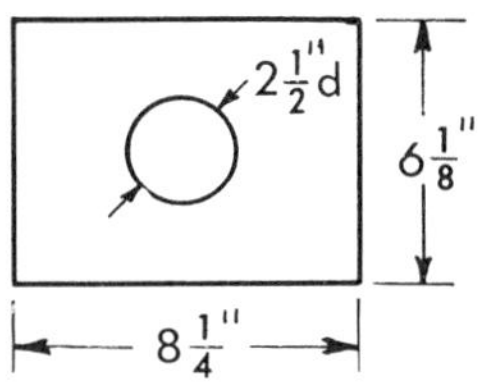

3. Flange

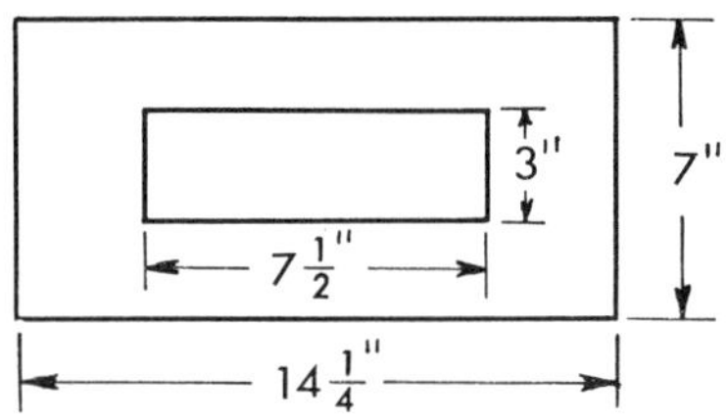

4. Plate

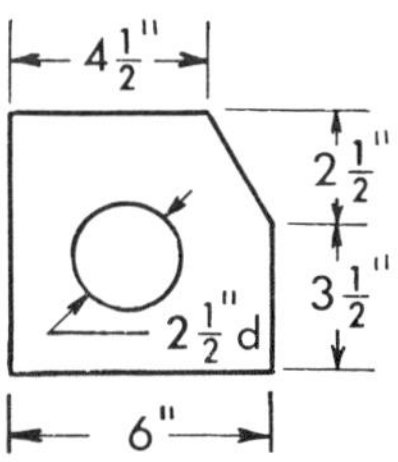

5. Bracket

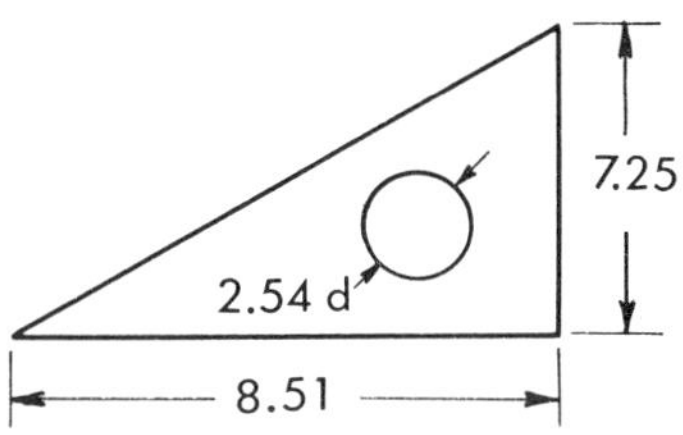

6. Plate

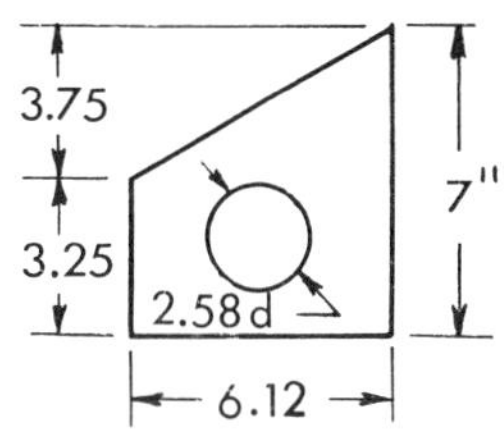

C. Weight Problems: Find the weight of each job in pounds. Refer to the table on page 149.

1. Ship Plate

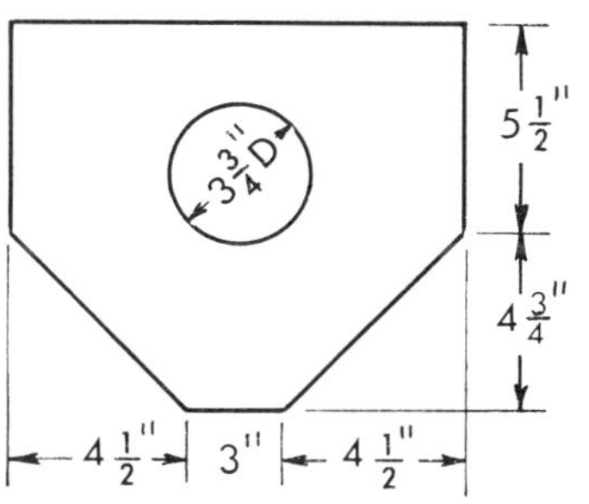

10-GAGE BLK. IRON

2. Motor Guard

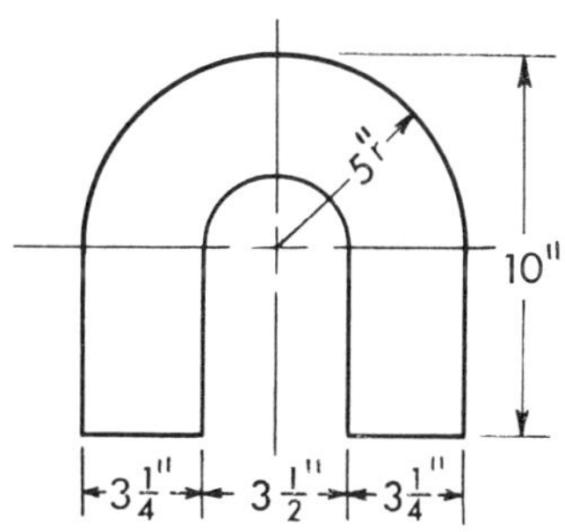

26-GAGE GALV. IRON

D. Waste Problems

Compute the amount of waste in sq. in. and the percent of waste for each of the problems on this page. To find the amount of waste, find the area of the sheet and the area of the piece, and then subtract one area from the other. To find the percent of waste, divide the amount of waste by the area of the sheet and change to a percent.

1. Triangular pieces

a.

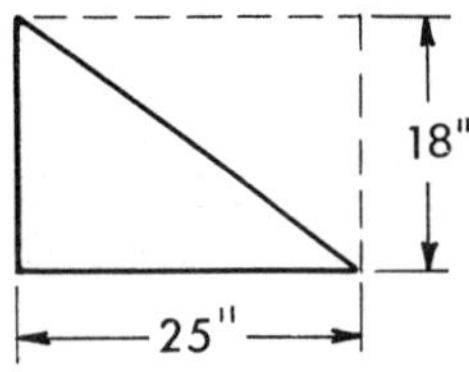

b.

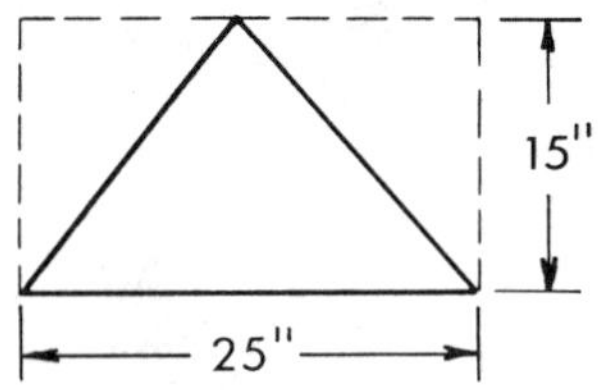

c.

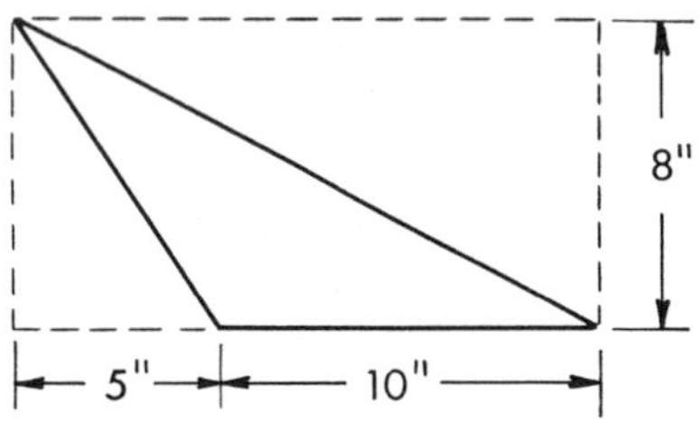

d.

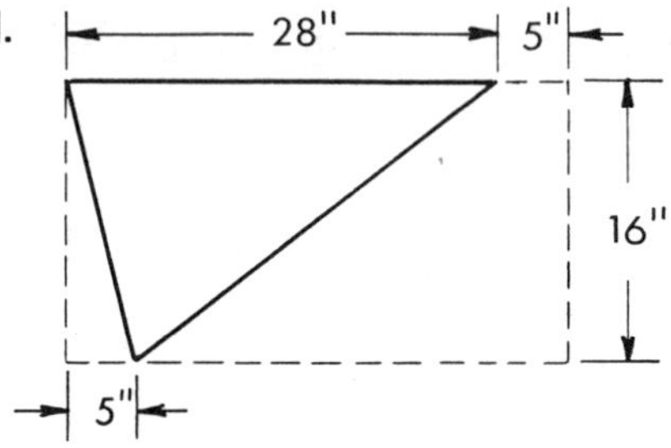

2. Trapezoidal pieces

a.

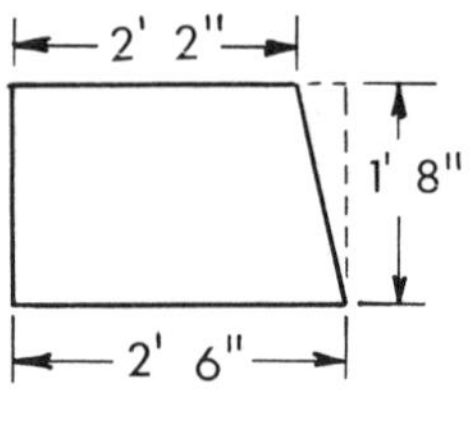

b.

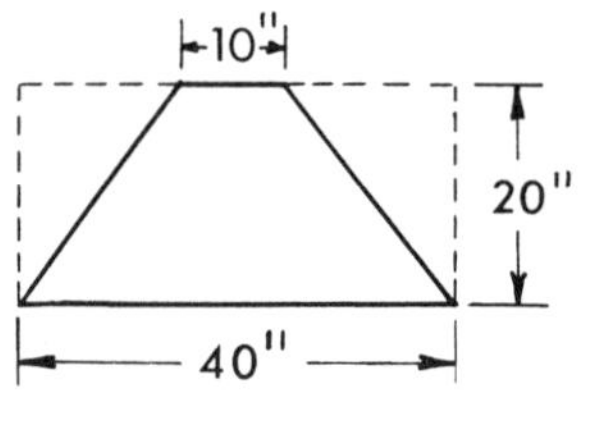

c.

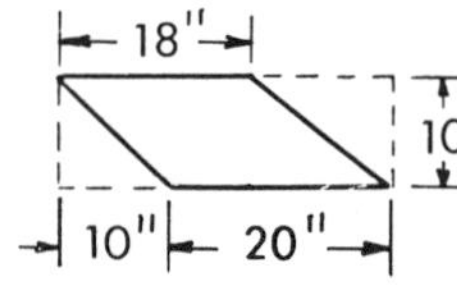

3. Circular pieces

a.

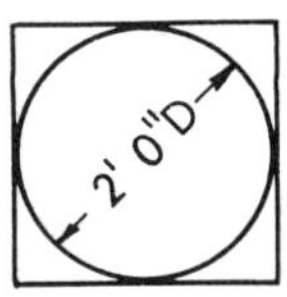

b.

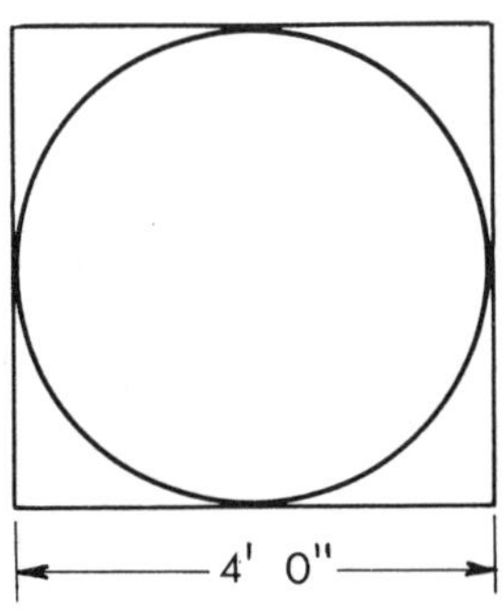

c.

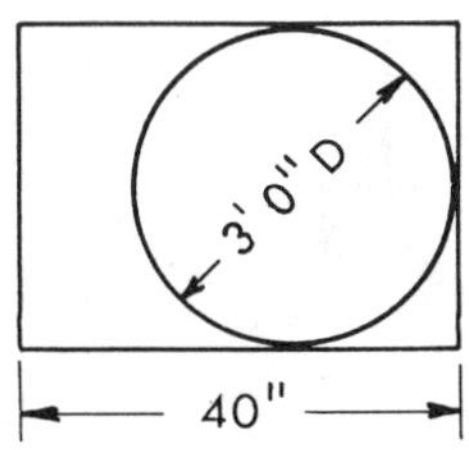

d.

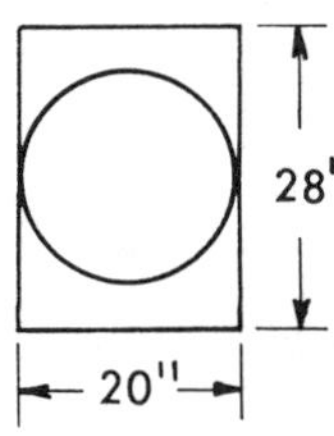

UNIT 65 AREA OF SEMICIRCULAR-SIDED FIGURES AND SOLIDS

The computation of the area of the semicircular-sided figure has the same applications to sheet metal work as the computation of the area of the circle does. This area is found by the method of adding areas explained in the previous unit.

RULE: The area of a semicircular-sided figure is equal to the sum of the areas of the two semicircles and the rectangle.

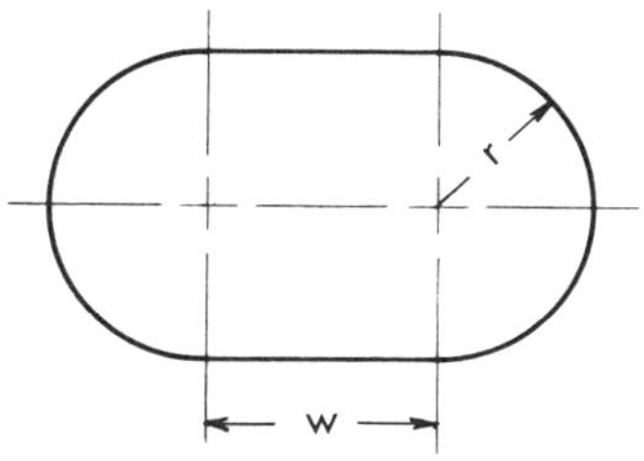

Formula: $A = \pi r^2 + 2rw$

Symbols: A = area

r = radius of the semicircles

w = distance between centers of semicircles

EXPLANATION OF RULE AND FORMULA

The semicircular-sided figure can be broken up into two semicircles and a rectangle:

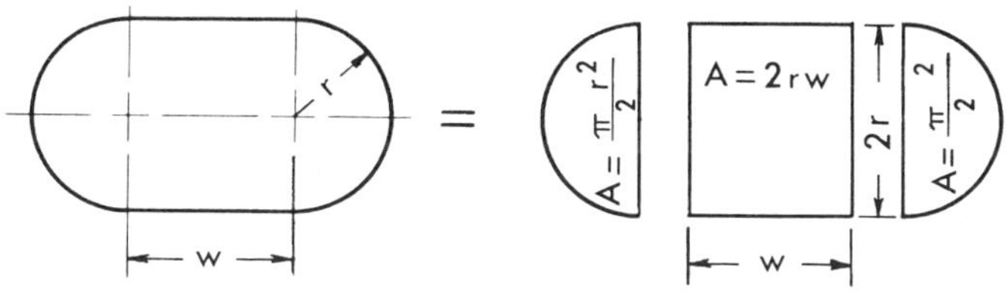

$$A \text{ (total)} = \frac{\pi r^2}{2} + 2r \times w + \frac{\pi r^2}{2}$$

$$\therefore A = \pi r^2 + 2rw \text{ (combining)}$$

The radius of each of the semicircles of a semicircular-sided job is 3 in. and the distance between the centers of semicircles is 5 1/2 in. Compute the area in sq. in.

1. Formula $A = \pi r^2 + 2rw$

$$\begin{array}{r} 3 \\ \times 3 \\ \hline 9 \end{array} \qquad \begin{array}{r} 2 \\ \times 3 \\ \hline 6 \end{array}$$

2. Substitution $A = (3.14 \times 3^2) + (2 \times 3 \times 5\ 1/2)$

3. Evaluation $A = (3.14 \times 3 \times 3) + (2 \times 3 \times 5.5)$

$A = (3.14 \times 9) + (6 \times 5.5)$

$$\begin{array}{r} 3.14 \\ \times 9 \\ \hline 28.26 \text{ sq. in.} \end{array} \qquad \begin{array}{r} 5.5 \\ \times 6 \\ \hline 33.0 \text{ sq. in.} \end{array}$$

$A = 28.26 + 33$

$A = 61.26$ sq. in. Ans.

$$\begin{array}{r} 28.26 \\ 33.00 \\ \hline 61.26 \text{ sq. in.} \end{array}$$

AREA OF A SPHERICAL-ENDED SOLID

RULE: The surface area of a spherical-ended solid is equal to the sum of the areas of two hemispheres and the cylinder.

Formula: $A = \pi D \ell$ or $2\pi r \ell$

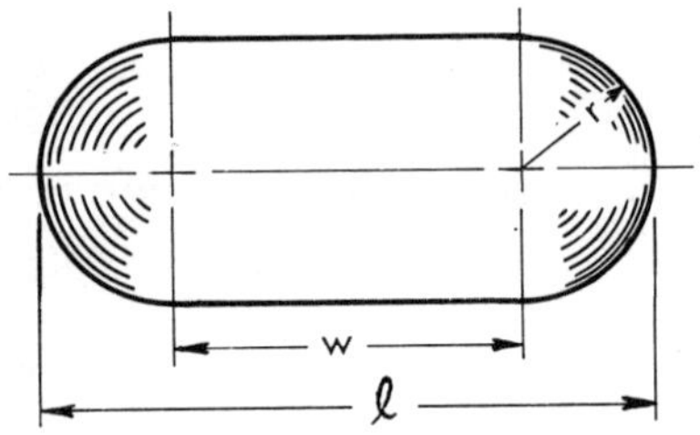

Symbols: A = area

r = radius of hemispheres

w = distance between the centers of the hemispheres, or $(\ell - 2r)$

Explanation of Rule and Formula

The spherical-ended solid can be broken up into two hemispheres and a cylinder.

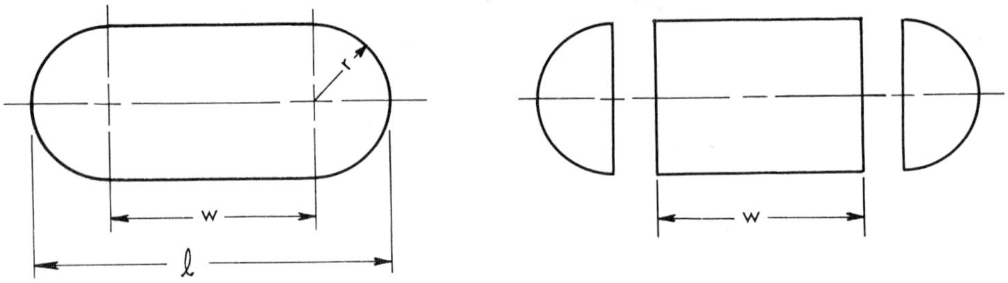

The area of the surface of a sphere equals four times the area of one of the great circles, or $A = 4\pi r^2$.

The surface area of the cylinder is equal to the circumference of the circle times the length of the cylinder, or $A = \pi D \ell$.

Therefore, combining these areas:

$$A = 4\pi r^2 + \pi D w$$

$$A = 4\pi \left(\frac{D}{2}\right)^2 + \pi D (\ell - 2r)$$

$$A = 4\pi \frac{D^2}{4} + \pi D \left(\ell - 2\frac{D}{2}\right)$$

OR

$$A = \pi D^2 + \pi D \ell - \pi D^2$$

$$A = \pi D \ell \quad \text{substituting } D = 2r$$

$$A = 2\pi r \ell$$

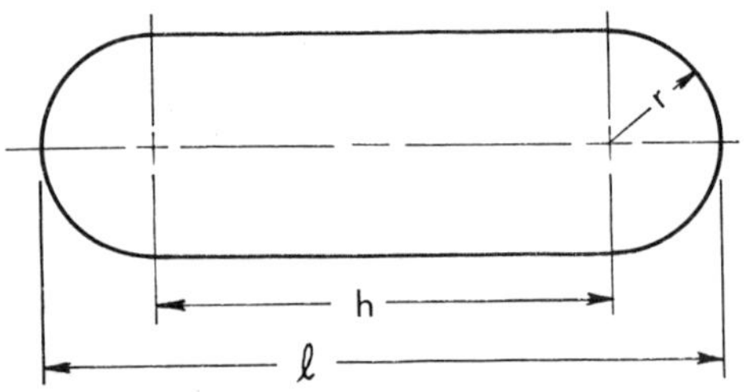

The radius of each of the hemispheres of a spherical-ended tank is 2 in. and the distance between the centers of the hemispheres is 8 in. Compute the area in square in.

$r = 2$

$w = 8$

$\ell = 8 + 4 = 12$

1. Formula $\quad A = 2\pi r \ell$
2. Substitution $\quad A = 2\pi(2) \times 12$
3. Evaluation $\quad A = 6.28 \times 24$

$A = 150.72$ sq. in. Ans.

3.14	12	6.28
2	2	24
6.28	24	150.72 sq. in.

ASSIGNMENT

1. Find the cross-section area in sq. in. of each pipe:

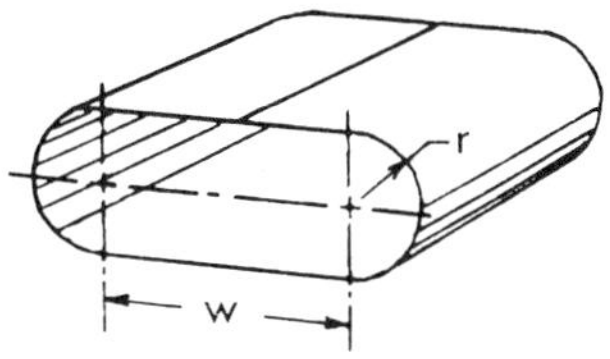

a. r = 10″,	w = 30″	e. r = 1 5/8″,	w = 3 1/2″
b. r = 5″,	w = 12″	f. r = 6 3/4″,	w = 2 1/8″
c. r = 8″,	w = 10″	g. r = 7′6″,	w = 1′ 9″
d. r = 4″,	w = 5″	h. r = 2′ 3″,	w = 2′ 5″

2. What is the weight in lb. of each of the following pieces?

a.

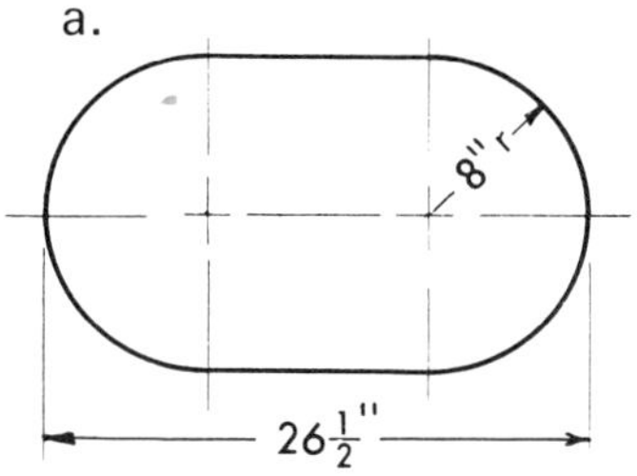

28-GAGE GALV. IRON

b.

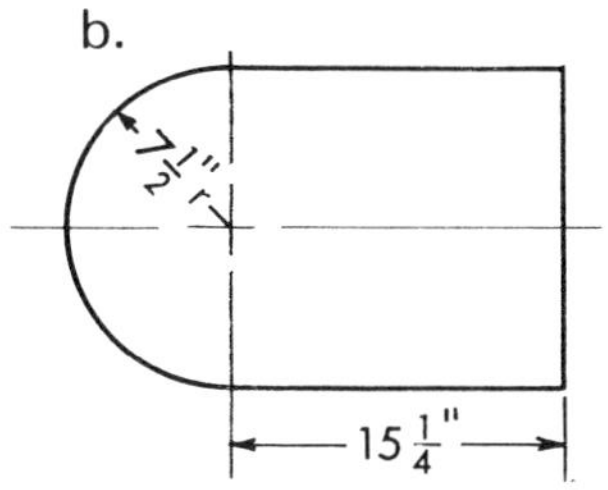

26-GAGE GALV. IRON

c.

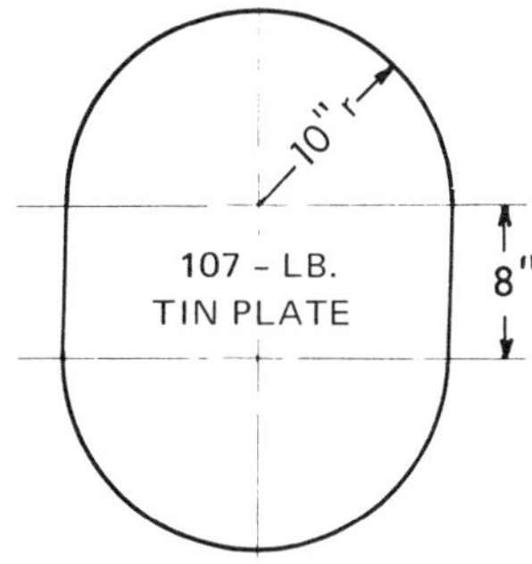

3. Compute the amount and the percentage of waste in each problem:

a.

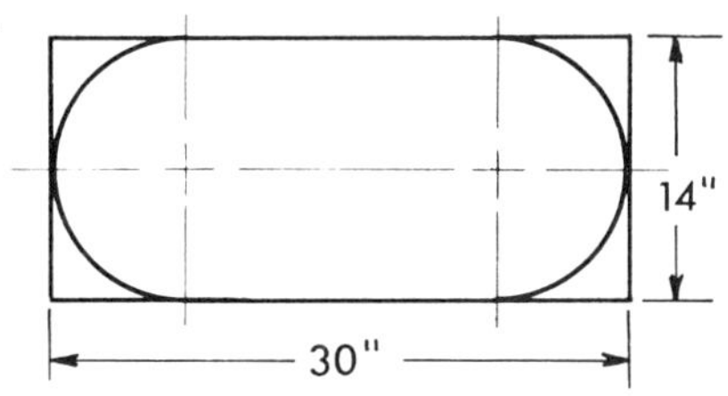

b.

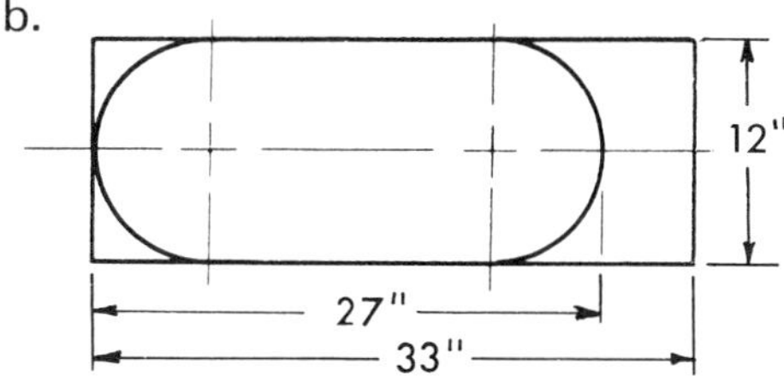

c.

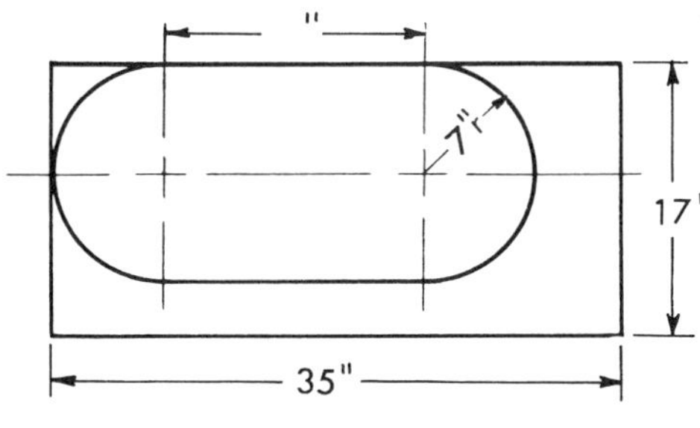

d.

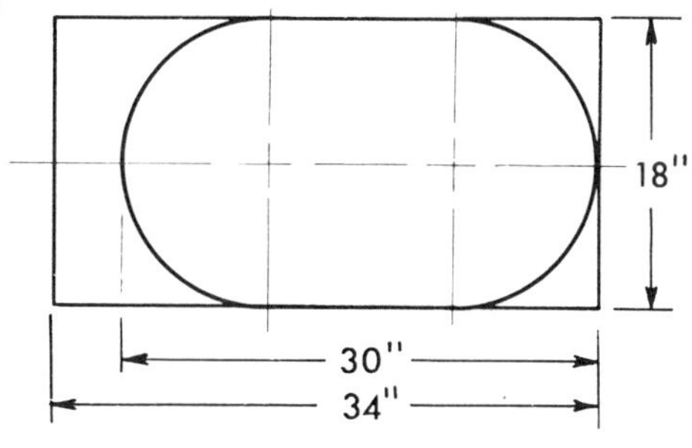

4. Draw a semicircular-sided plate and find its area in sq. in.; the distance between centers of the semicircles is 2 in.; the radius of each semicircle is 1 1/4 in.

5. Find the weight saved in pounds if a semicircular-sided hole (distance between semicircles 20 in.; radius of semicircles 15 in.) is cut from an 8-gage plate of galvanized iron. Refer to the table on page 149.

6. Compute the weight in lb. of the link shown at the right. The link is made of 10-gage metal. Refer to the table on page 149.

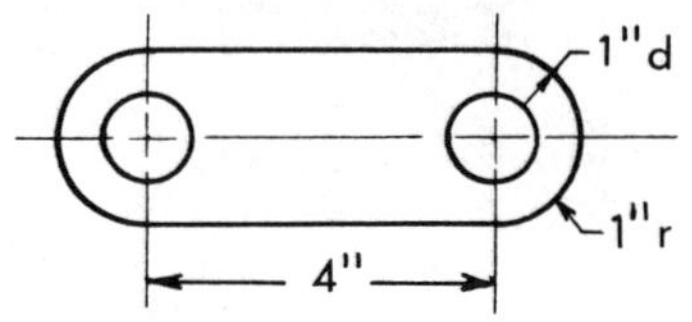

7. Calculate the surface area of the following spherical-ended tanks.

a. D = 7", ℓ = 2'

b. D = 28", ℓ = 6'

c. D = 63", ℓ = 12'

d. r = 7', ℓ = 16'

e. r = 3.5', ℓ = 24'

f. r = 4', ℓ = 21'

SECTION 9 — VOLUME MEASURE

UNIT 66 DESCRIPTION OF PARALLEL-SIDED SOLIDS

Most of the jobs a sheet metal worker must lay out, fabricate, or assemble, are shaped like solids. Solids can be divided into three types: the parallel-sided, the tapered, and the double-curved. Many jobs are a combination of two or more of these solids.

Parallel-sided solids may be prism-shaped or cylinder-shaped. (See the drawing at the right.) *Prisms* are solids whose upper and lower bases are equal, parallel polygons, and whose faces are parallelograms. *Cylinders* are solids with parallel, curved bases of the same shape and size.

The altitude, or height, of either a prism or a cylinder is the perpendicular distance between the bases. The *axis* of a cylinder is the line connecting the centers of the bases.

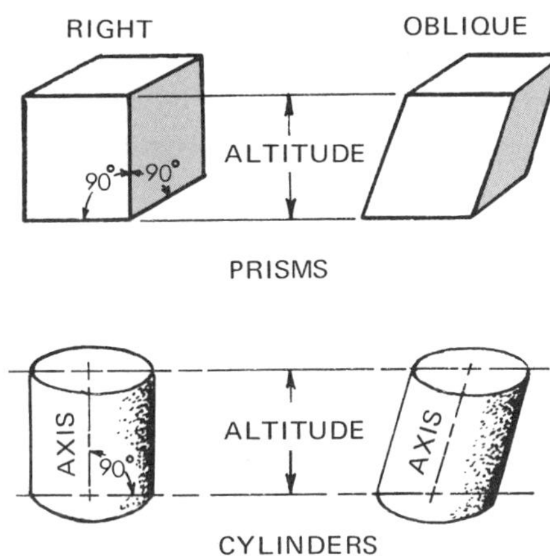

TYPES OF PRISMS AND CYLINDERS

Prisms or cylinders are called *right* or *oblique*, depending upon whether the sides of the prisms or the axes of the cylinders are at right angles or oblique to the bases. See the drawing above.

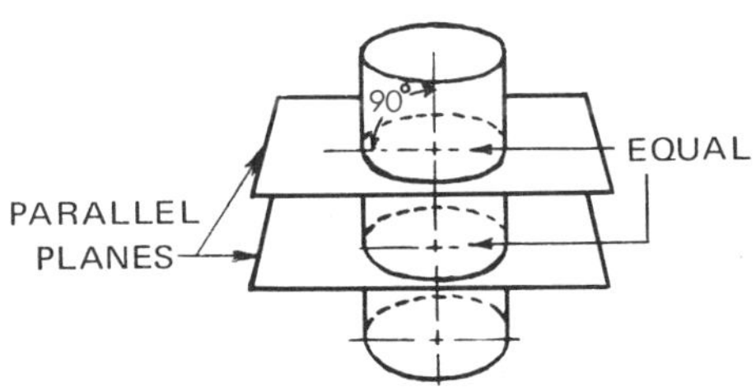

The cross sections of parallel-sided solids are equal. A cross section is a right section (a plane figure formed by a plane cutting the solid at right angles to the edge or the axis). Both bases of a right prism or right cylinder are the same size and shape as a cross section of that solid.

Prisms or cylinders cut by a plane at an angle to the base are called *truncated* prisms, or *truncated* cylinders. The sections or pieces of a pieced, circular elbow are truncated cylinders.

The shape of the base determines the kind of prism or cylinder. Common types of right prisms and right cylinders are described below:

TRUNCATED

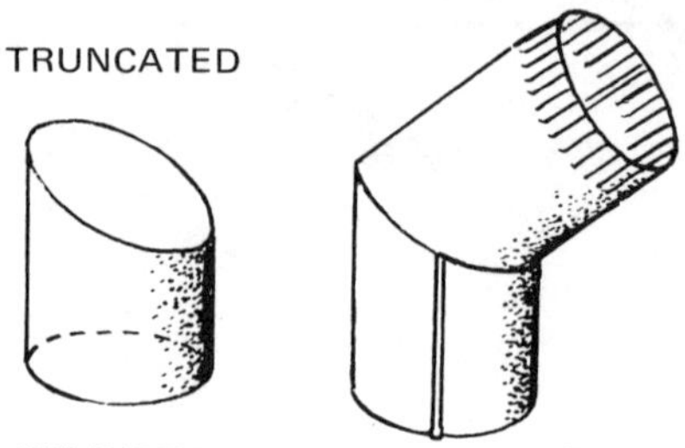

CYLINDER ELBOW

- A *cube* is a right prism with six equal square faces. The edges of a cube are equal. The opposite faces are parallel, and the adjacent faces are at right angles to each other. The length of the sides or the edges is often designated by the letter *s.*
- A *square prism* is a right prism with a square base. The opposite faces are parallel, and the adjacent faces are at right angles to each other and the bases. The side of the base is designated by *s* and the height of the prism by *h.* Pans, tanks, etc. may have this shape.

CUBE

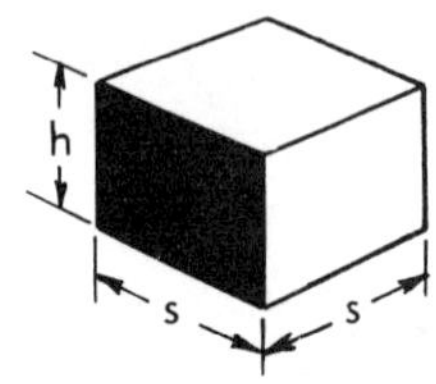

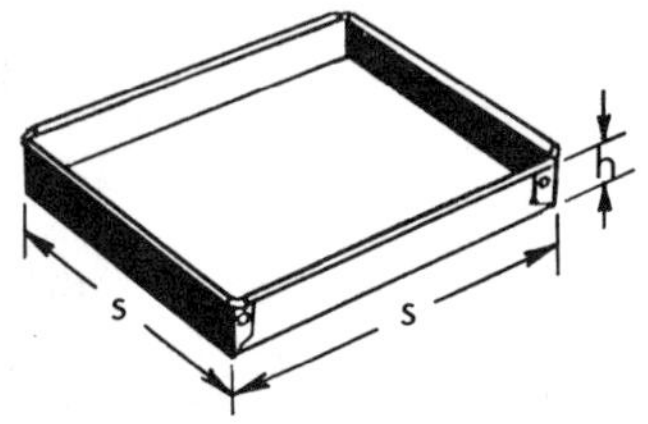

- A *rectangular solid* is a right prism whose bases are rectangles. The opposite faces are parallel, and the adjacent faces are at right angles to each other and the bases. The length is designated by ℓ, the width by *w*, and the height by *h.* Typical jobs are pans, tanks, boxes, and pipes.

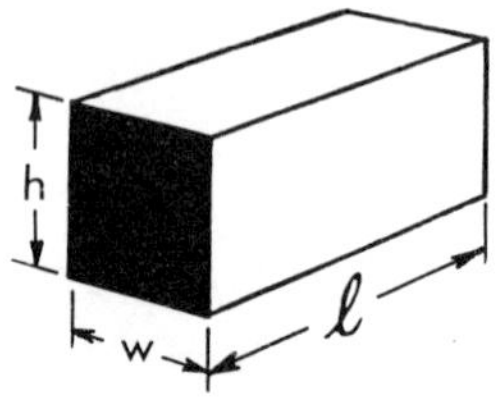

RECTANGLE

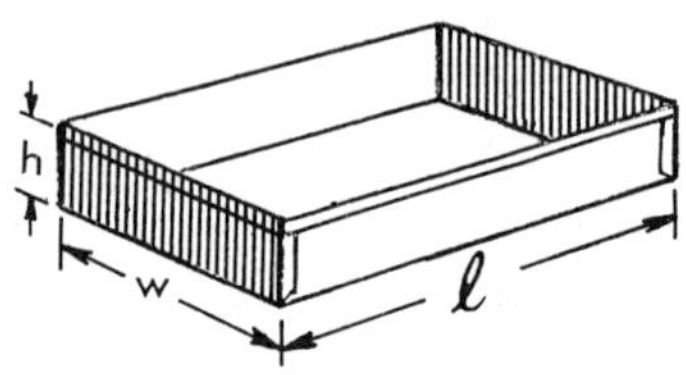

- A *right circular cylinder* is a right cylinder whose bases are circles. Often such solids are called cylinders. The height is designated by *h*, and the diameter by *d*. Tanks and pipes are jobs of this shape.

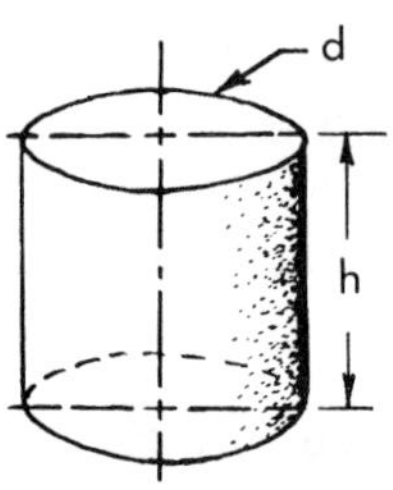

CYLINDER

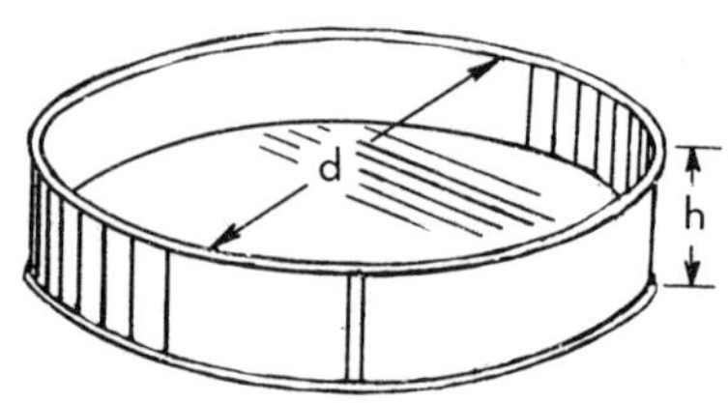

- *Rings* may be square or rectangular in cross section. Such rings have two diameters which may be indicated as *D* and *d*. Square or rectangular elbows are parts of such rings.

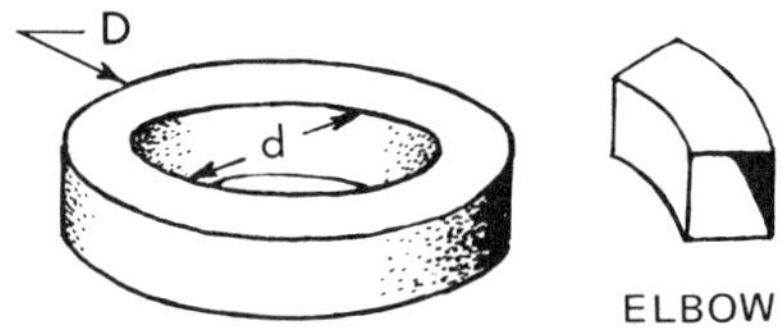

- A *semicircular-sided solid* is a type of cylinder whose base is a semicircular-sided figure. See the drawing below for the dimensions usually indicated for this solid. Tanks, containers, and pipes may be of this shape.

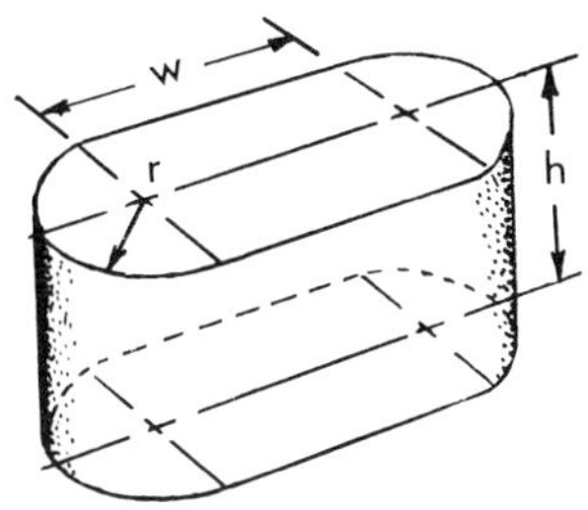

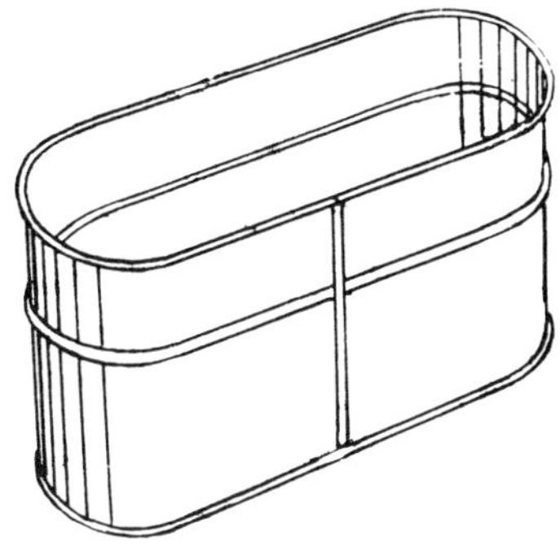

- A *spherical-ended solid* is a type of cylinder whose ends are hemispherical. Tanks and containers are sometimes made in this shape.

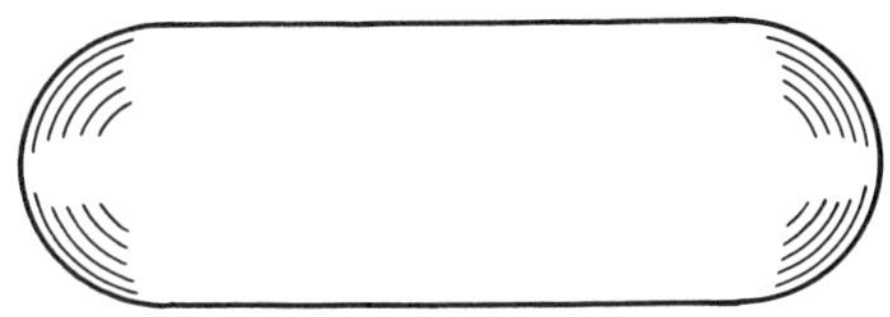

ASSIGNMENT

A. Identify with the proper mathematical terms each of the following jobs or parts of jobs:

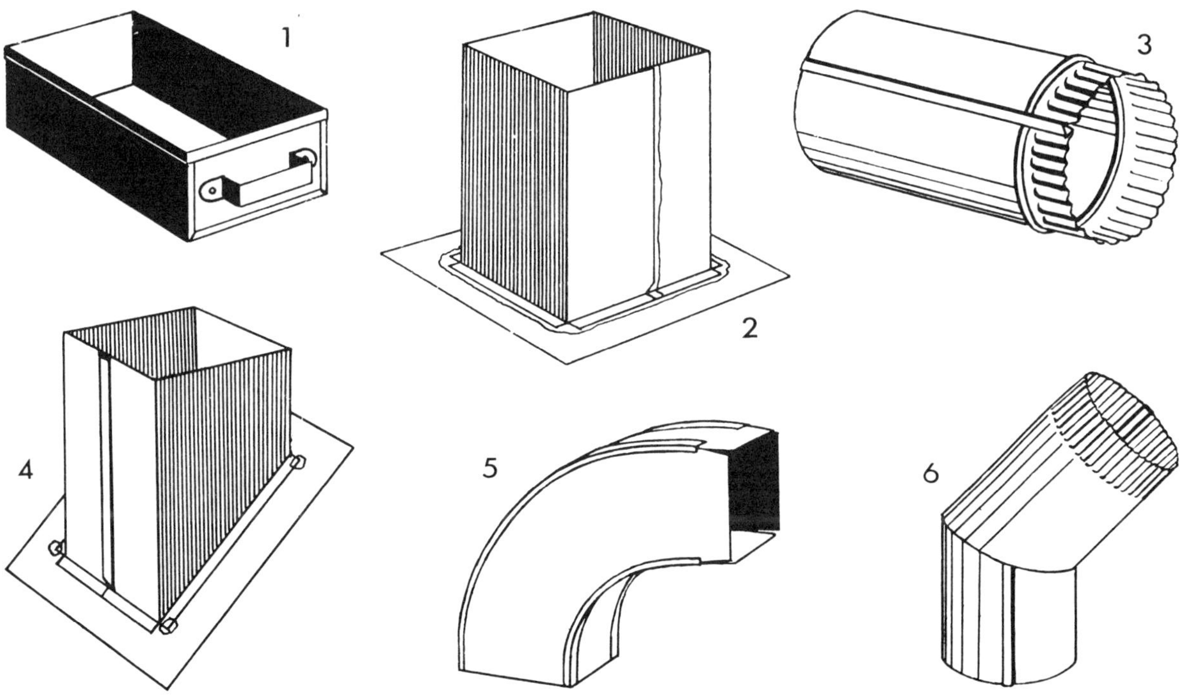

B. Answer the following questions:

1. What shape are the lateral faces of an oblique prism?
2. What is another name for the altitude of a prism or a cylinder?
3. What is the relationship of the lengths of the edges of a cube?
4. What is the essential difference between a cube and a square prism; between a cube and a rectangular solid?
5. If the diameter of the lower base of a right circular cylinder is 10 in., what is the diameter of the upper base; of the cross section?
6. What is the relationship between the lengths of the axis and the altitude of a right circular cylinder; of an oblique circular cylinder?
7. Name two shapes of the cross sections of rings.
8. How is a truncated square prism formed? Are the sides of equal height?
9. Name five jobs that you have made in shop which are shaped like prisms or cylinders, and designate the type of each job with the proper mathematical term.
10. What is the difference between a semicircular-sided solid and a hemispherical-ended solid?
11. Give several applications of a hemispherical-ended container. What would be the advantage of this shape?
12. Would a hemispherical-ended container ever be used as an open tank? Explain your answer.

UNIT 67 THE UNITS OF VOLUME MEASURE

Lengths are measured by units of *linear* measure; areas, by units of *square* measure; volumes, by units of *cubic* measure. The volume of a solid is the number of cubic units of measure contained in that solid. The cubic unit of measure is a cube whose length of side is one linear unit of measure.

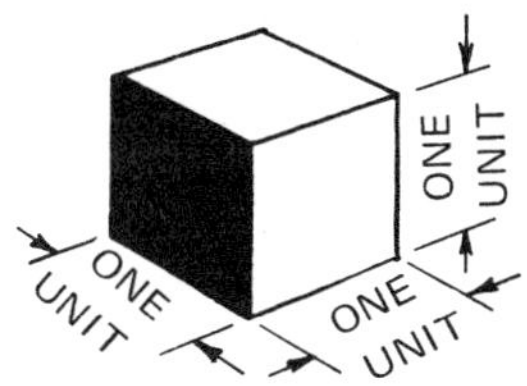

Volume measure is used when the capacity or contents of tanks, containers, etc. is to be found. Heating and ventilating problems also involve the mathematics of volume measure.

The formula for the volume of parallel-sided solids is $V = Ah$; A is the area of the base and h is the average height (altitude).

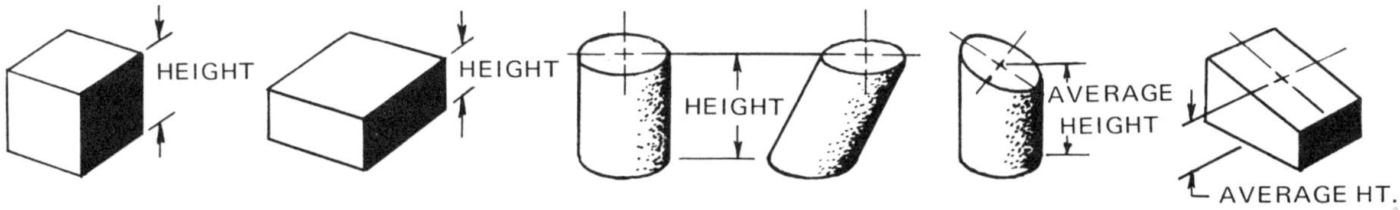

CUBIC MEASURE

TABLE OF CUBIC MEASURE

1728 cubic inches	=	1 cubic foot (cu. ft.)
27 cubic feet	=	1 cubic yard (cu. yd.)

To find the number of cubic inches contained in one cubic foot, multiply 12 x 12 x 12. The answer, 1728, proves that there are 1728 one-inch cubes in one cubic foot. **Note:** in. x in. x in. = $in.^3$, or cu. in.

The number 27 (cu. ft.) is obtained by multiplying 3 x 3 x 3 which equals 27 cu. ft. Note: ft. x ft. x ft. = $ft.^3$, or cu. ft.

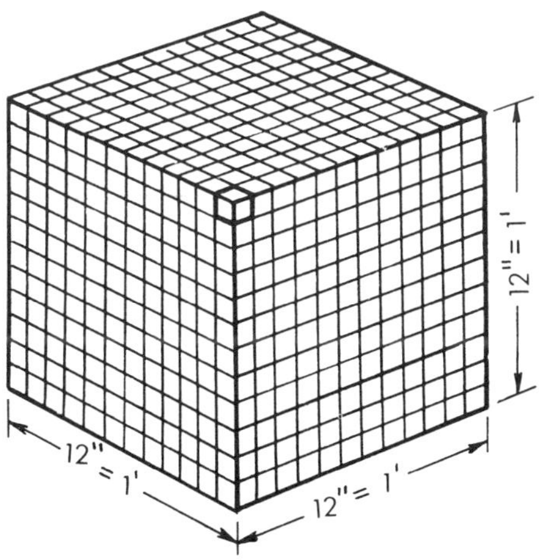

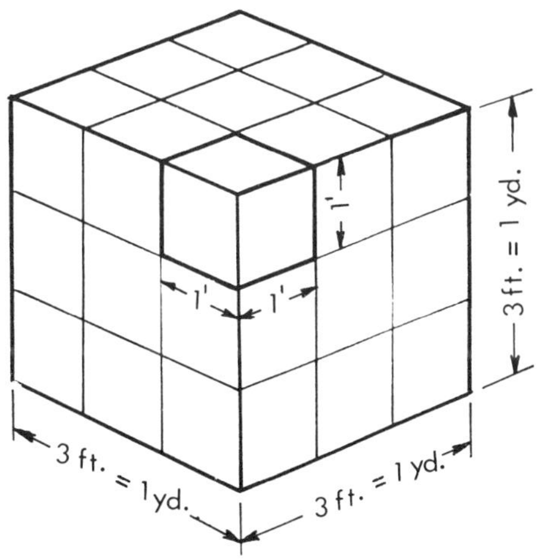

Conversion Rules and Examples

RULE 1: To change cu. in. to cu. ft., divide the given cu. in. by 1728.

RULE 2: To change cu. ft. to cu. yd., divide the given cu. ft. by 27.

Example: Change 182,021 cu. in. to cu. ft. and then to cu. yd.

1. Divide the given cu. in. (182,021) by 1728 to obtain cu. ft.

$$1728\,\overline{)\,182021.000} = 105.336 \text{ cu. ft.}$$

2. Divide this result (105.336) by 27. The answer is 3.90 cu. yd.

$$27\,\overline{)\,105.336} = 3.901 \text{ cu. yd.}$$

RULE 3: To change cu. yd. to ft., multiply the given cu. yd. by 27.

RULE 4: To change cu. ft. to cu. in., multiply given cu. ft. by 1728.

Example: Change 2 cu. yd. to cu. ft. and then to cu. in.

1. Multiply the given cu. yd. (2) by 27 to obtain cu. ft.
2. Multiply this result (54) by 1728. The answer is 93,312 cu. in.

```
27             1728
x2             x 54
54 cu. ft.     6912
              8640
              93312 cu. in.
```

DRY MEASURE

TABLE OF DRY MEASURE

2 pints	=	1 quart (qt.)
8 quarts	=	1 peck (pk.)
4 pecks	=	1 bushel (bu.)

CONVERSION CONSTANTS

1 bushel	=	2150.42 cu. in.
1 bushel	=	1.24446 cu. ft.

1 CU. IN.

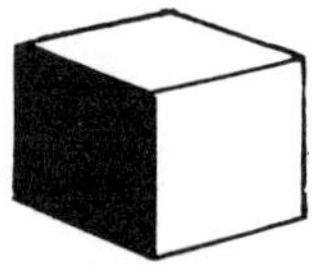

1 CU. FT.

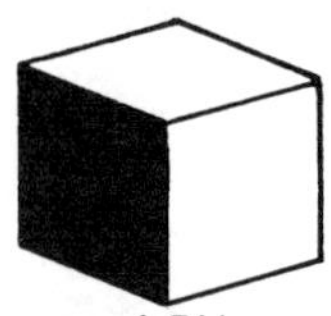

1 BU.

Conversion Rules and Examples

RULE 1: To change cu. in. to bu., divide the given cu. in. by 2150.42.

RULE 2: To change cu. ft. to bu., divide the given cu. ft. by 1.24446.

RULE 3: To change bu. to cu. in., multiply the given bu. by 2150.42.

RULE 4: To change bu. to cu. ft., multiply the given bu. by 1.24446.

Example: Change 10,000 cu. in. to bushels.

Divide the given cu. in. (10,000) by 2150.42. The answer is 4.65 bu.

$$2150.42_{\wedge}\,\overline{)\,10000.00_{\wedge}000} = 4.650 \text{ bu.}$$

Example: Change 10 1/2 bushels to cubic feet.

Multiply the given bu. (10 1/2) by 1.24446. The answer is 13.07 cu. ft.

```
     1.2 4 4 4 6
       x  1 0.5
     -----------
     6 2 2 2 3 0
   1 2 4 4 4 6 0
   -------------
   1 3.0 6 6 8 3 0  cu. ft.
```

LIQUID MEASURE

TABLE OF U.S. LIQUID MEASURE

4 gills	=	1 pt. (pt.)
2 pints	=	1 quart (qt.)
4 quarts	=	1 gallon (gal.)

U.S. LIQUID MEASURE

1 cu. ft.	=	7.48 gallons
231 cu. ft.	=	1 gallon
31 1/2 gallon	=	1 barrel (bbl.)

TABLE OF IMPERIAL LIQUID MEASURE

1 cu. ft.	=	6.25 gallons
277 cu. in.	=	1 gallon
45 gallons	=	1 barrel

CONVERSION CONSTANTS

1 Imp. gal. = 1.2 U.S. gal.

1 U.S. gal. = 5/6 Imp. gal.

1 Imp. gal. weighs 10 lb.
1 U.S. gal. weighs 8.33 lb. } distilled water at 68 deg. F.

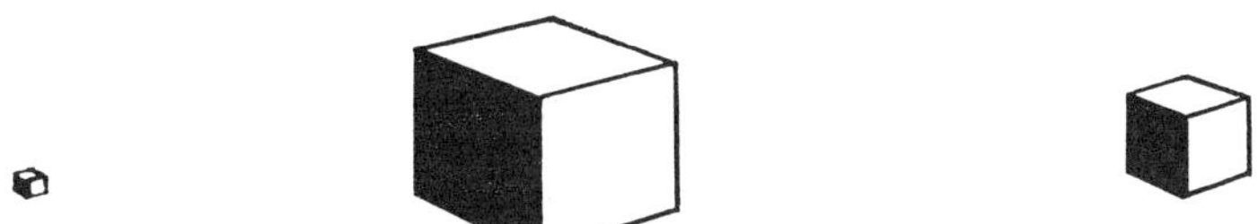

.1 CU. IN. 1 CU. FT. 1 GAL.

Conversion Rules and Examples

	U. S. MEASURE	IMP. MEASURE
RULE 1: To change cu. in. to gallons,	*divide by 231*	*divide by 277*
RULE 2: To change cu. ft. to gallons,	*multiply by 7.48*	*multiply by 6.25*
RULE 3: To change gallons to cu. in.,	*multiply by 231*	*multiply by 277*
RULE 4: To change gallons to cu. ft.	*divide by 7.48*	*divide by 6.25*

Example: Change 2000 cu. in. into gallons. Divide the volume by 231.

$$\frac{2000}{231} = 8.658 \text{ U.S. gal.}$$

Divide the volume by 277

$$\frac{2000}{277} = 7.22 \text{ Imp. gal.}$$

Example: Change 100 cu. ft. into gallons. Multiply the volume by 7.48.

100 x 7.48 = 748 U.S. gal.

Example: Change 100 cu. ft. into gallons. Multiply the volume by 6.25

100 x 6.25 = 625 Imp. gal.

Note: The ratio of pints and quarts in Imperial measure to U.S. measure are the same as the gallons.

1 Imp. qt. = 1.2 U.S. qt.

1 Imp. pt. = 1.2 U.S. pt.

ASSIGNMENT

A. Cubic Measure

Using the tables given, make the following conversions:

1. 5184 cu. in. to cu. ft.
2. 10,000 cu. in. to cu. ft.
3. 2015.23 cu. in. to cu. ft.

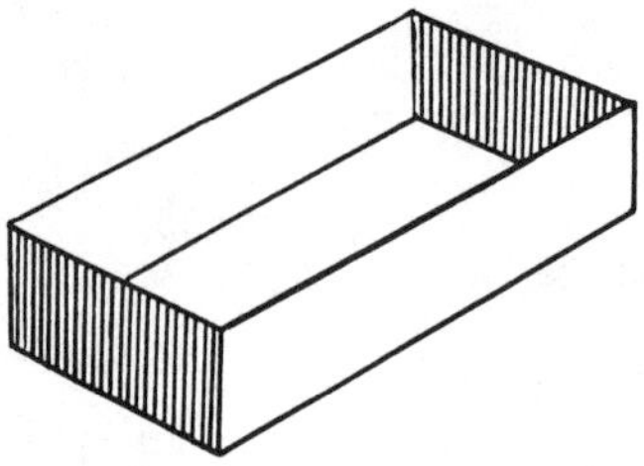

4. 10 3/4 cu. ft. to cu. in.
5. 5 cu. yd. to cu. in.
6. 28.96 cu. yd. to cu. ft.

B. Dry Measure

Using the tables given, make the following conversions.

1. 5300.84 cu. in. to bu.
2. 10 cu. ft. to bu.
3. 4 bu. to cu. ft.

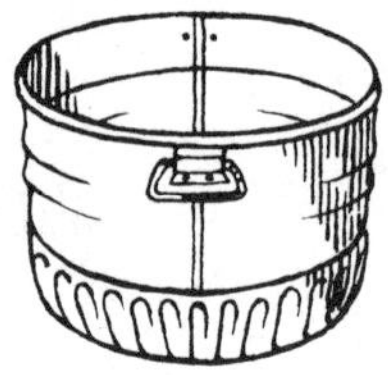

4. 3 bu., 3 pk. to cu. in.
5. 1 bu. to qt.
6. 22 pk. to bu.

C. Liquid Measure

Using the tables given, make the following conversions.

1. 2000 cu. in. to U.S. gal, to Imp. gal.
2. 10,521 cu. in. to U.S. gal., to Imp. gal.
3. 11 1/2 cu. ft. to U.S. gal., to Imp. gal.
4. 10 gal. to cu. in.

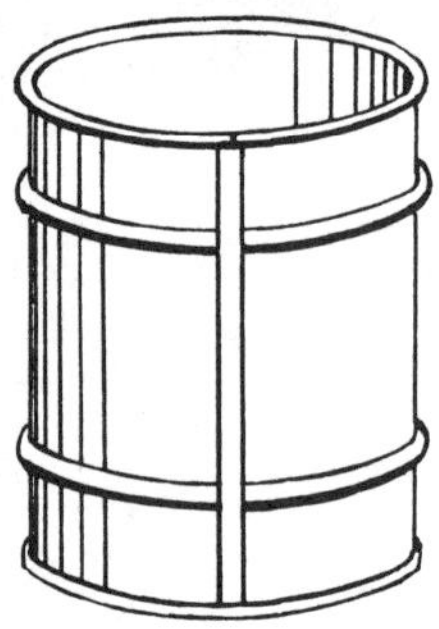

5. 15 1/4 U.S. gal., 15 1/4 Imp. gal. to cu. ft.
6. 20 bbl. to U.S. gal, to Imp. gal.
7. 100 U.S. gal., 100 Imp. gal. to bbl.
8. 21.5 U.S. gal. to qt., pt.
9. A tank holds 500 Imp. gal., how many U.S. gal. will it hold?
10. How many Imp. gal. will a chemical tank with a capacity of 2500 U.S. gal. hold?
11. A 1000 Imp. gal. tank is being filled with a pump rated at 125 U.S. gal. per hour. How long, to the closest minute will it take to fill it?
12. A car from Oregon gets 15 miles per U.S. gal. How many miles per Imp. gal. will the car get on a trip in British Columbia?

UNIT 68 VOLUMES: THE CUBE, THE SQUARE PRISM, AND THE RECTANGULAR SOLID

Tanks, containers, and pans of various kinds are often constructed in the shape of a cube, a square prism, or a rectangular solid. Rooms which are to be heated and ventilated may be of these shapes.

CUBE

CUBE

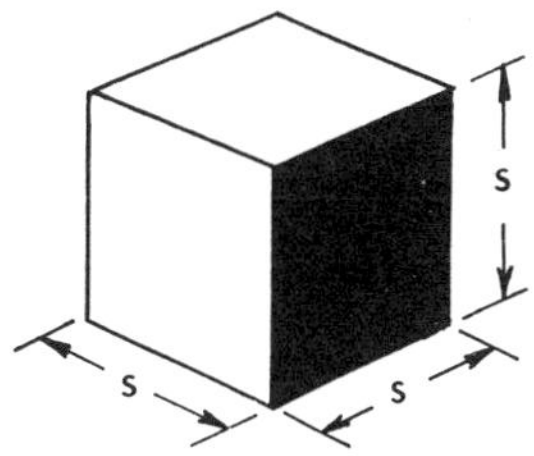

RULE: The volume of a cube is equal to the length of one side cubed.

Formula: $V = s^3$

V = volume

s = length of one side

Example: The side of a cube is 11½ in. Find its volume in cu. in.

1. Formula $V = s$
2. Substitution $V = (11\frac{1}{2})^3$
3. Evaluation $V = (11.5)^3$

$V = 11.5 \times 11.5 \times 11.5$

$V = 1520.875$ cu. in., or 1520.88 cu. in. Ans.

SQUARE PRISM

SQUARE PRISM

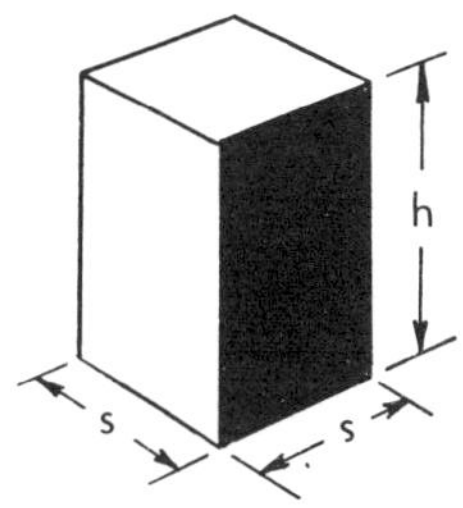

RULE: The volume of a square prism is equal to the area of the base times the height.

Formula: $V = Ah$, or

$V = s^2h$ (since $A = s^2$)

Symbols: V = volume

A = area of base

s = length of one side of base

h = height of solid

Example: The length of each side of a container is 10 in. and its height is 20 in. Find the volume in cu. in.

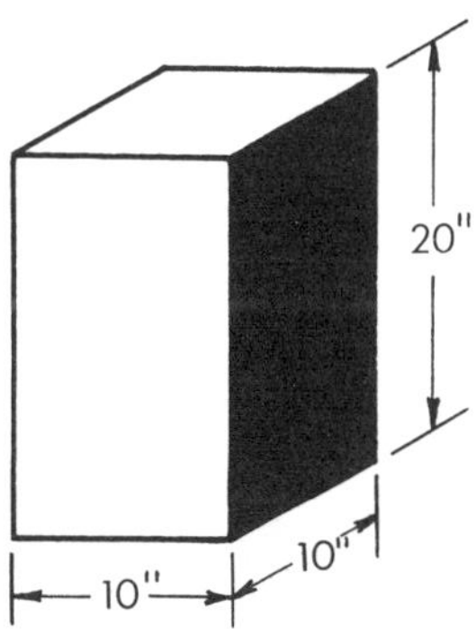

1. Formula: $V = s^2h$
2. Substitution $V = 10^2 \times 20$
3. Evaluation $V = 10 \times 10 \times 20$

10″	100 sq. in.
×10″	×20″
100 sq. in.	2000 cu. in.

$V = 2000$ cu. in. Ans.

RECTANGULAR SOLID

RULE: The volume of a rectangular solid is equal to the area of the base multiplied by the height.

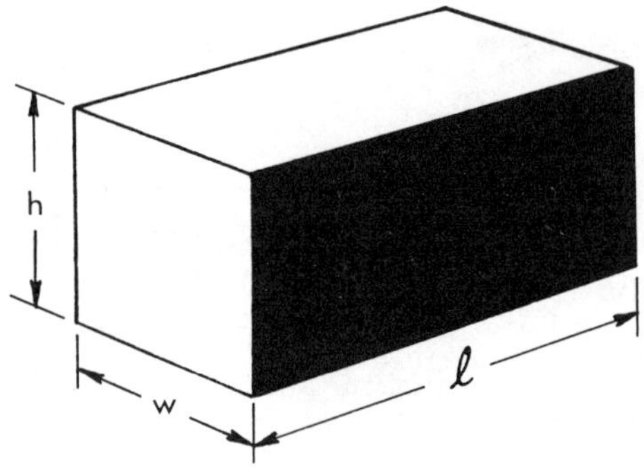

Formula: $V = Ah$, or

$V = \ell wh$ (since $A = \ell w$)

Symbol:

V	=	volume
A	=	area of base
ℓ	=	length of base
w	=	width of base
h	=	height of solid

Example: The length of a container is 22½ in., the width 14 in., and the height 6 in. Find the volume in cu. in.

1. Formula $V = \ell wh$
2. Substitution $V = 22\frac{1}{2} \times 14 \times 6$
3. Evaluation $V = 22.5 \times 14 \times 6$

$V = 1890$ cu. in. Ans.

```
  2 2.5"
 x 1 4"
 ------
  9 0 0
2 2 5
-------
3 1 5.0□"

  3 1 5□"
     x 6"
 --------
 1 8 9 0 in.³
```

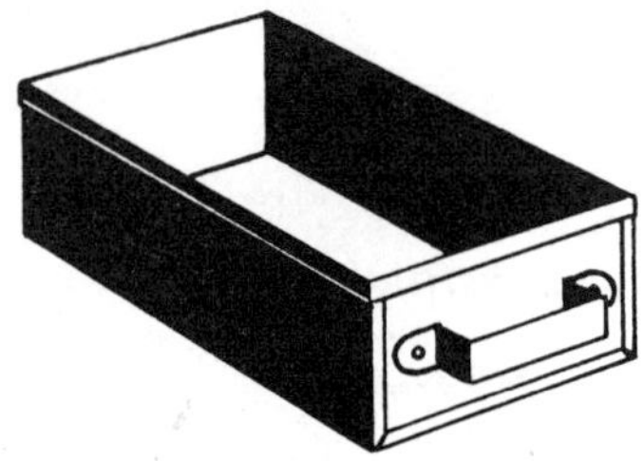

ASSIGNMENT

A. Cube

1. side = 16″, volume in cu. in. = ?
2. side = 19½″, volume in cu. ft. = ?
3. side = 28″, capacity in gal. = ?
4. side = 49″, capacity in gal. = ?

B. Square Prism

1. s = 12″, h = 10″, vol. in. in.^{3} = ?
2. s = 18″, h = 20″, cap. in gal. = ?
3. s = 8¾″, h = 15½″, cap. in qt. = ?
4. s = 3′ 6″, h = 6′ 3″, cap. in bu. = ?

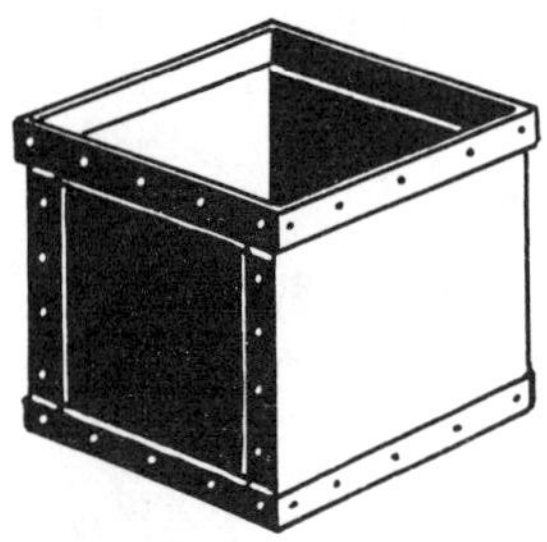

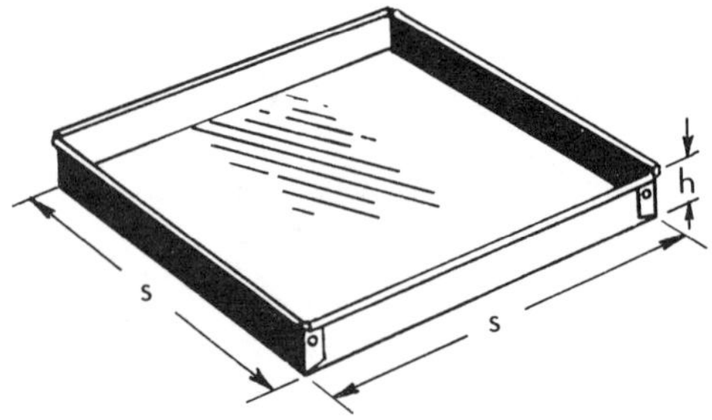

C. Rectangular Solid

1. ℓ = 24″, w = 18″, h = 6″ cap. in gal. = ?
2. ℓ = 21″, w = 15½″, h = 9″, cap. in gal. = ?
3. ℓ = 20½″, w = 18¼″, h = 9″, cap. in bu. = ?
4. ℓ = 5′ 6″, w = 4′ 3″, h = 6′, cap. in gal. = ?

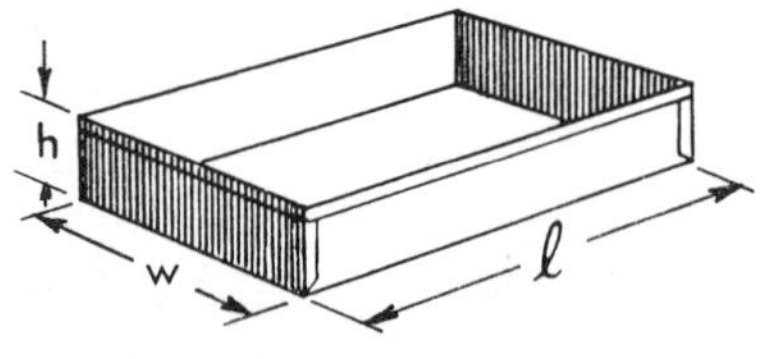

D. Additional Problems

1. Find the volume of a square room in cu. ft. if the length of one side is 12′ 6″, and the height is 8′ 6″.
2. Compute the volume of a room in cu. ft. if the length is 12′ 6″, the width 10′ 3″, and the height 8′ 9″.
3. The length of a room is 14′ 4″, the width 12′ 6″, and the height 8′ 4″. What is the volume of the room in cu. ft.?
4. Find the oil capacity of a square pan in gallons if the length of each side is 20″, and the height is 2″.
5. The length of a gasoline tank is 25½″, the width 19¼″, and the height 17″. Compute the gasoline capacity in gallons.
6. The length of the side of a square-based tank is 2′ 6″, and the height is 4′. How many gallons of milk will the tank hold?

UNIT 69 VOLUMES OF CYLINDRICAL, SEMICIRCULAR-SIDED, AND SPHERICAL-ENDED SOLIDS

Sheet metal workers are often called upon to make tanks similar in shape to circular cylinders and semicircular-sided solids. It is therefore important to know how to find the volume of such solids.

CIRCULAR CYLINDER

RULE: The volume of a cylinder is equal to the area of the end times the length.

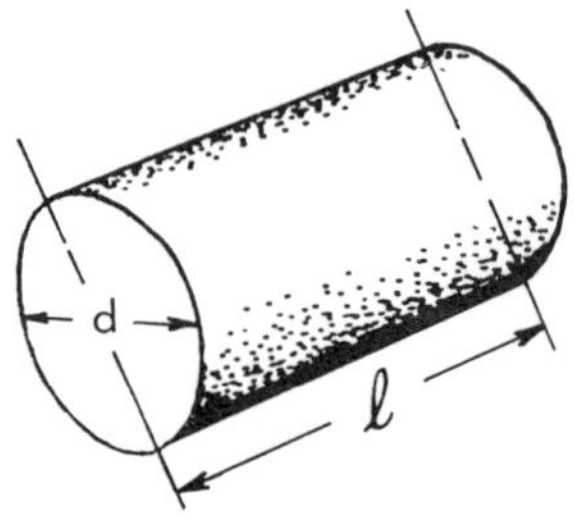

Formula: $V = A\ell$, or $\pi r^2 \ell$

Symbols:
V = volume
A = area of end
π = 3.14
r = radius
ℓ = length

Example: If the diameter of a circular cylindrical tank is 24″, and the height is 8″, find the volume in cu. in.

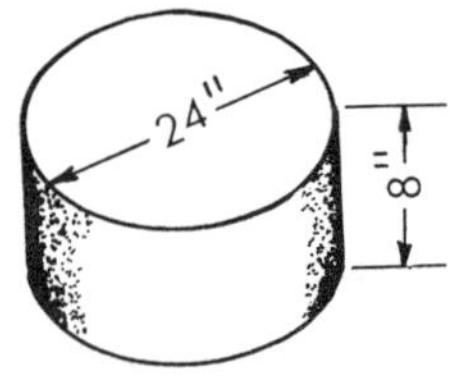

1. Formula $V = \pi r^2 \ell$
2. Substitution $V = 3.14 \times 12^2 \times 8$
3. Evaluation $V = 3.14 \times 12 \times 12 \times 8$
$V = 3617.28$ cu. in. Ans.

SEMICIRCULAR-SIDED SOLIDS

RULE: The volume is equal to the area of the base times the height.

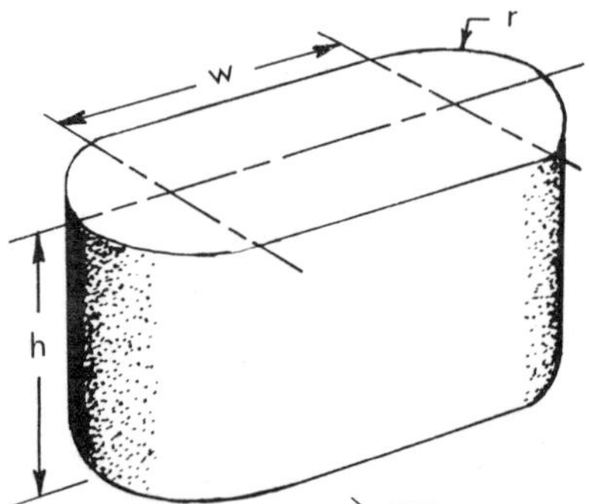

Formula: $V = Ah$, or $(\pi r^2 + 2rw)h$

Symbols:
V = volume A = area of base
π = 3.14 r = radius
w = distance between semicircles
h = height

Example: Find the volume in cu. in. of a semicircular-sided tank with the following dimensions: r = 10″, w = 15″; h = 25″.

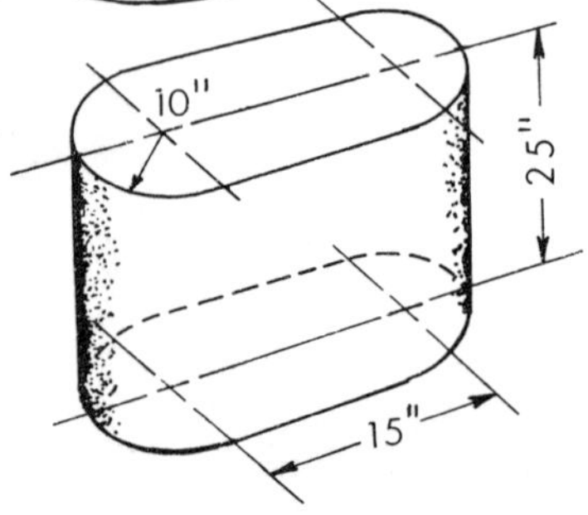

1. Formula $V = (\pi r^2 + 2rw)h$
2. Substitution $V = (3.14 \times 10^2 + 2 \times 10 \times 15) \times 25$
3. Evaluation $V = (314 + 300) \times 25$
$V = 614 \times 25$
$V = 15{,}350$ cu. in. Ans.

SPHERICAL-ENDED SOLIDS

RULE: The volume of a spherical-ended solid is equal to the sum of the volumes of two hemispheres and the volume of the cylinder.

Formula: $V = 4/3\,\pi r^3 + \pi r^2 h$

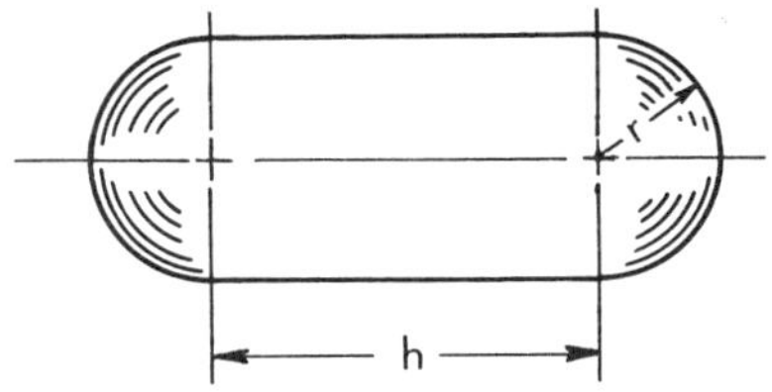

Symbols:
V = volume
r = radius of hemisphere
h = distance between the centers of the hemispheres

EXPLANATION OF RULE AND FORMULA

The spherical-ended solid can be broken up into two hemispheres and a cylinder. thus, the volume of a sphere is $4/3\pi r^3$ and the volume of a cylinder is $\pi r^2 h$. Therefore, combining the two separate volumes, the total volume becomes $V = 4/3\pi r^3 + \pi r^2 h$.

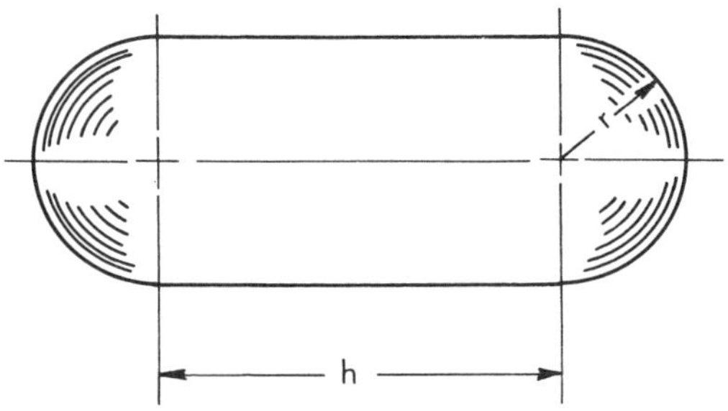

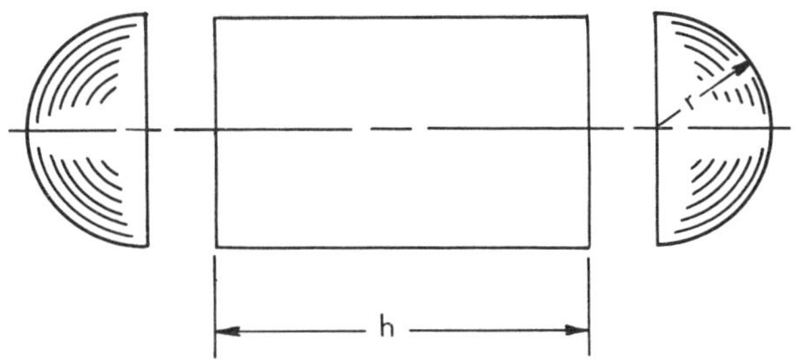

This formula may be reduced to:

$$V = 4/3\pi r^3 + \pi r^2 h$$
$$V = \pi r^2 (4/3 r + h)$$

When the diameter (d) is given, the formula may be written as:

$$V = \pi r^2 (4/3 r + h)$$
$$V = \pi(\frac{d}{2})^2 (\frac{\cancel{4}^2}{3} (\frac{d}{\cancel{2}}) + h)$$
$$V = \frac{\pi}{4} d^2 (\frac{2}{3} d + h)$$
$$V = .7854 d^2 (\frac{2}{3} d + h)$$

Example: The radius of each of the hemispheres of a spherical-ended tank is 6 in. and the distance between the centers of the hemispheres is 8 in. Compute the volume in cu. in.

1. Formula $V = 4/3\pi r^3 + \pi r^2 h$, or
   $V = \pi r^2\ (4/3 r + h)$
2. Substitution $V = \pi(6)^2\ (4/3\ (6) + 8)$
3. Evaluation $V = \pi 36\ (8 + 8)$
   $V = \pi \times 36 \times 16$
   $V = 3.14 \times 576$
   $V = 1808.64$ cu. in. Ans.

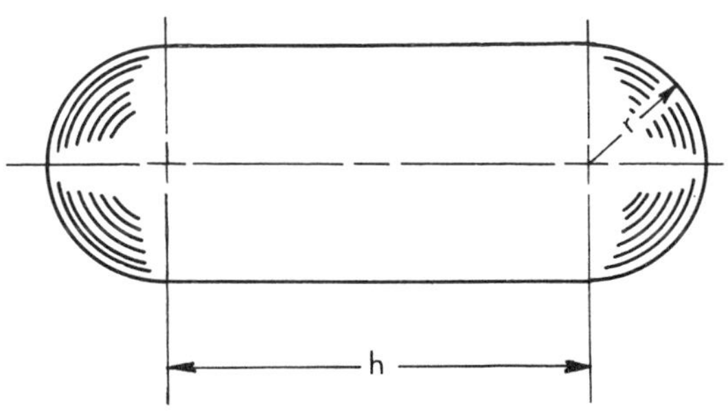

36	8	113.04
3.14	+8	16
113.04	16	1808.64 cu. in.

ASSIGNMENT

A. Cylindrical Containers

Find the volumes or the capacities in the units of measure indicated:

1. d = 10″ h = 15″; vol. in cu. in.
2. d = 12″, h = 28″; vol. in cu. ft.
3. d = 15″, ℓ = 7″; vol. in cu. in.
4. d = 19″, h = 25½″; cap. in gal.
5. d = 6″, h = 4¼″; cap. in. qt.
6. r = 25½″, h = 30¼″; cap. in gal.
7. d = 25′, ℓ = 3′ 6″; cap. in gal.
8. d = 4′ 3″, h = 7′ 9″; cap. in bu.

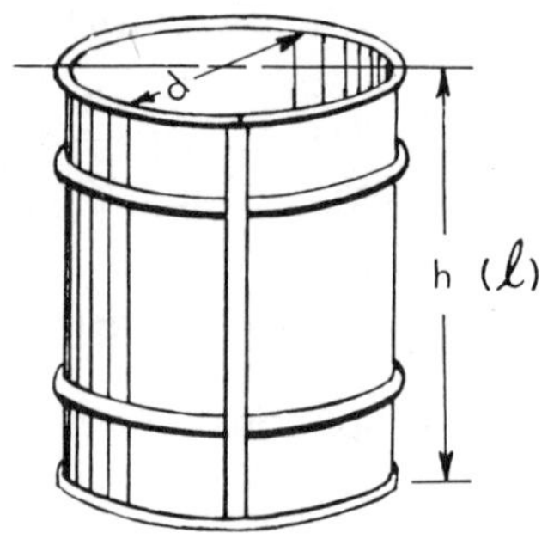

B. Semicircular-Sided Containers

Find the volumes or the capacities in the units indicated:

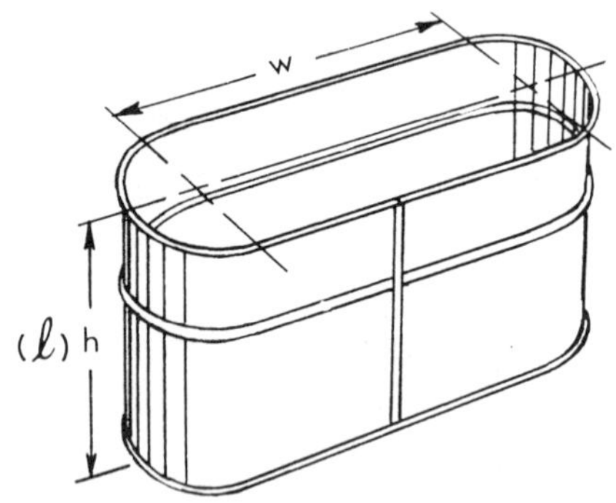

1. r = 4″, w = 10″, h = 15″; vol. in cu. in.
2. r = 3″, w = 5″, h = 10″; vol. in cu. in.
3. r = 5″, w = 7″, ℓ = 20″; vol. in cu. ft.
4. r = 10″, w = 15½″, ℓ = 18″; cap. in gal.
5. r = 12″, w = 30″, ℓ = 25½″; cap. in bu.
6. r = 6″, w = 8½″; ℓ = 10¼″; cap. in qt.
7. r = 1′ 3″, w = 1′ 6″, ℓ = 2′ 3″; cap. in gal.

C. Spherical-Ended Containers

1. Calculate the volume of the following spherical-ended tanks.

   a. r = 7′ h = 10′
   b. r = 14′ h = 16′
   c. r = 11′ h = 20′

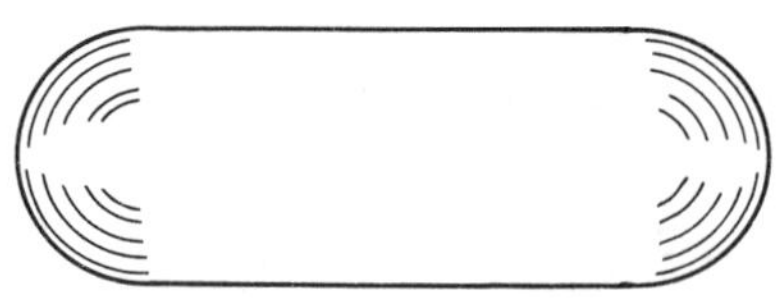

   d. d = 20′ h = 21′
   e. d = 24″ h = 35″
   f. d = 5′ h = 9′

D. Additional Problems

1. What is the capacity of a circular pan in quarts if the diameter is 6″, and the height is 3½″?
2. Compute the oil capacity in gallons of a semicircular-sided tank. Radii of semicircles is 7½″, distance between centers of semicircles is 10″; height of tank is 21″.
3. Find the capacity in gallons of a cylindrical oil tank; diameter is 11′; height is 15′ 6″.
4. Compute the capacity of a tank in gallons. Radius of semicircles is 5′ 6″; distance between centers of semicircles is 4′ 3″; length of tank is 6′.

UNIT 70 VOLUMES OF CONES AND PYRAMIDS

Parallel-sided solids have been covered in units 66 and 68. It will be valuable to review the terminology given in these units before beginning the study of tapered solids.

TAPERED SOLIDS

Tapered solids may be pyramid-shaped or cone-shaped.

Pyramid

A pyramid is a tapered solid whose base is a polygon and whose sides form triangles. The triangular sides meet in a common point to form the vertex.

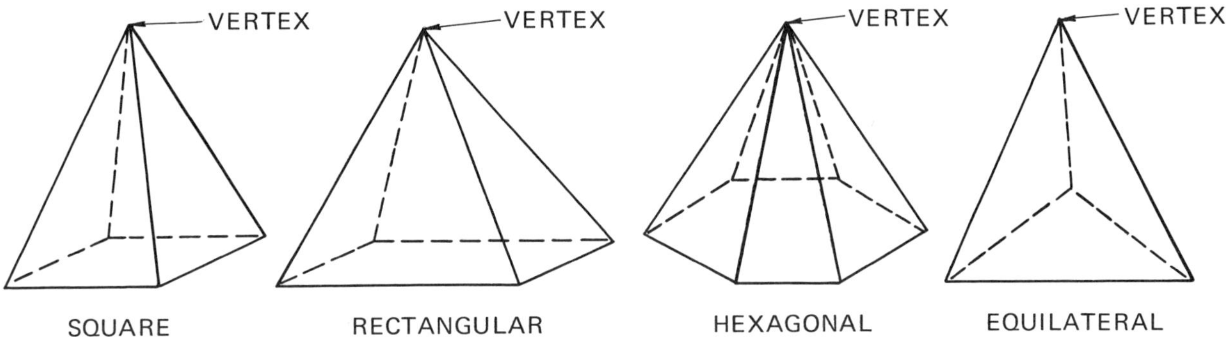

Cone

A cone is a tapered solid whose base is a single circle and whose lateral surface tapers uniformly to a point which forms the vertex.

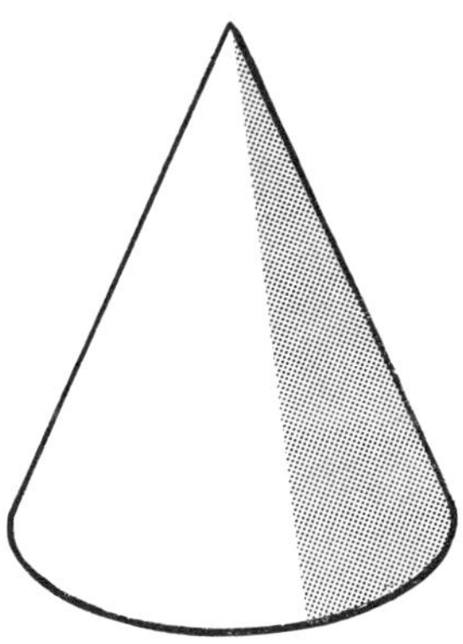

CIRCULAR CONE

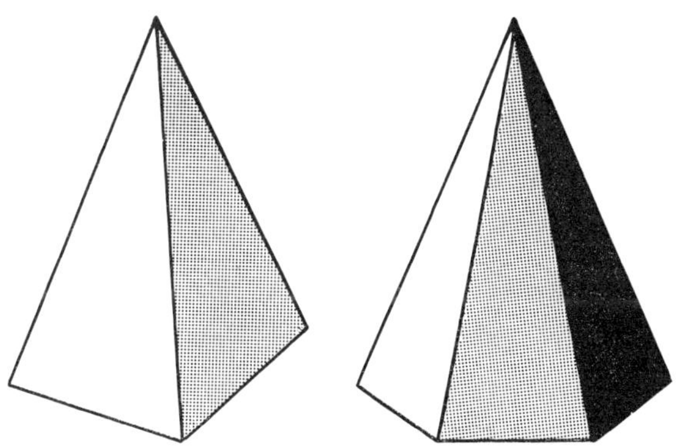

PYRAMIDS

The shape of the base and the angle of the axis determines the kind of pyramid or cone. They are classified as right or oblique, depending upon whether the axes are at right angles or oblique to the base.

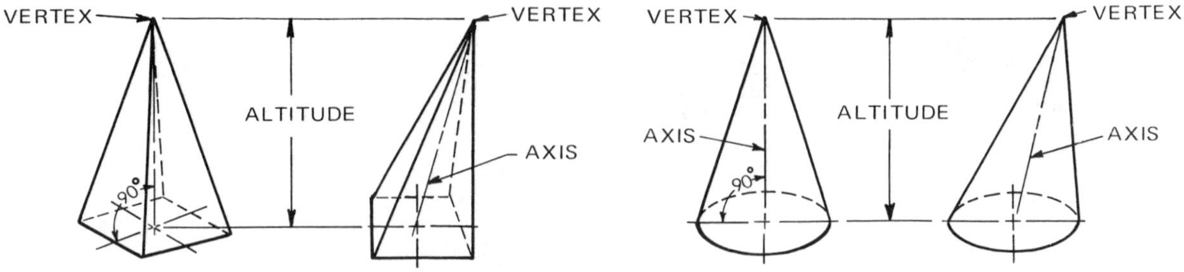

RIGHT OBLIQUE
PYRAMIDS

RIGHT OBLIQUE
CONES

To find the volume of any tapered solid, it is necessary to first find the area of the base. Next, multiply the area of the base by the vertical height or altitude of the solid and divide the product by 3.

$$\text{Volume} = \frac{\text{Area of Base} \times \text{Vertical Height}}{3}$$

Right Quadrangular Pyramid

A right quadrangular pyramid is a pyramid with a square or rectangular base whose vertex lies at a point in a line erected perpendicular to the center of the base.

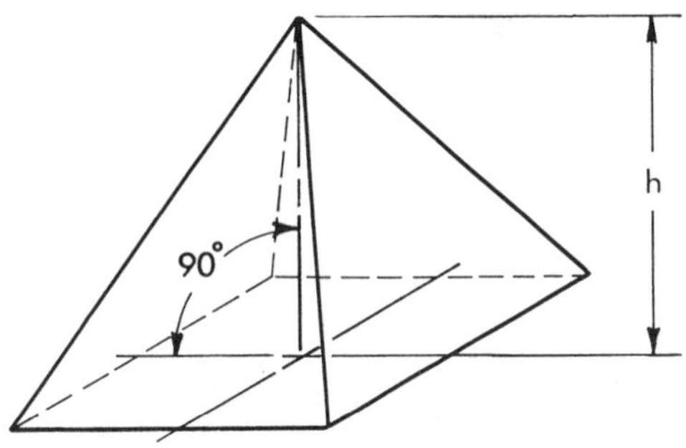

SQUARE RIGHT PYRAMID

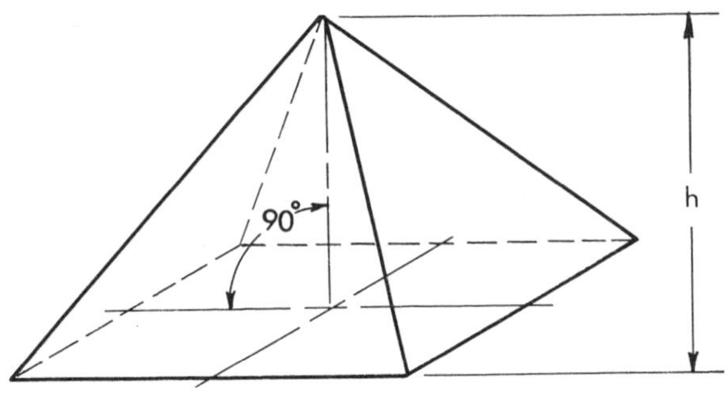

RECTANGULAR RIGHT PYRAMID

Finding the Volume of a Right Square Pyramid

RULE: The volume of a right square pyramid is equal to the area of the square base ($A = s^2$), multiplied by the vertical height, and divided by 3.

Formula: $V = \dfrac{s^2 h}{3}$

Symbols: V = volume

s = length of base sides

h = height of the pyramid

EXPLANATION OF RULE AND FORMULA

The base of a right square pyramid is a square. In order to find its area, the length of the sides is squared. ($A = s^2$). This area is then multiplied by the height to give us the volume of a square prism. Since the pyramid is tapered, we must divide the product by 3. (See drawing below.)

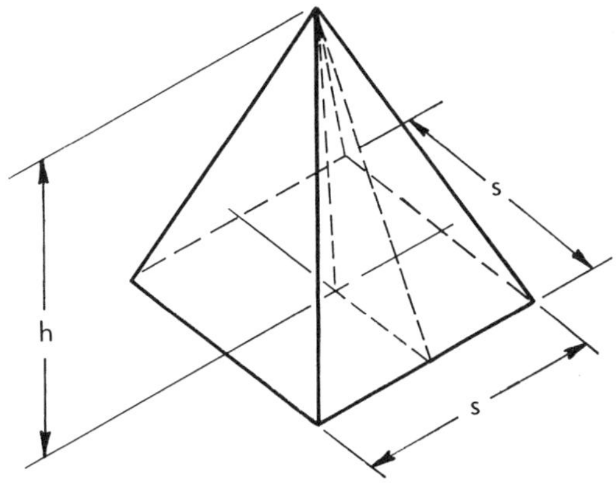

RIGHT SQUARE PYRAMID

Example: A right square pyramid has a base which is 3 in. long and a height which is 5 in. Compute its volume in cu. in.

1. Formula $V = \frac{s^2 h}{3}$ $\quad 3^2 = 3 \times 3 \quad 3 \times 3 = 9$
2. Substitution $V = \frac{3^2 \times 5}{3}$ $\quad 9 \times 5 = 45$
3. Evaluation $V = \frac{45}{3}$ $\quad 45 \div 3 = 15$ cu. in.

   $V = 15$ cu. in.

Right Rectangular Pyramid

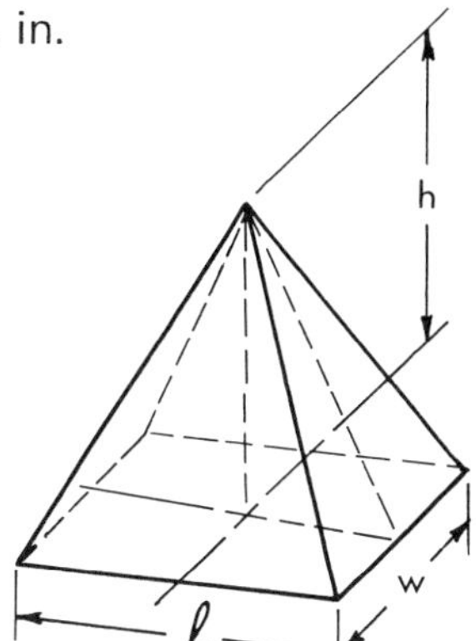

A right rectangular pyramid is a right pyramid with a rectangular-shaped base. The altitude or the height is designated by *a*, or *h*, the edges of the base by ℓ and *w*. The slant heights of the end faces and of the side faces differ in length.

Finding the Volume of a Right Rectangular Pyramid

RULE: The volume of a right rectangular pyramid is equal to the area of the rectangular base ($A = \ell w$) multiplied by the vertical height and divided by 3.

Formula: $V = \frac{(\ell w)h}{3}$

Symbols:
V = volume
ℓ = length of base
w = width of base
h = height

EXPLANATION OF RULE AND FORMULA

Since the base of the right rectangular pyramid is a rectangle, the area of the rectangular base must be found first and then multiplied by the vertical height to give the volume of a rectangular solid. Since the pyramid is tapered, we must divide the product by 3. (See drawing at right.)

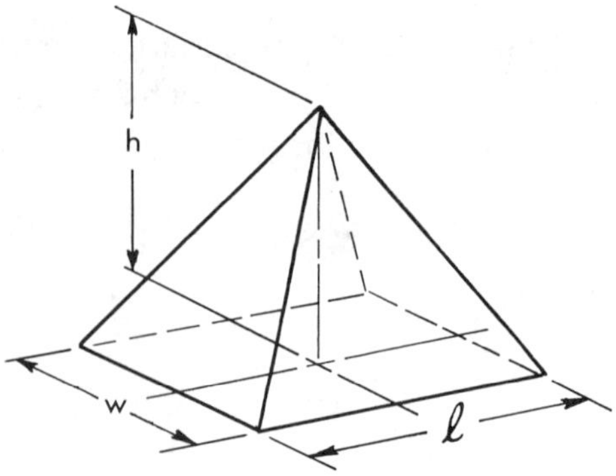

RECTANGULAR PYRAMID

Example: A right rectangular pyramid has a base which is 4 ft. long by 3 ft. wide and 8 ft. high. Compute its volume in cu. ft.

1. Formula $V = \frac{(\ell w)h}{3}$

2. Substitution $V = \frac{(4 \times 3) \times 8}{3}$

3. Evaluation $V = \frac{96}{3}$

$V = 32$ cu. ft.

$4 \times 3 = 12$

$12 \times 8 = 96$

$96 \div 3 = 32$ cu. ft. Ans.

Finding the Volume of an Equilateral Triangle Pyramid

RULE: The volume of an equilateral triangle pyramid is equal to the area of the base (A = 1/2 A x h) multiplied by the perpendicular height, and divided by 3.

Note: An equilateral triangle is one in which all sides and all angles are equal. Thus, side (a) = side (b) = side (c) and ∠ a = ∠ b = ∠ c.

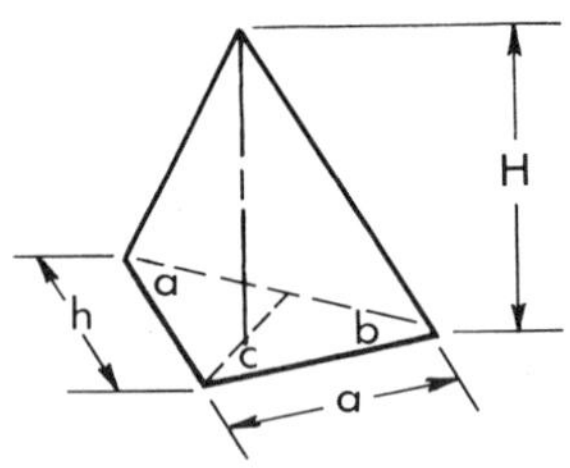

EQUILATERAL TRIANGLE PYRAMID

Formula: $V = \frac{\text{area of base} \times H}{3}$

$V = \frac{(1/2\, a \times h) \times H}{3}$

Symbols: V = volume
a = length of side (a)
h = perpendicular height of base triangle
H = perpendicular height of pyramid

EXPLANATION OF RULE AND FORMULA

The base of the equilateral triangle pyramid is an equilateral triangle. The area of the triangle, therefore, is 1/2 base x height of the triangle, or (1/2 a x h). This area is then multiplied by the height of the pyramid (H) and divided by 3 to give the volume of the pyramid.

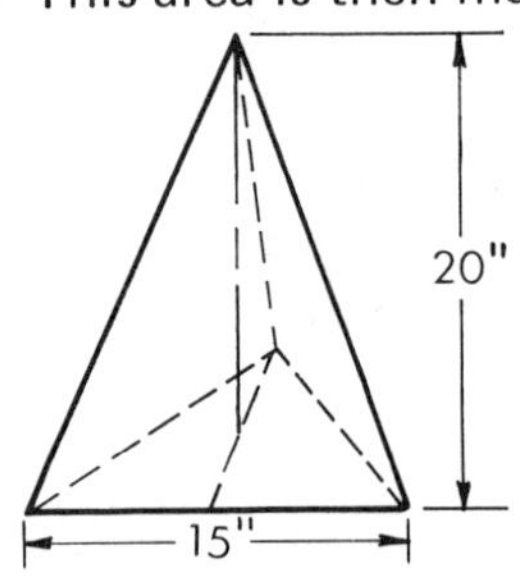

Example: The base of an equilateral triangle pyramid is 15 in. long on each side. The perpendicular height of an equilateral triangle pyramid is 20 in. Find its volume in cu. in.

Note: First find the area of the base triangle.

1. Formula : $A = 1/2\, a \times h$

Note: *h* is the unknown and must be found. Draw *h* perpendicular to side (a). Since all the angles are equal, (h) will bisect line (a), or 7.5″. Solve for *h* by use of the formula:

$$h = \sqrt{c^2 - (\frac{a}{2})^2}$$

$$h = \sqrt{(15)^2 - (7.5)^2}$$

$$h = \sqrt{225 - 56.25}$$

$$h = \sqrt{168.75}$$

$$h = 12.99''$$

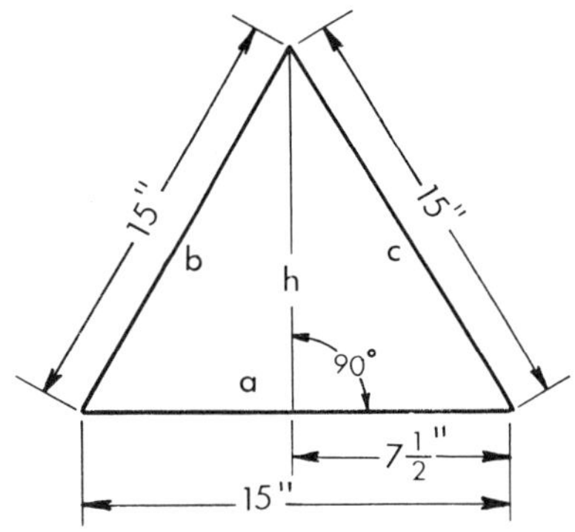

2. Substitution $A = 1/2a \times h$

$A = 1/2 \times 15 \times 12.99$

3. Evaluation $A = 7.5 \times 12.99 = 97.43$ cu. in.

Note: To find the volume of the equilateral triangle pyramid, substitute in the formula:

1. Formula $V = \frac{\text{area of base} \times H}{3}$

2. Substitution $V = \frac{97.43 \times 20}{3}$

3. Evaluation $V = \frac{1948.60}{3}$

$V = 649.53$ cu. in.

Right Circular Cone

A right circular cone is a right cone whose base is a circle. Right circular cones are usually identified by the word cone. The vertex of a right cone lies at a point in a line erected perpendicular to the center of the base.

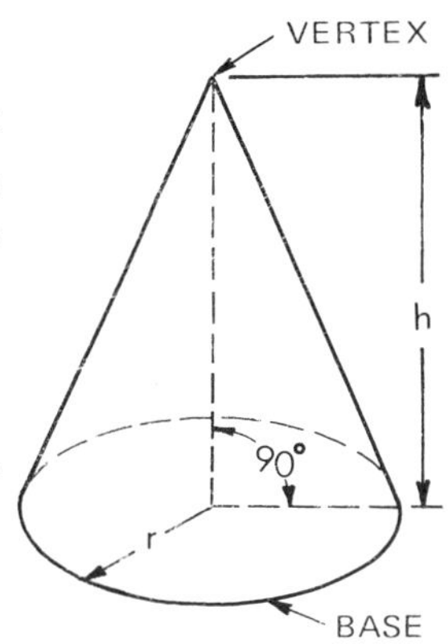

RIGHT CIRCULAR CONE

Finding the Volume of a Right Circular Cone

RULE: The volume of a right circular cone is equal to area of circular base ($A = \pi r^2$), multiplied by vertical height, and divided by 3.

Formula: $V = \frac{(\pi r^2)h}{3}$

Symbols: V = volume r = radius of base

π = 3.14 h = height of cone

EXPLANATION OF RULE AND FORMULA

The base of a right circular cone is a circle. In order to find its area, the radius is squared and multiplied by 3.14. This area is then multiplied by the vertical height to give us the volume of a circular cylinder. Since the cone is tapered, we must divide the product by 3. (See the drawing at right.)

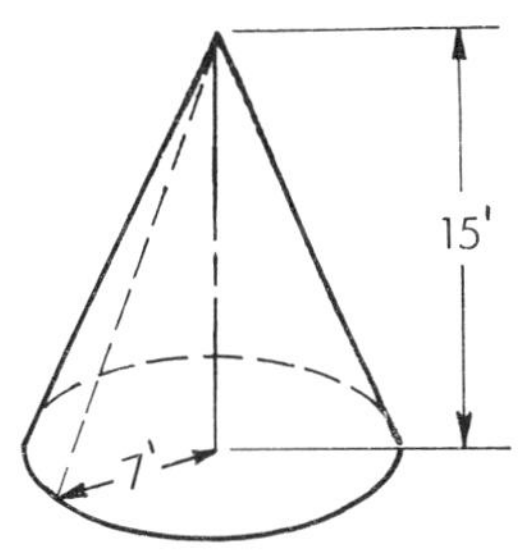

Example: A right circular cone has a radius of 7 ft. and a height of 15 ft. Compute its volume in cu. ft.

1. Formula $V = \dfrac{\pi r^2 h}{3}$

2. Substitution $V = \dfrac{3.14 \times 7 \times 7 \times 15}{3}$

3. Evaluation $V = \dfrac{2307.9}{3}$

$V = 769.3$ cu. ft.

$7 \times 7 = 49$

```
  3.14          153.86
  x 49            x 15
  2826           76930
 1256           15386
 153.86        2307.90

                 769.3 cu. ft.
              3)2307.9
```

ASSIGNMENT

A. Volumes of Cones

Compute the volumes of the following right cones:

1. r = 7′ h = 16′
2. r = 14′ h = 28′
3. r = 3′ h = 7′
4. r = 5′ h = 11′
5. D = 6′ h = 13′
6. D = 8′ h = 17′

B. Volumes of Pyramids

Compute the volumes of the following right pyramids:

1. Base = 3′ x 4′; height = 3′ 6″
2. Base = 4′ x 4′ 6″; height = 5′ 3″
3. Base = 22″ x 18″; height = 16″
4. Base = 3′ 4″ x 20″; height = 2′ 7″
5. Base is triangular (20″ each side); 20″ high
6. Base is rectangular (8½″ x 14½″); 10½″ high

C. Calculate the volume of the following pyramids and cones.

1. S = 4′ h = 9′
2. L = 4′ w = 6′ h = 12′
3. r = 7″ h = 21″
4. D = 3′ h = 7′

D. Calculate the capacity in gal. of a right circular cone, radius 35″, height 21″.

SECTION 10 — FORMULAS: SPECIAL PROBLEMS

UNIT 71 ALLOWANCES FOR EDGES

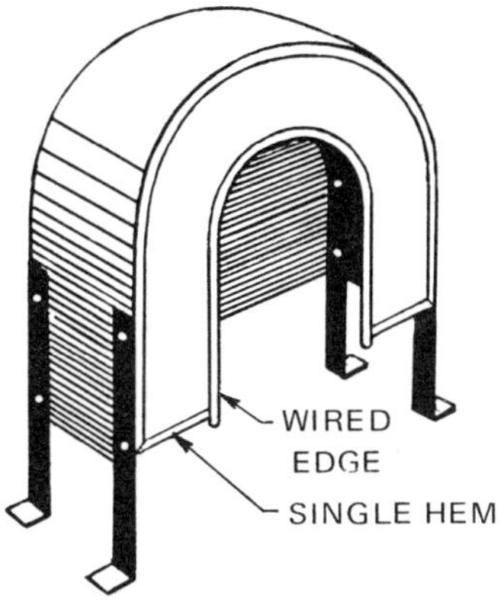

Edges of various kinds are needed on the job to eliminate raw edges, to improve the appearance, and to add strength. There are three methods of forming these edges: (1) by using the metal itself, (2) by inserting wire or rod, and (3) by attaching band or angle iron.

The addition of the seams and edges to a job requires that additional metal be allowed for them on the pattern. The allowances given in this unit are for 24-gage metal and lighter.

SINGLE HEM

RULE: The allowance for a single hem is equal to the width of the seam.

Formula: Allowance = W (W = width of seam)

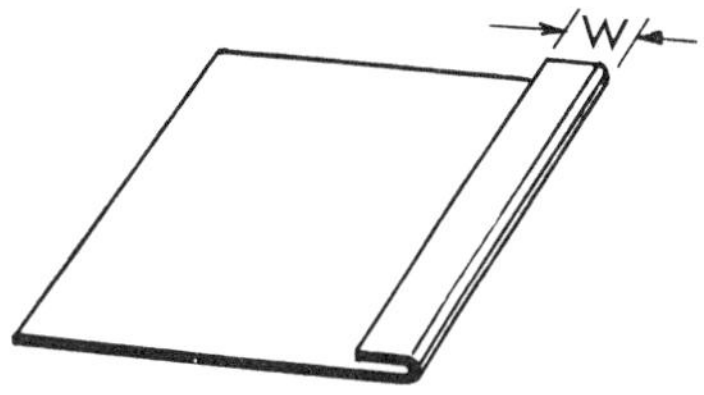

Example: The allowance for a single hem, width 3/16 in., is 3/16 in.

DOUBLE HEM

RULE: The allowance for a double hem is twice the width of the seam.

Formula: Allowance = 2W (W = width of seam)

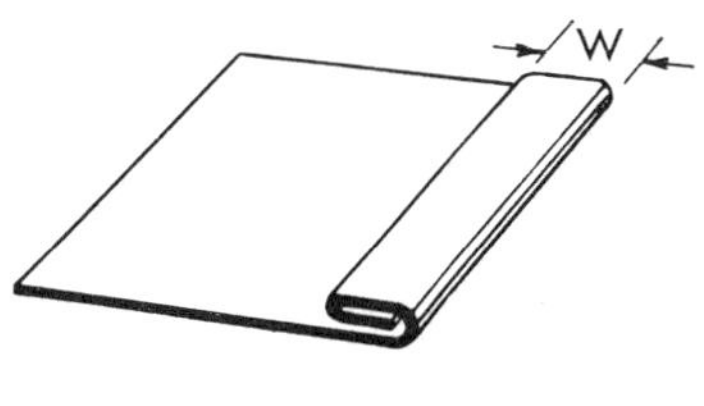

Example: What is the allowance for a double hem, width 5/32 in.?

Allowance is $2 \times \frac{5''}{32} = \frac{10''}{32}$, or $\frac{5''}{16}$ Ans.

WIRED EDGE

RULE: The allowance for a wired edge is 2 1/2 times the diameter of the wire.

Formula: Allowance = 2 1/2D (D = dia. of wire)

Example: Compute the alowance for a wired edge, diameter of wire 5/32 in.

Allowance is $2\frac{1}{2} \times \frac{5''}{32}$, or $\frac{5}{2} \times \frac{5''}{32} = \frac{25''}{64}$ Ans.

ASSIGNMENT

Find the allowances of material needed for one side or two sides of the jobs according to the dimensions indicated in the following lists.

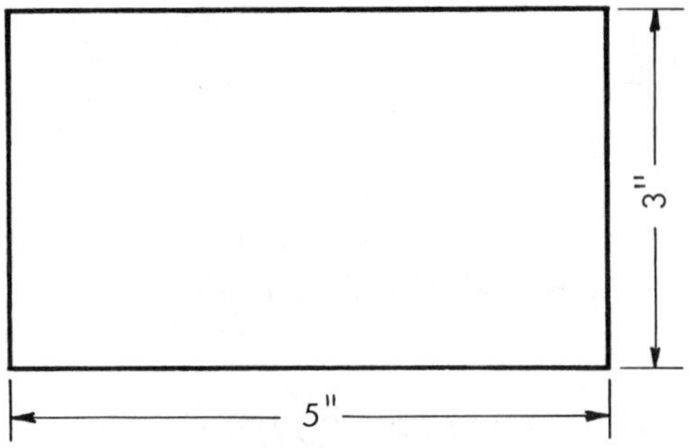

A. Single Hem

1. Width of hem, 3/16", (one side)
2. Width of hem, 1/4", (one side)
3. Width of hem, 5/32", (two sides)

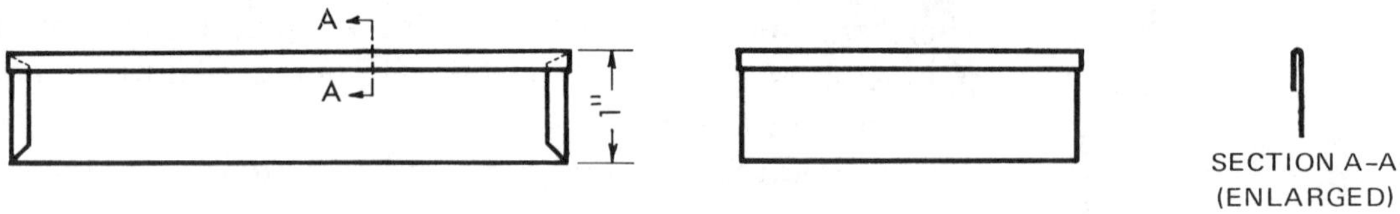

B. Double Hem

1. Width of hem, 1/4", (one side)
2. Width of hem, 3/32", (one side)
3. Width of hem, 5/16", (one side)
4. Width of hem, 1/8", (two sides)
5. Width of hem, 5/32", (two sides)

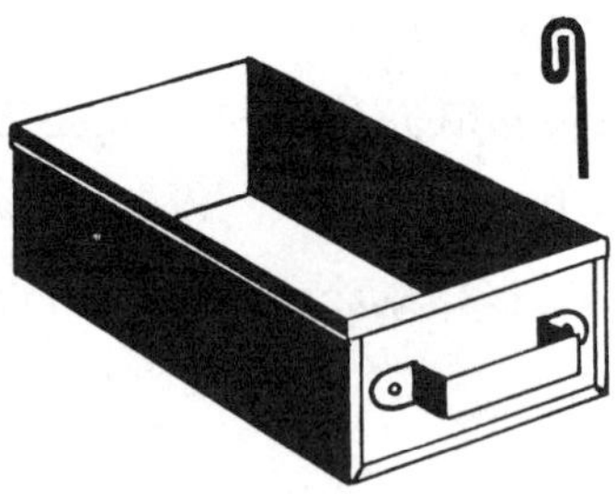

C. Wired Edge

1. Diameter of wire, 1/8", (one side)
2. Diameter of wire, 1/4", (one side)
3. Diameter of wire, 3/16", (one side)
4. Diameter of wire, 5/16", (two sides)
5. Diameter of wire, 3/32", (two sides)
6. Diameter of wire, 5/32", (two sides)
7. Diameter of wire, 7/32", (one side)

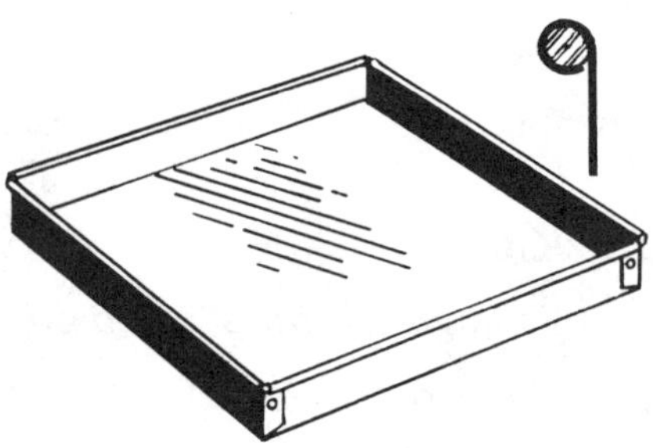

UNIT 72 ALLOWANCES FOR SEAMS

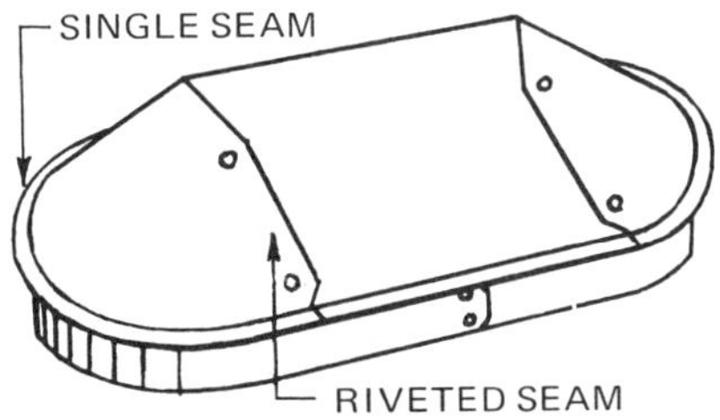

Parts of sheet metal jobs can be joined by many different kinds of seams. On light metal, these can be formed of the metal itself; on heavy metals and on jobs where strength is needed, rivets, metal screws, and bolts can be used. The drawing shows two kinds of seams, a riveted seam and a seam formed of the metal.

The addition of seams to a job requires that additional metal be allowed on the pattern. The allowances given in this unit are for 24-gage metal and lighter.

LAP SEAMS

RULE: The allowance for lap seams is equal to the seam width.

Formula: Allowance = W (W = width of lap seam)

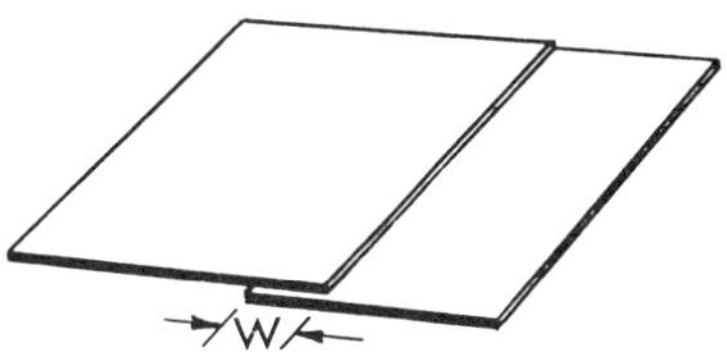

LAP SEAM

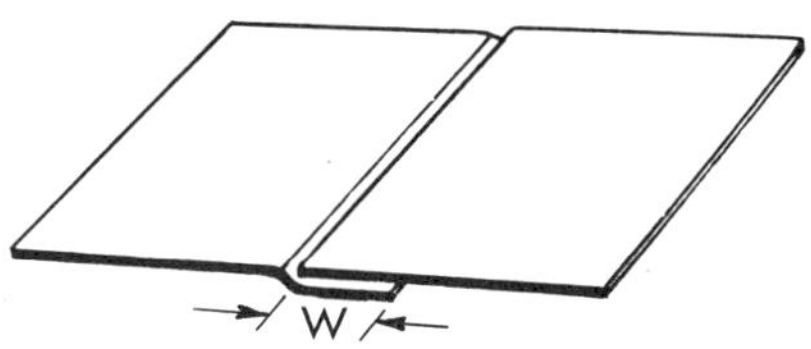

OFFSET SEAM

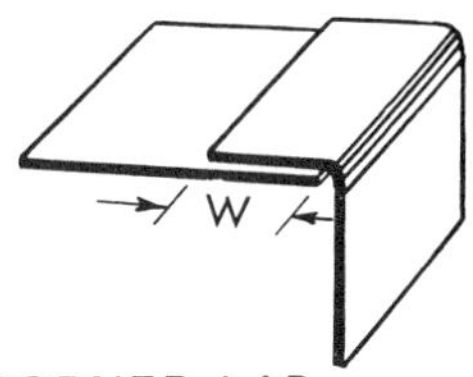

CORNER LAP

Example: The allowance for a lap seam, width 1/8″, is 1/8″.

Grooved Lock Seams

RULE: The allowance for both edges of an outside or an inside grooved seam is three times the width of the seam, one-half of this being added to each edge of the pattern.

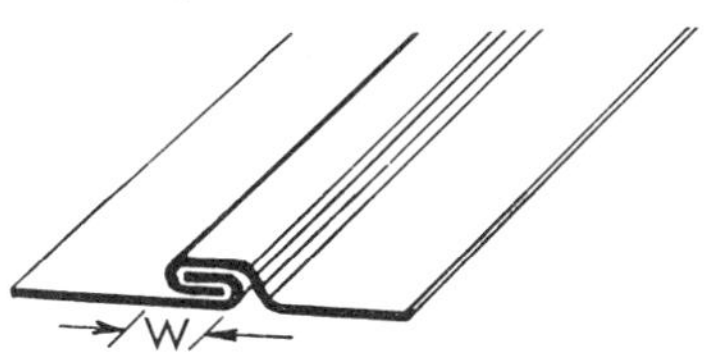

OUTSIDE GROOVED SEAM

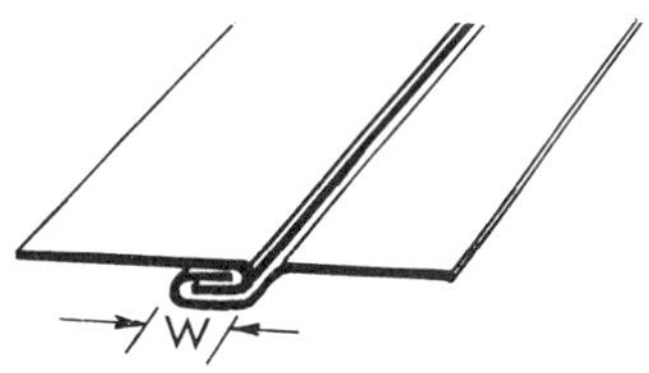

INSIDE GROOVED SEAM

Formula: Allowance (total) = 3W (W = width of grooved seam)

Example: Find (1) the total allowance, and (2) the allowance for each edge of a grooved seam, width 3/16″.

1. Allowance (total) is 3 x 3/16″ = 9/16″ Ans.
2. Allowance (one edge) is 1/2 x 9/16″ = 9/32″ Ans.

RIVETED LAP SEAM

RULE: The allowance for each edge of a riveted seam is twice the diameter of the rivet.

Formula: Allowance = 2D (D = dia. of rivet)

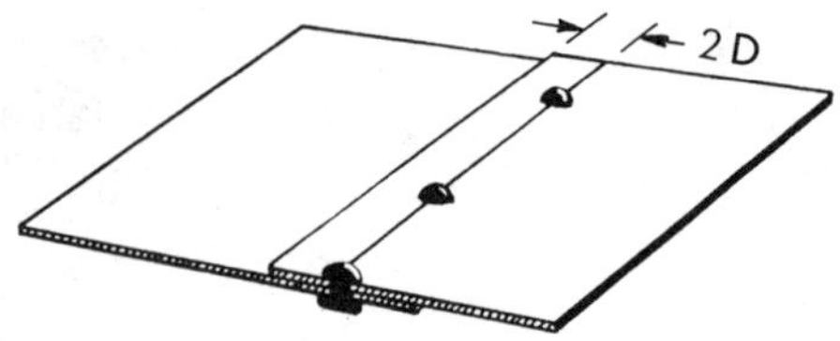

Example: Find the allowance for one edge of a riveted seam; the rivet size is 14 oz.

1. The dia. of a 14-oz. rivet is .109″. (See rivet table, page 265.)
2. Allowance is 2 x .109″ = .218″ or 7/32″ Ans.

DOVETAILED SEAM

RULE: The allowance for a dovetailed seam is equal to the width of the seam.

Formula: Allowance = W (W = width of seam)

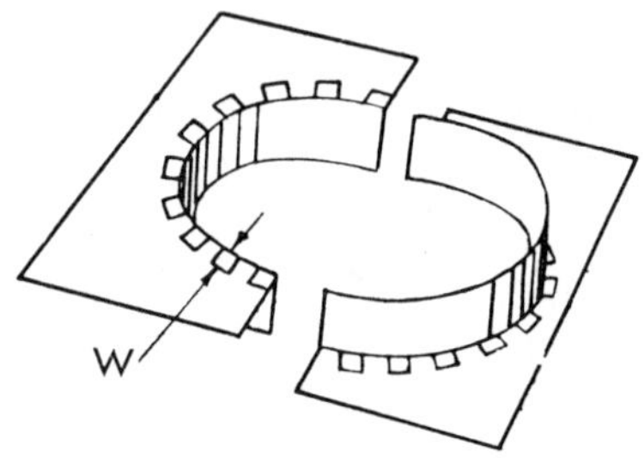

Example: The allowance for a 5/16″ dovetailed seam is 5/16″.

BURRED BOTTOM SEAM

RULE: The allowance for one side of a job with a burred bottom seam is equal to the width of the lap or burr.

Formula: Allowance = W (W = width of lap or burr)

Example: The allowance for a burred bottom seam, width 1/8″, is 1/8″.

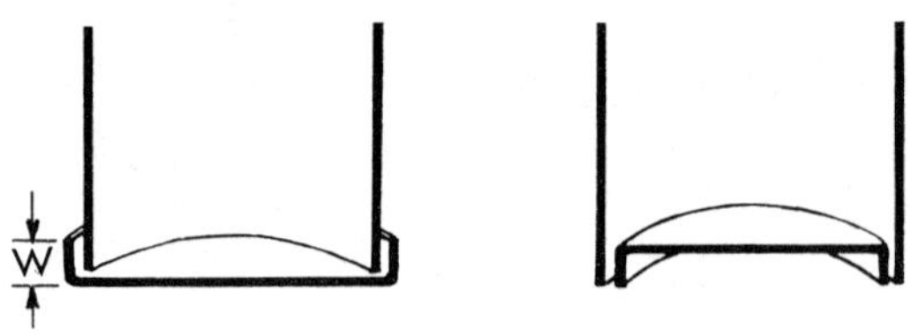

SET-IN BOTTOM SEAM

RULE: The allowance on one side of the body for a set-in bottom seam is equal to twice the width of the seam; the allowance for one side of the bottom is equal to the width of the seam.

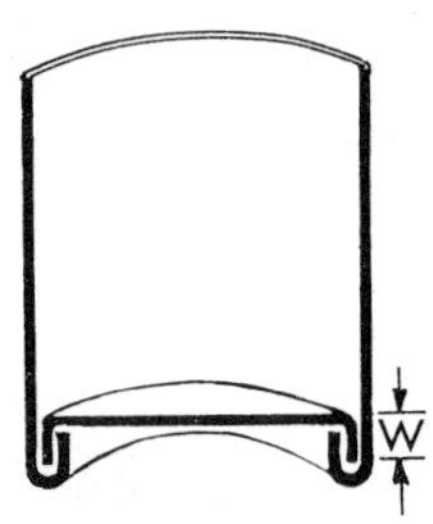

Formulas: Allowance (one side of body) = 2W

Allowance (one side of bottom) = W

SINGLE AND DOUBLE SEAMS

RULE: The allowance on one side of the bottom for either the single seam or the double seam is equal to twice the width of the flange on the body, but the seam should be made so that the outer allowance is equal to the width minus 1/32″, and the inner allowance is equal to the width plus 1/32″.

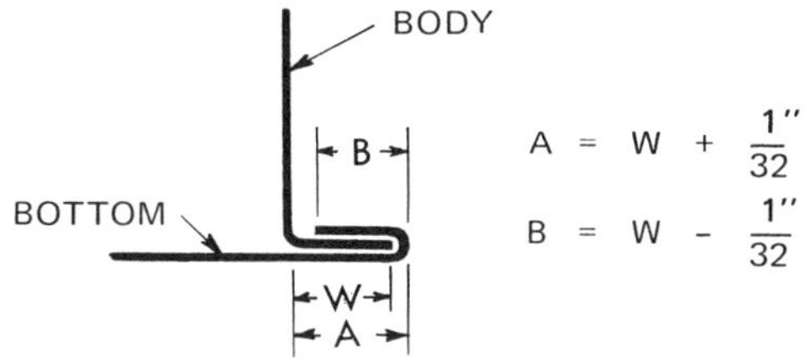

Formulas: Allowance (one side of body) = W

Allowance (one side of bottom) = 2W

Distribution of allowance (bottom): A = W + 1/32″

B = W - 1/32″

Example: Find the allowances on one side of the body and the bottom of a job with a single seam, width of flange 5/32″. Compute the distribution for the bottom.

Allowance (one side of body) = 5/32″ Ans.

Allowance (one side of bottom) = 2 x 5/32″, or 5/16″ Ans.

Distribution (bottom): A = 5/32″ + 1/32″, or 3/16″ Ans.

B = 5/32″ - 1/32″, or 1/8″ Ans.

PITTSBURGH LOCK

RULE: The allowance for the pocket part of one side of a job is equal to twice the width of the pocket plus 3/16″.

Formula: Allowance (for pocket of one side) = 2W + 3/16″.

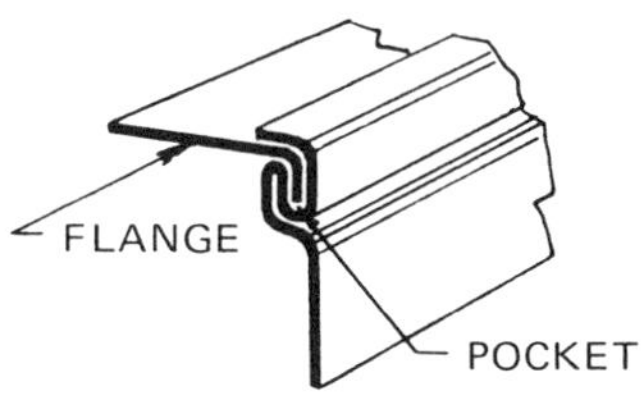

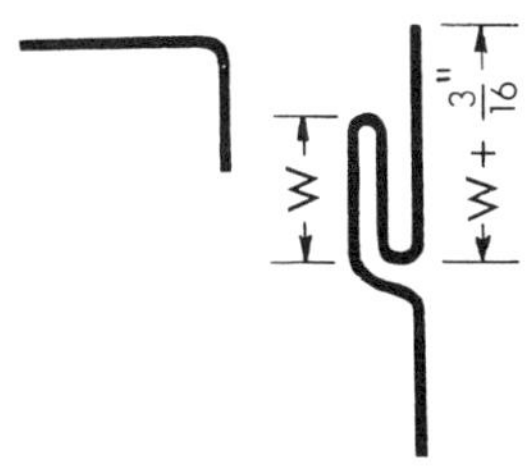

Example: Find the allowance for the pocket part of one side of a job with a Pittsburgh lock, width of pocket 1/2″.

Allowance for pocket (one side) = 2 x 1/2″ + 3/16″

= 1″ + 3/16″

= 1 3/16″ Ans.

ASSIGNMENT

Find the allowances for one side or two sides of the jobs according to the dimensions indicated in the following lists.

A. Lap Seam

1. Width 1/8″

2. Width 3/16″

B. Grooved Lock Seams

1. Width 1/8″ (one side)
2. Width 3/16″ (one side)

3. Width 1/8″ (two sides)
4. Width 5/16″ (two sides)

C. Riveted Lap Seam (See unit 17 for dia. of rivets.)

1. Rivet size 3/16″ (one side)
2. Rivet size .144″ (one side)

3. Rivet size 3 lb. (one side)
4. Rivet size 2 lb. (two sides)

D. Dovetailed Seam

1. Width 1/8″ (one side)

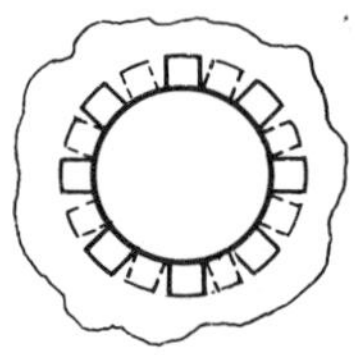

2. Width 5/16″ (one side)

E. Burred Bottom Seam

1. Width 3/32″ (one side)
2. Width 1/8″ (one side)

3. Width 3/16″ (two sides)
4. Width 5/32″ (two sides)

F. Set-In Bottom Seam (Allowances on the body and the bottom)

1. Width 3/16″ (one side)

2. Width 1/8″ (two sides)

G. Single and Double Seams (Allowances on the body and the bottom)

1. Width 3/32″ (one side)
2. Width 1/8″ (one side)
3. Width 3/16″ (one side)

4. Width 1/4″ (two sides)
5. Width 5/32″ (two sides)
6. Width 5/16″ (two sides)

H. Pittsburgh Lock (Allowances for the pocket and the flange)

1. Width 1/4″ (one side)
2. Width 1/2″ (one side)

3. Width 1/4″ (two sides)
4. Width 1/2″ (two sides)

UNIT 73 STRETCHOUTS OF SQUARE PIPES

When the size of the material needed for a square pipe must be determined, it is necessary to compute the size of the stretchout (pattern) for such a job. The length of the stretchout is found by using the formula for the perimeter and adding the allowance for the seam. The width of the stretchout is equal to the length of the pipe plus the allowance for the edge or seam. The effect of the thickness of metal is disregarded for square pipe. Use the letters L.S. and W.S. to designate the length and the width of the stretchout.

BUTT OR WELDED SEAM

Formulas:

L.S. = 4s

s = length of end

W.S. = ℓ

ℓ = length of job

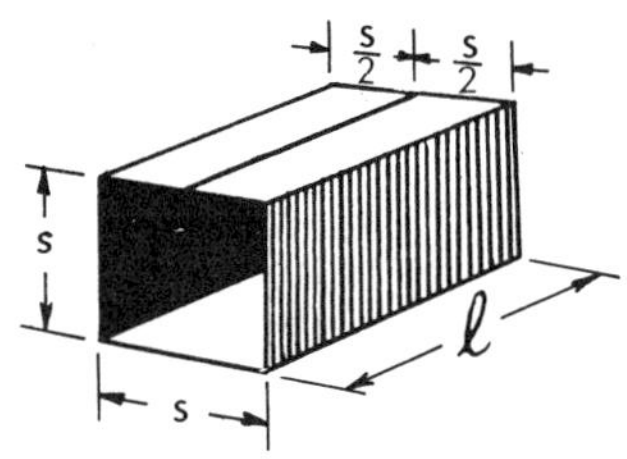

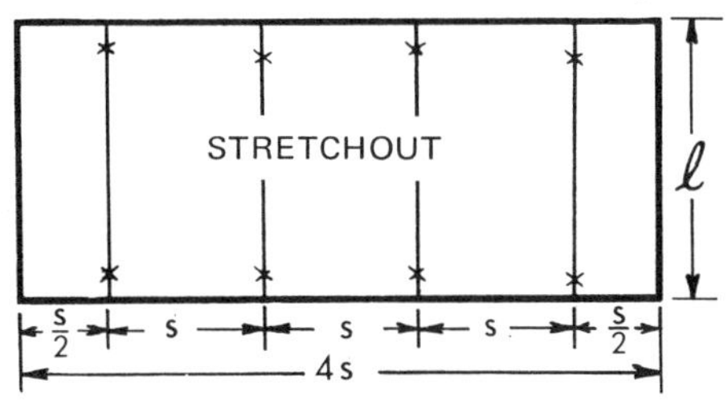

Example: A square pipe has the following dimensions: end, 6.5 in.; length, 12 in. Compute the size of the stretchout.

Length: L.S. = 4s

L.S. = 4 x 6.5

L.S. = 26″ Ans.

Width: W.S. = ℓ

W.S. = 12″ Ans.

OTHER SEAMS

Lap Seam

Formulas:

L.S. = 4s + W

s = length of end

W = width of seam

W.S. = ℓ

ℓ = length of job

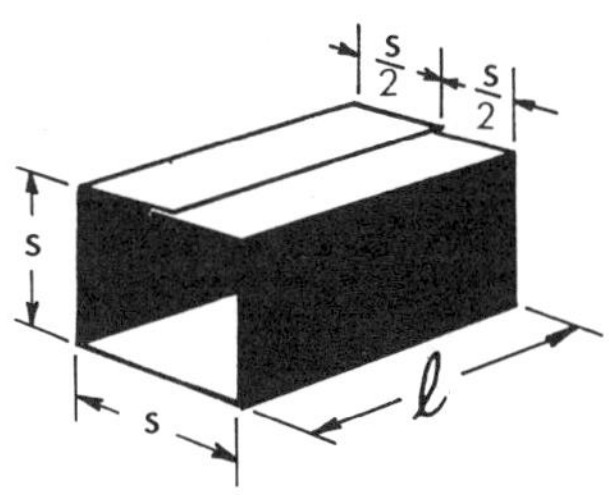

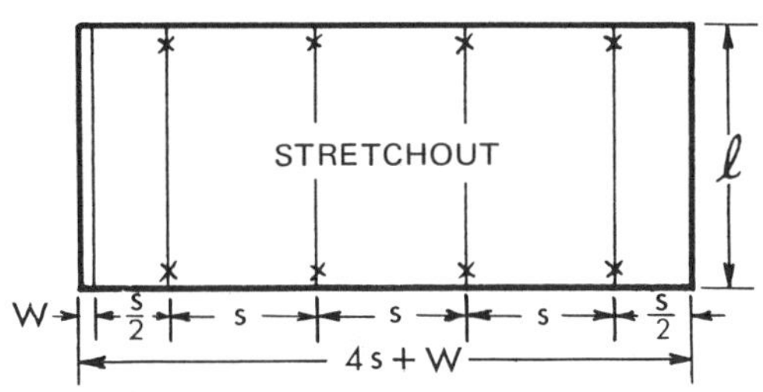

Example: A square pipe has the following dimensions: end, 3″; length, 18″; lap seam 1/8″. Compute size of stretchout.

Length: L.S. = 4s + W

L.S. = 4 x 3 + 1/8

L.S. = 12 + 1/8

L.S. = 12 1/8″ Ans.

Width: W.S. = ℓ

W.S. = 18″

Grooved Seam

Formulas:

L.S. = 4s + 3W

s = length of end

W = width of lock

W.S. = ℓ

ℓ = length of job

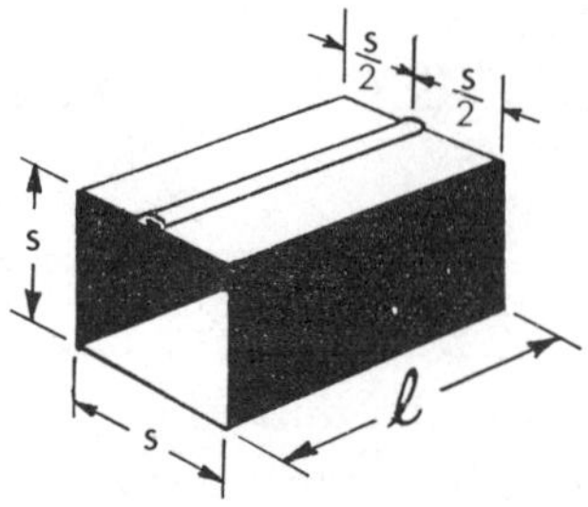

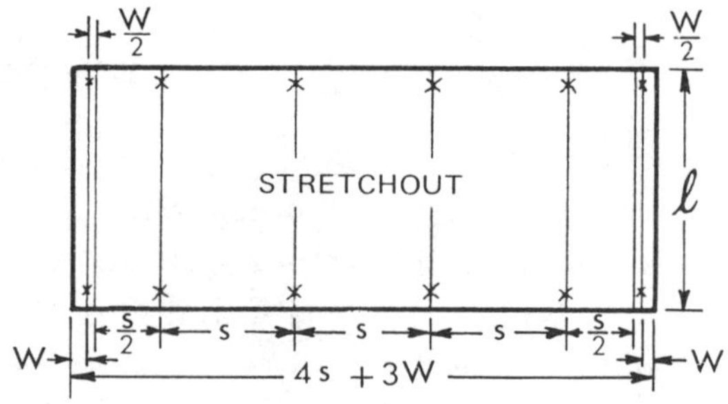

Example: A square pipe with a 1/8-in. grooved seam has the following dimensions: end, 6 in.; length, 14 in. Compute the size of the stretchout.

Length: L.S. = 4s + 3W

L.S. = 4 x 6 + 3 x 1/8

L.S. = 24 3/8″ Ans.

Width: W.S. = ℓ

W.S. = 14″ Ans.

Riveted Seam

Formulas:

L.S. = 4s + 2D + 2D

s = length of end

D = dia. of rivet

W.S. = ℓ

ℓ = length of job

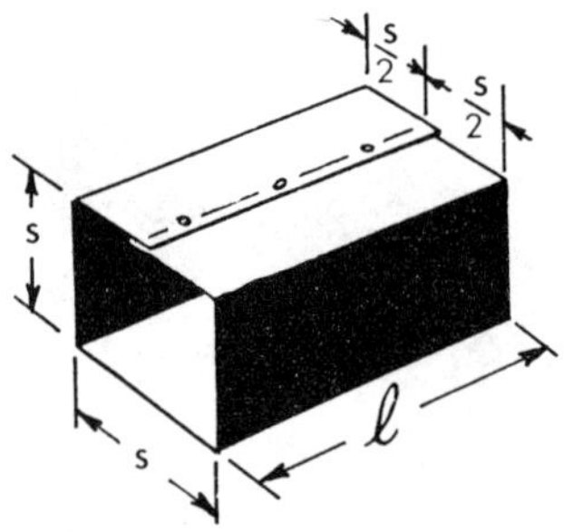

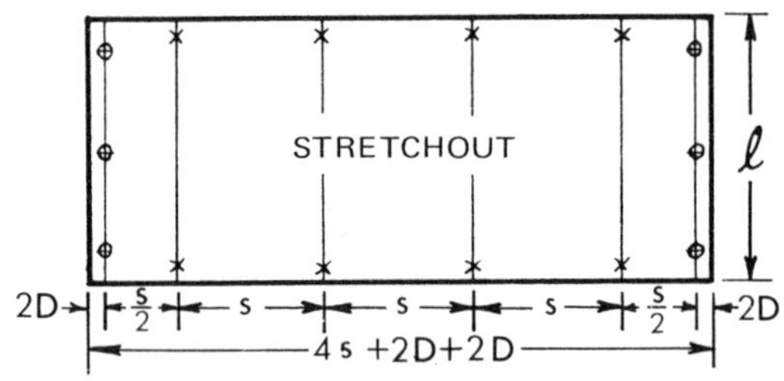

Example: A 12 in. length of 5-in. square pipe is to be made of 30-gage black iron with a riveted seam. Compute the size of the stretchout.

Length: L.S. = 4s + 2D + 2D

2D = .095″ x 2, or .190″, or 3/16″

Width: W.S. = ℓ

W.S. = 12″ Ans.

Note: See unit 17 for dia. of rivet

L.S. = 4 x 5 + 3/16 + 3/16

L.S. = 20 + 3/16 + 3/16, or 20 3/8″ Ans.

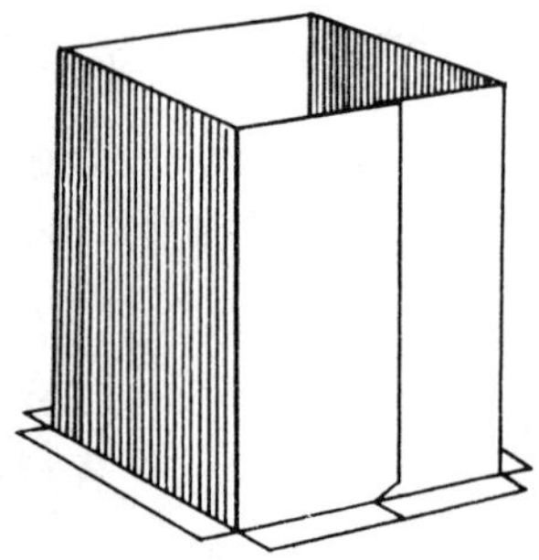

End Seams

If the pipe has an end seam or end seams, add the allowances for these to the width of the stretchout.

ASSIGNMENT

Compute the length and width of the stretchouts in the problems listed below. Draw stretchouts and dimension all parts of stretchouts in the problems starred. Check the length and width of these stretchouts by actual measurement.

A. Butt or Welded Seam

*1. s = 1 1/2″, ℓ = 2″

*2. s = 1 1/4″, ℓ = 2 1/2″

*3. s = 1 5/8″, ℓ = 3″

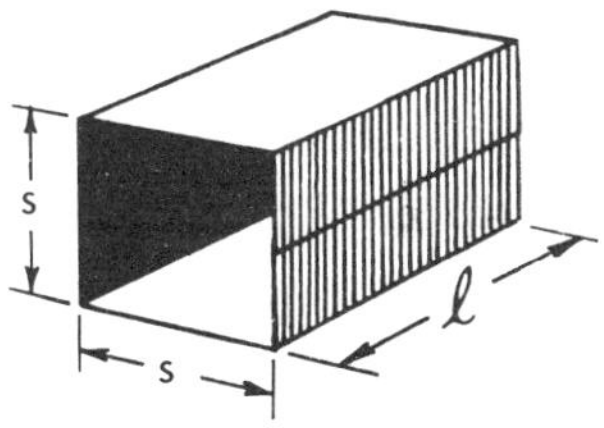

4. s = 10 1/2″, ℓ = 18″

5. s = 2′ 6″, ℓ = 2′

6. s = 3′ 3″, ℓ = 2′ 6″

B. Lap Seam (W = Width of Seam)

*1. s = 1 5/8″, ℓ = 2″, W = 3/16″

*2. s = 1 3/8″, ℓ = 3″, W = 3/16″

3. s = 4 1/4″, ℓ = 9″, W = 3/16″

4. s = 3 7/8″, ℓ = 12″, W = 5/32″

C. Grooved Seam (W = Width of Seam)

*1. s = 1 1/2″, ℓ = 2″, W = 1/8″

*2. s = 1 3/4″, ℓ = 2 1/2″, W = 3/16″

*3. s = 1 3/8″, ℓ = 3″, W = 1/8″

4. s = 10″, ℓ = 12″, W = 1/4″

5. s = 20″, ℓ = 18″, W = 5/16″

6. s = 2′ 4″, ℓ = 2′6″, W = 3/8″

D. Riveted Seam (D = Dia. of Rivets)

Note: See unit 17 for dia. of rivets.

*1. s = 1 1/4″, ℓ = 2″, D = .095″

*2. s = 1 1/8″, ℓ = 3″, D = .105″

*3. s = 1 7/8″, ℓ = 2″, D = .095″

4. s = 12″, ℓ = 18″, 26-gage

5. s = 10 1/2″ ℓ = 18″, 26 gage

6. s = 20″, ℓ = 30″, 24-gage

E. End Seams

1. Compute the length and width of the stretchout of the pipe part of the job shown at the right.

2. If the pipe in the preceding problem had a riveted seam instead of the grooved seam, a lap seam of 1/2 in. width instead of 3/16 in. width, and were made of 26-gage galvanized iron, compute the size of the stretchout.

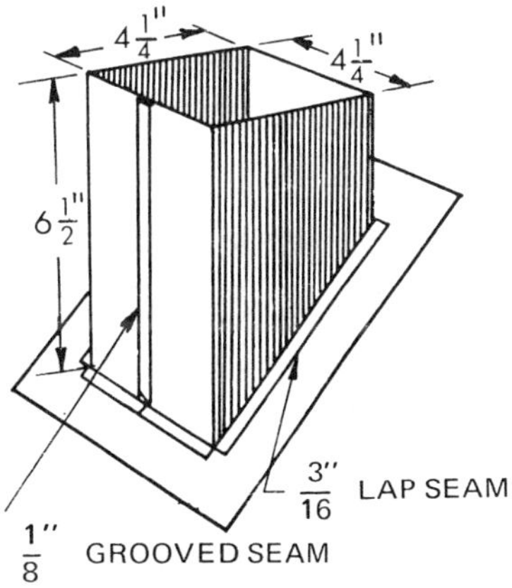

UNIT 74 STRETCHOUTS OF RECTANGULAR PIPES

The size of the stretchout of a rectangular pipe is computed in a manner similar to that used for the stretchout of a square pipe, that is, by adding the allowance for the seams to the perimeter of the end.

BUTT OR WELDED SEAM

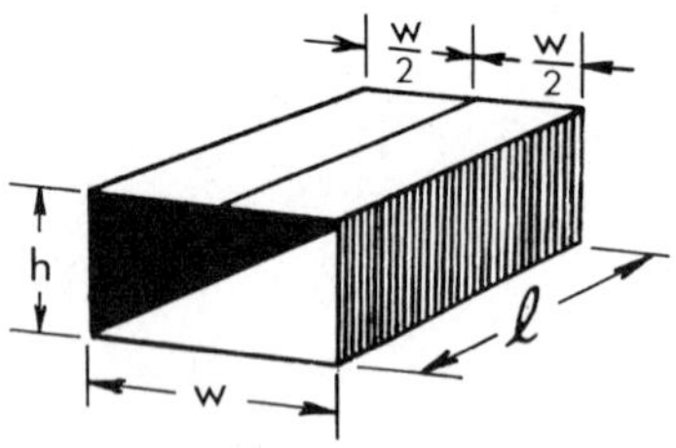

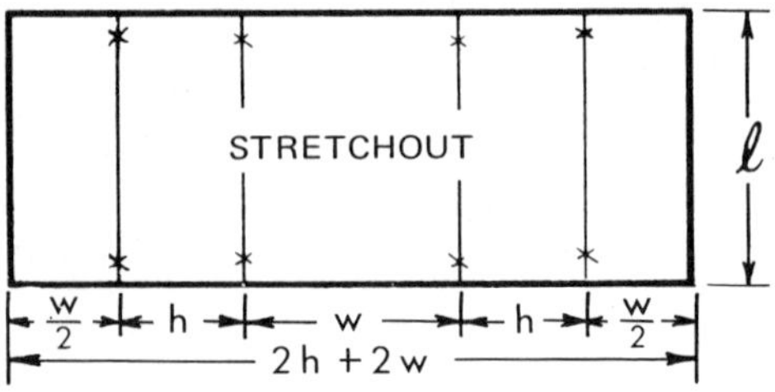

Formulas: Length of Stretchout

L.S. = 2h + 2w

h = height of end

w = width of end

Width of Stretchout

W.S. = ℓ

ℓ = length of job

OTHER SEAMS

Lap Seam

Formulas: L.S. = 2h + 2w + W (W = width of lap seam)

W.S. = ℓ (ℓ = length of job)

Grooved Seam

Formulas: L.S. = 2h + 2w + 3W (W = width of grooved seam)

W.S. = ℓ (ℓ = length of job)

Riveted Seam

Formulas: L.S. = 2h + 2w + 2D + 2D (D = diameter of rivet)

W.S. = ℓ (ℓ = length of job)

Note: If a rectangular pipe has an edge or a seam at the end, the allowances for the edge or the seam are added to the width of the stretchout.

ASSIGNMENT

Compute the length and width of stretchouts in the problems listed. Draw the stretchouts in the problems listed. Draw the stretchouts and dimension all parts of stretchouts in the problems starred. Check the length and width of these stretchouts by actual measurement.

A. Butt or Welded Seam

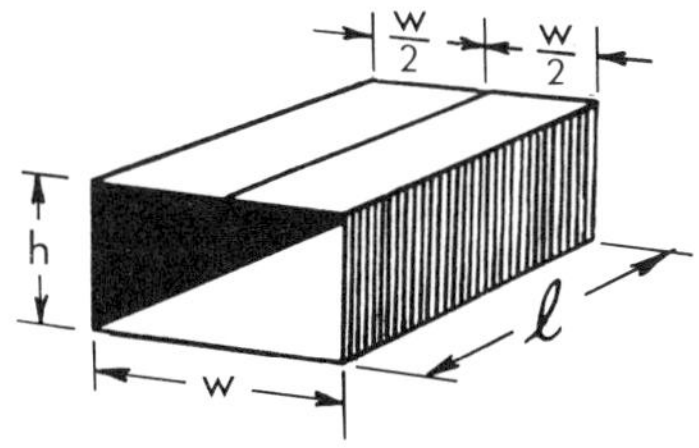

*1. w = 2″, h = 1″, ℓ = 2 1/2″

*2. w = 1 3/4″, h = 1 1/4″ ℓ = 3″

*3. w = 1 7/8″, h = 1 1/8″, ℓ = 2″

B. Lap Seam

(W = width of seam; seam is at center of shorter side of pipe.)

*1. w = 1 1/2″, h = 1″ ℓ = 2″ W = 1/8″

*2. w = 1 5/8″, h = 3/4″, ℓ = 2 1/2″, W = 5/32″

C. Grooved Seam

(W = width of seam; seam is at center of shorter side of pipe.)

*1. w = 1 3/4″, h = 1 1/4″, ℓ = 2″, W = 1/8″

*2. w = 1 1/2″, h = 1 1/8″, ℓ = 2 1/4″, W = 3/16″

3. w = 5″, h = 20″, ℓ = 18″, W = 1/4″

D. Riveted Seam

(D = dia. of rivet; seam is at center of longer side of pipe.)

*1. w = 1 1/4″, h = 2″, ℓ = 2″, D = .095″

*2. w = 1 1/8″, h = 1 7/8″, ℓ = 2″, D = .105″

*3. w = 3/4″, h = 2 1/8″, ℓ = 2 1/2″, D = .095″

4. w = 12 3/4″, h = 6 5/8″, ℓ = 30″, 26-gage metal

E. Additional Problem

Compute the size of stretchout of a one-piece elbow, 6″ x 4″, with the outside heights as indicated. The width of the inside grooved seam is 3/16″.

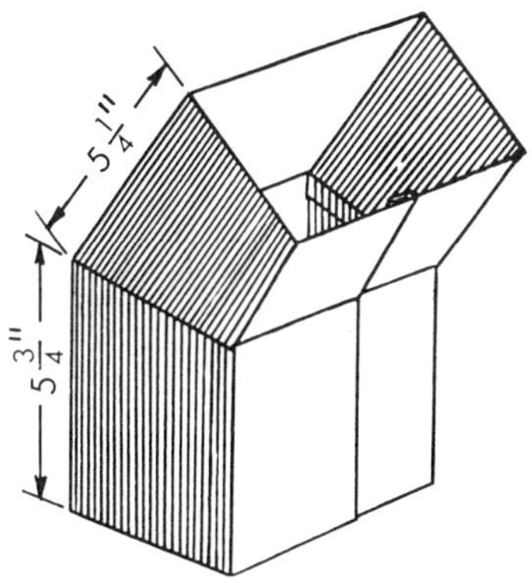

Note: No allowance is to be made for the seam between the two pieces.

UNIT 75 STRETCHOUTS OF CIRCULAR JOBS

To find the length of the stretchout of a circular job, the circumference of the end is found and the allowance for the seam added. The thickness of the metal is disregarded for circular pipes of 24 gage or lighter in problems of this kind.

BUTT OR WELDED SEAM

Formulas:

L.S. $= \pi d$
W.S. $= \ell$

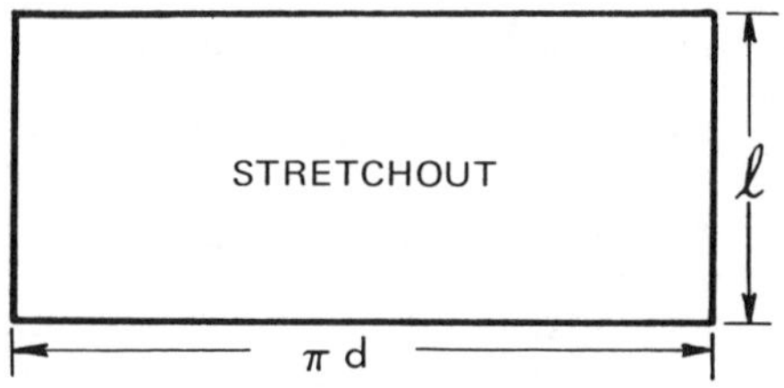

Example: Compute the size of stretchout of a 6-in. pipe (nearest 32nd of an inch). The length of the pipe is 12 in.

Length: L.S. $= \pi d$

L.S. = 3.14 x 6

L.S. = 18.84, or 18 27/32″ Ans.

Width: W.S. $= \ell$

W.S. = 12″ Ans.

OTHER SEAMS

Formulas: Grooved Seam – L.S. $= \pi d + 3W$ (W = width of seam)

Riveted Seam – L.S. $= \pi d + 2D + 2D$ (D = dia. of rivet)

EDGES AND SEAMS OF CIRCULAR JOBS

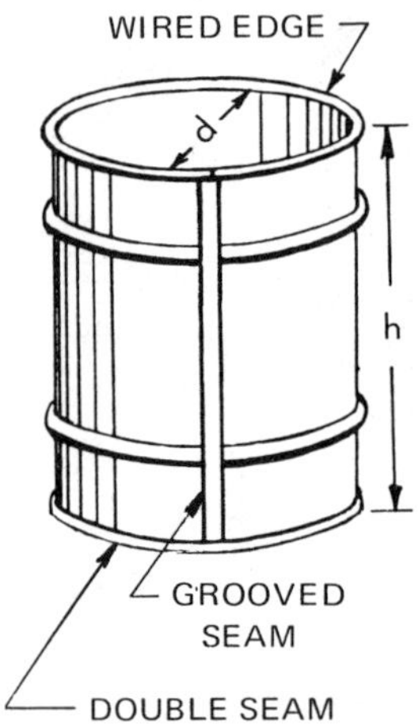

Body – Formulas:

L.S. $= \pi d + 3W$

d = dia. of job

W = width of grooved seams

W.S. $= h + W + 2\ 1/2D$

h = height of job

W = width of double seam

D = dia. of wire

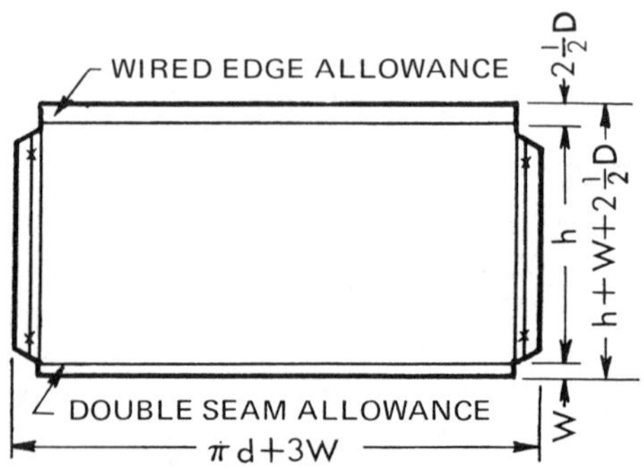

Bottom – Formula:

OD $= d + 2W + 2W$, or $d + 4W$

OD = dia. of stretchout

d = dia. of job

W = width of double seam

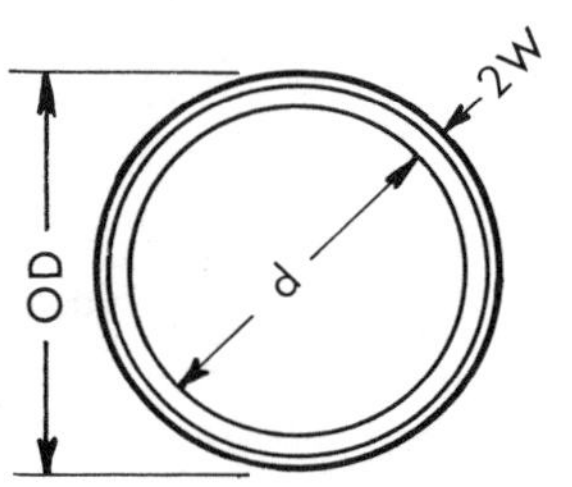

ASSIGNMENT

Compute, to the nearest 32nd of an inch, the length and width of the stretchouts in the problems listed below. Draw the stretchouts and dimension all parts of the stretchouts in the problems starred. Check the size of these stretchouts by measurement.

A. Butt or Welded Seam

*1. d = 1 1/2″, ℓ = 2″

*2. d = 2″, ℓ = 3″

3. d = 12″, ℓ = 18″

4. d = 2′ 3″, ℓ = 2′ 6″

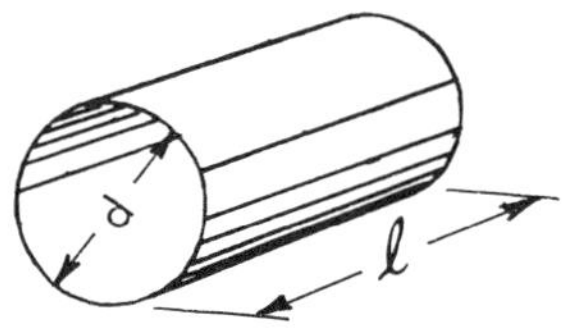

B. Other Seams

*1. dia. = 1″,	length = 2″,	1/8″ grooved seam
2. dia. = 10 1/2″,	length = 24″,	3/16″ grooved seam
*3. dia. = 1 1/4″,	length = 2″,	diameter of rivet = .095″
4. dia. = 22″,	length = 18″,	riveted seam, 24-gage metal

C. Circular Jobs

1. Compute the size of the stretchout of the pipe.

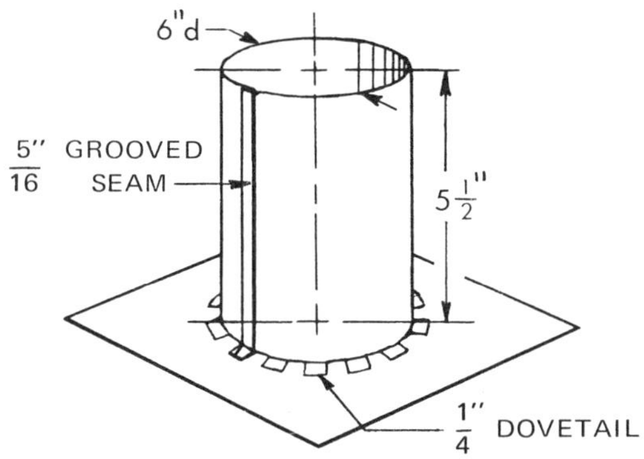

2. Compute the size of the stretchout of the pipe.

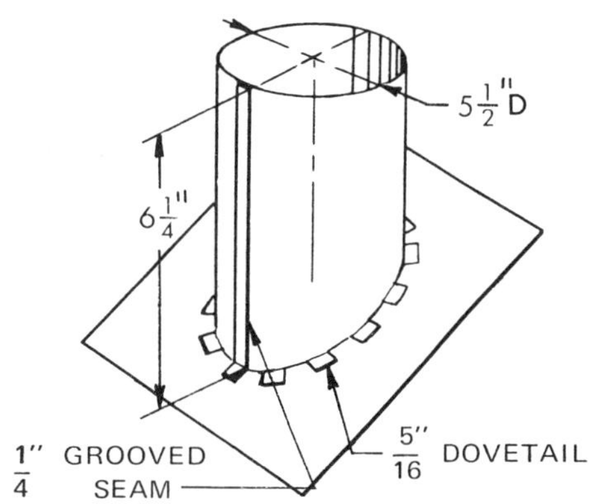

3. What is the size of the stretchout of the collar?

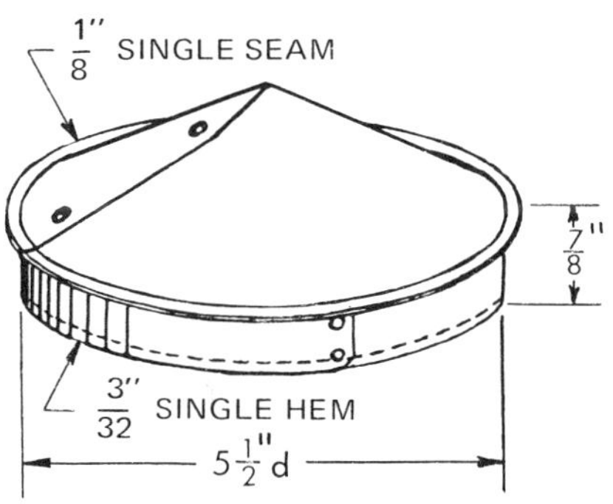

4. What is the actual size of the sheet metal needed for the side and for the bottom?

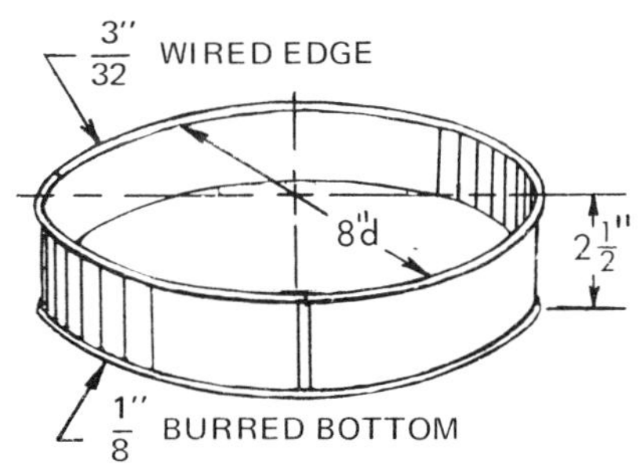

UNIT 76 STRETCHOUTS OF SEMICIRCULAR-SIDED JOBS

The method of computing the size of the stretchout for a semicircular-sided job is somewhat similar to that used in computing the stretchout of a circular job. In this case, however, the straight parts must be added to the circular parts to find the length of the stretchout. Disregard thickness for 24-gage and lighter.

BUTT OR WELDED SEAM

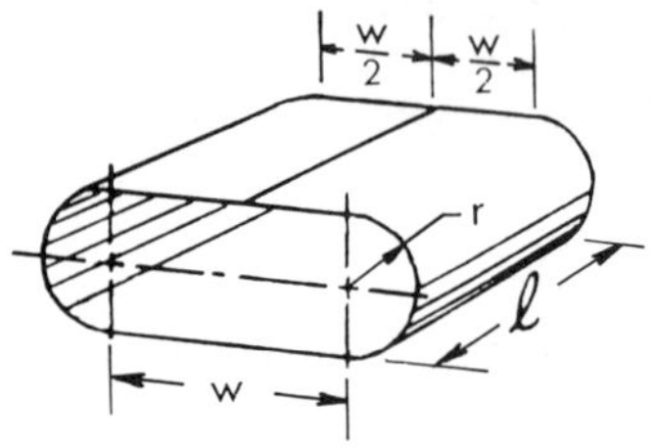

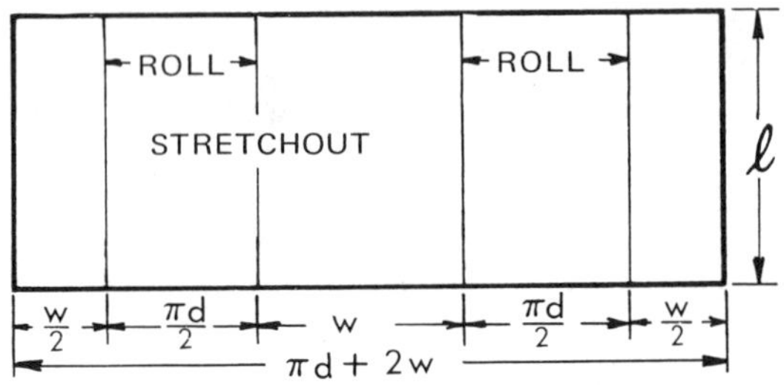

Formulas:

Length of Stretchout

L.S. = $\pi d + 2w$

d = diameter, or 2r

w = distance between centers of semicircles

Roll

Roll = $\frac{\pi d}{2}$

Width of Stretchout

W.S. = ℓ

ℓ = length of job

Example: A semicircular-sided pipe has the following dimensions: radius of semicircles 2 1/2 in., distance between centers of semicircles 6 in., length of pipe 18 in. Compute (closest 32nd) the size of the stretchout, and the length and the location of each part to be rolled

Length

L.S. = $\pi d + 2w$

L.S. = $3.14 \times 5 + 2 \times 6$

L.S. = $15.7 + 12$

L.S. = 27.7″, or 27 23/32″ Ans.

Width

W.S. = ℓ

W.S. = 18″ Ans.

Roll: $\frac{\pi d}{2} = \frac{15.7''}{2}$, or 7.85″, or 7 27/32″ Ans.

Location of Roll: $\frac{w}{2} = \frac{6''}{2}$, or 3″ from each end of the stretchout

OTHER SEAMS

Grooved Seam

L.S. = $\pi d + 2w + 3W$

W = width of grooved seam

W.S. = ℓ

ℓ = length of job

Riveted Seam

L.S. = $\pi d + 2w + 2D + 2D$ W.S. = ℓ

D = diameter of rivet ℓ = length of job

ASSIGNMENT

Compute, to the nearest 32nd of an inch, the length and width of the stretchouts in the problems listed below. Draw the stretchouts and dimension all parts (including the roll) in the problems starred. Check the length and width of these stretchouts by the actual measurement.

A. Butt or Welded Seam

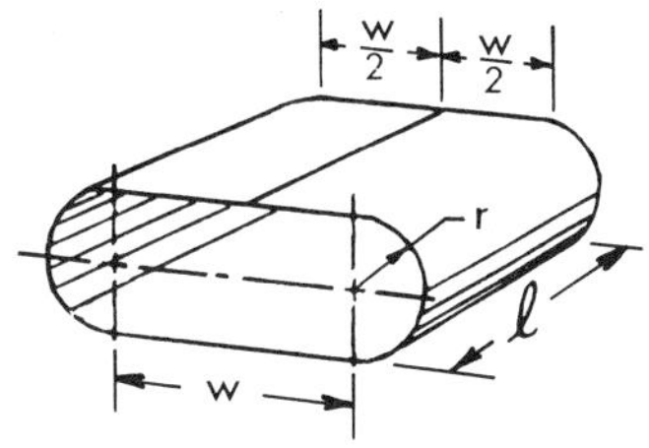

*1. r = 1/2″, w = 1 1/2″, ℓ = 2″

*2. r = 5/8″, w = 1 1/4″, ℓ = 2 1/2″

3. r = 4 1/2″, w = 3″, ℓ = 18″

B. Other Seams

*1. r = 3/8″, w = 1 1/2″, ℓ = 2″, 1/8″ grooved seam

*2. r = 1/2″, w = 1 1/4″, ℓ = 2 1/2″, 1/8″ grooved seam

3. r = 3″, w = 4″, ℓ = 12″, 3/16″ grooved seam

4. r = 5″, w = 2″, ℓ = 8″, 3/16″ grooved seam

*5. r = 7/16″, w = 1 1/2″, ℓ = 3″, dia. of rivet = .105″

6. r = 6″, w = 8″, ℓ = 14″, dia. of rivet = .105″

7. r = 10″, w = 12″, ℓ = 15″, 26-gage metal

C. Semicircular-Sided Jobs

1. Find the size of the stretchout of the collar part of the cover shown at the right. The collar is made of 28-gage metal.

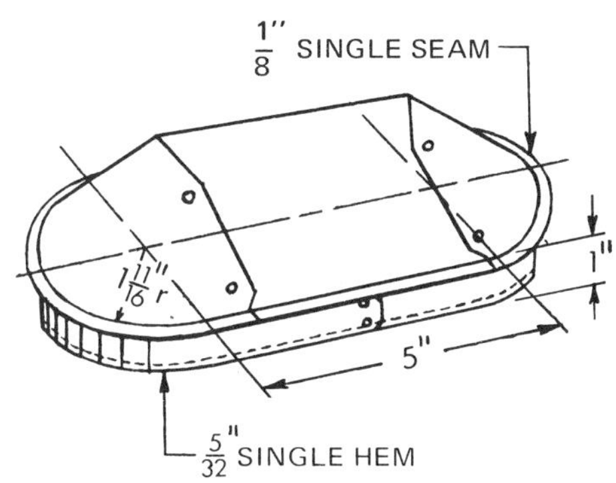

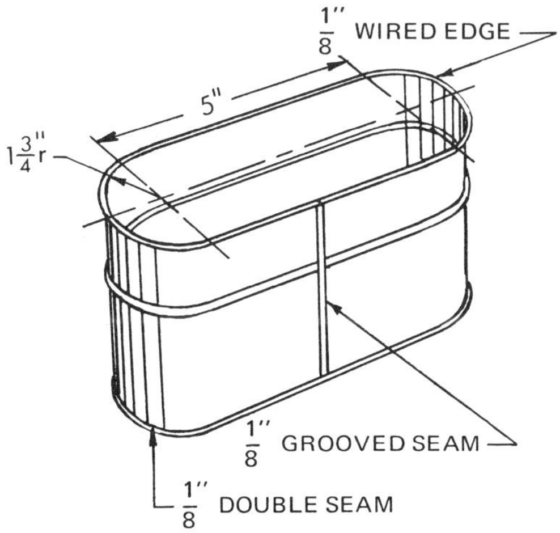

2. Find the size of the stretchout of the body part and the stretchout of the bottom part of the container shown at the left.

UNIT 77 STRETCHOUTS OF BOXES

Box-shaped jobs are common in sheet metal work. When it is necessary to compute the size of stretchout of such a job, the allowances for the edges must be added to the stretchout without the edges.

BOXES WITHOUT EDGES

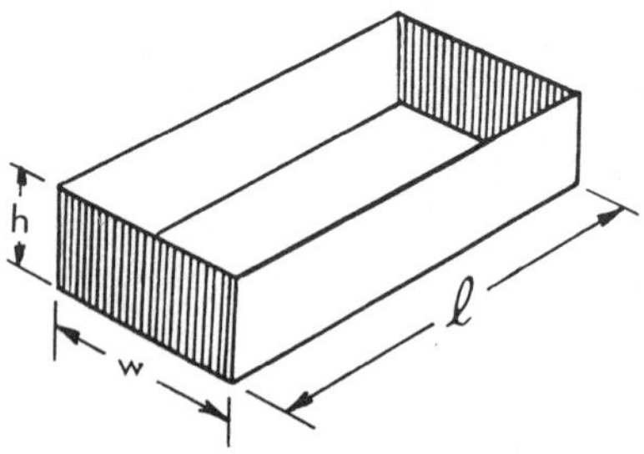

Formulas:

$L.S. = \ell + 2h$

ℓ = length

h = height

$W.S. = w + 2h$

w = width

h = height

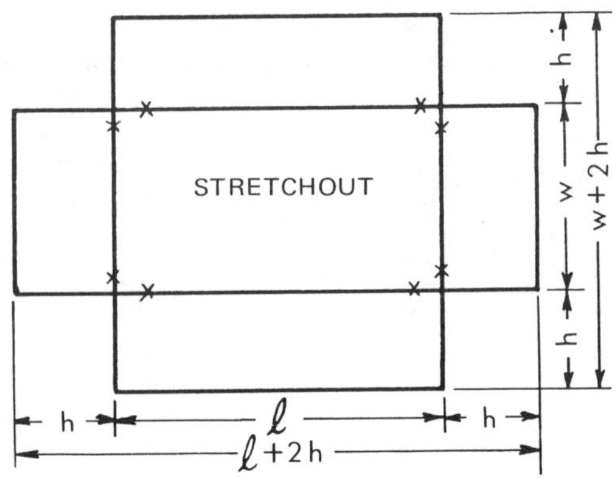

Example: If the length of a box is 10 in., the width 8 in., and the height 3 in., find the size of the stretchout.

Length

$L.S. = \ell + 2h$

$L.S. = 10 + 2 \times 3$

$L.S. = 10 + 6 = 16''$ Ans.

Width

$W.S. = w + 2h$

$W.S. = 8 + 2 \times 3$

$W.S. = 8 + 6 = 14''$ Ans.

BOXES WITH EDGES

Single Hem

$L.S. = \ell + 2h + 2W$

$W.S. = w + 2h + 2W$

W = width of single hem

Example: The length of a box is 5 in., width 4 in., height 3 in.; the width of the single hem is 1/8″. Find the size of the stretchout.

Length

$L.S. = \ell + 2h + 2W$

$L.S. = 5 + 2 \times 3 + 2 \times 1/8$

$L.S. = 5 + 6 + 1/4$

$L.S. = 11\ 1/4''$ Ans.

Width

$W.S. = w + 2h + 2W$

$W.S. = 4 + 2 \times 3 + 2 \times 1/8$

$W.S. = 4 + 6 + 1/4$

$W.S. = 10\ 1/4''$ Ans.

Double Hem

$L.S. = \ell + 2h + 4W$

$W.S. = w + 2h + 4W$

W = width of double hem

Wired Edge

$L.S. = \ell + 2h + 2\ 1/2D + 2\ 1/2D$

$W.S. = w + 2h + 2\ 1/2D + 2\ 1/2D$

D = diameter of wire

ASSIGNMENT

Compute the length and width of stretchouts in the problems listed below. Draw the stretchouts and label all parts of the stretchouts in problems starred. Check the length and width of these stretchouts by actual measurement.

A. Boxes Without Edges

*1. ℓ = 3″, w = 2″, h = 1″

*2. ℓ = 2″, w = 1 3/4″, h = 2″

*3. ℓ = 2 1/4″, w = 1 5/8″, h = 1 3/8″

4. ℓ = 10″, w = 6 1/2″, h = 2 1/4″

5. ℓ = 8 7/8″, w = 6″, h = 1 3/4″

6. ℓ = 9″, w = 6 7/8″, h = 3 3/4″

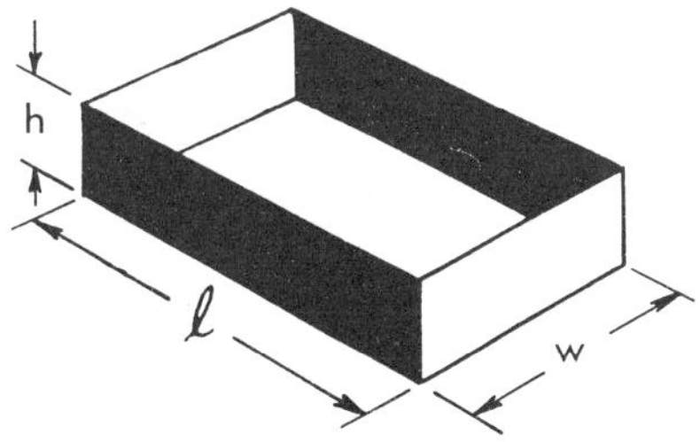

B. Boxes With Edges

*1. ℓ = 3″, w = 2″, h = 1″, width of single hem = 1/4″

*2. ℓ = 2 1/2″, w = 1 3/4″, h = 1 1/4″, width of single hem = 5/32″

3. ℓ = 12″, w = 9 3/4″, h = 2″, width of single hem = 3/16″

*4. ℓ = 2 3/4″, w = 2 1/4″, h = 1″, width of double hem = 1/8″

*5. ℓ = 3 1/2″, w = 2 7/8″, h = 1 1/4″, width of double hem = 5/32″

6. ℓ = 11 1/2″, w = 8″, h = 2 7/8″, width of double hem = 1/4″

*7. ℓ = 3″, w = 2″, h = 7/8″, dia. of wire = 1/8″

*8. ℓ = 3 1/4″, w = 2 5/8″, h = 1 1/8″, dia. of wire = 3/32″,

9. ℓ = 12″, w = 8″, h = 2 1/2″, dia. of wire = 7/32″

10. Compute the size of the stretchout of the container shown.

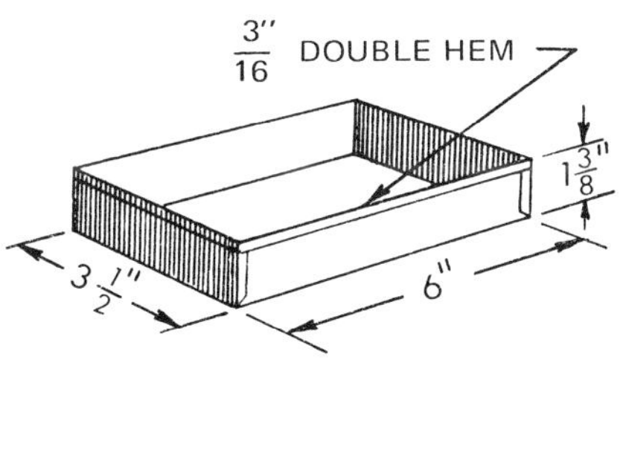

11. Compute the actual size of the sheet metal for the pan.

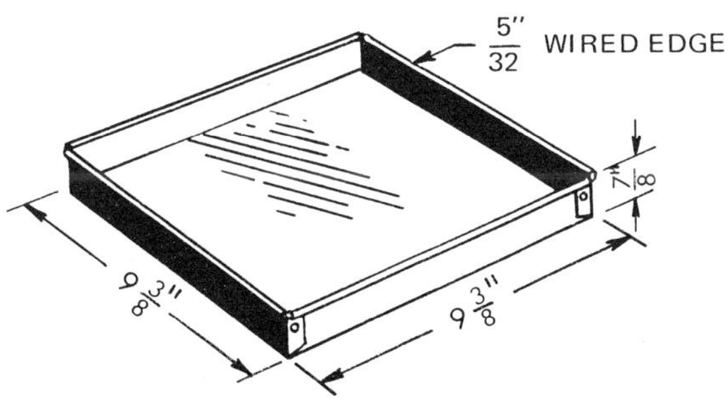

UNIT 78 NUMBER OF PIECES FROM A SHEET

In order to use material in the most economical manner, it is necessary to compute the the number of pieces which may be obtained from a sheet of metal. The steps in the procedure follow.

1. Sketch the sheet, and sketch the pattern in one corner.

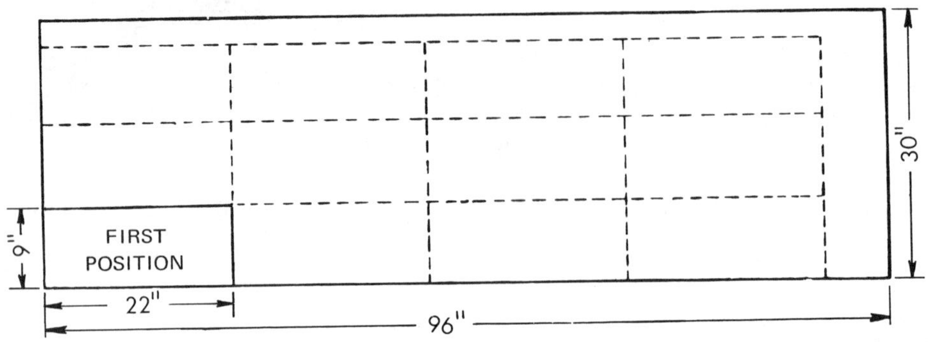

2. Divide width of sheet by width of pattern. Drop the fractional remainder. $\frac{W}{w} = \frac{30''}{9''}$, or $3\frac{1}{3}$, or 3 rows
3. Divide length of sheet by length of pattern. Drop the fractional remainder. $\frac{L}{\ell} = \frac{96''}{22''}$, or $4\frac{4}{11}$, or 4 pieces
4. Multiply these two results. $3 \times 4 = 12$ pieces (total)

   Note: If the strips left at the end of the sheet are wider than the pattern, they can be used. $4 \times 22'' = 88''$

   $96'' - 88'' = 8''$ (too narrow)

5. Rotate the sketch of the pattern 90 deg. and repeat the procedure.

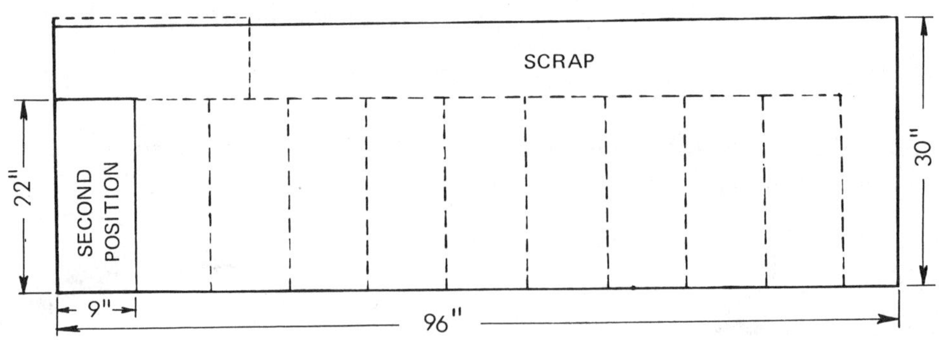

$\frac{W}{\ell} = \frac{30''}{22''}$, or 1 row $\frac{L}{w} = \frac{96''}{9''}$, or 10 pieces Total = 10 pieces

Note: Since $30'' - 22'' = 8''$, the top strip cannot be used.

Answer: The first position gives the greater number of pieces, 12.

ASSIGNMENT

A. Square, Circular, and Rectangular Pieces

Find the number of pieces of the shape and size indicated that can be obtained from the material listed below:

Kind of Piece	Dimensions		Sheet Size
1. Square		s = 4″	20″ x 28″
2. Circular		d = 6″	14″ x 20″
3. Rectangular	ℓ = 6″,	w = 5″	20″ x 28″
4. Rectangular	ℓ = 10″,	w = 8″	36″ x 120″
5. Rectangular	ℓ = 12″,	w = 10″	48″ x 120″
6. Rectangular	ℓ = 12 1/2″,	w = 10 1/2″	48″ x 120″
7. Rectangular	ℓ = 8″,	w = 4″	30″ x 120″
8. Rectangular	ℓ = 12″,	w = 9″	36″ x 120″
9. Rectangular	ℓ = 12″,	w = 4″	30″ x 120″

B. Stretchout Pieces From a Sheet

Note: Find the length and width of the stretchout before starting to compute the number of pieces obtainable from the sheet.

1. The side of a square pipe is 4 in., the length 18 in., and the lap seam 1/8″. How many pieces can be obtained from a sheet 36″ x 120″?
2. A rectangular pipe, 10″ x 3″, is 18 in. long and has a 3/16 in. grooved seam. How many pieces can be obtained from a sheet 48″ x 120″?
3. A 6-in. pipe with a 1/4-in. grooved seam is 12 in. long. Compute the number of pieces of this pipe which can be obtained from a sheet 48″ x 120″.

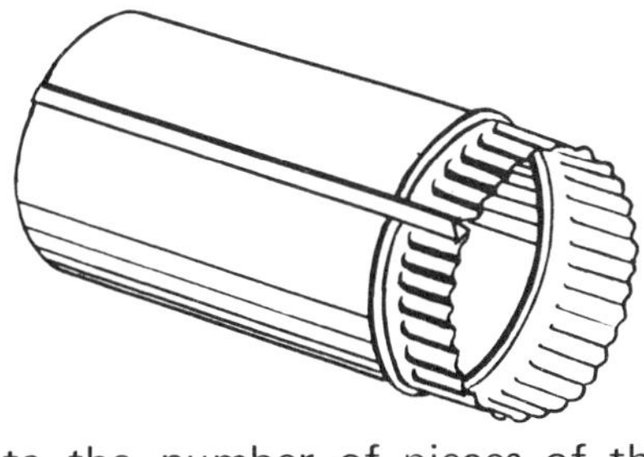

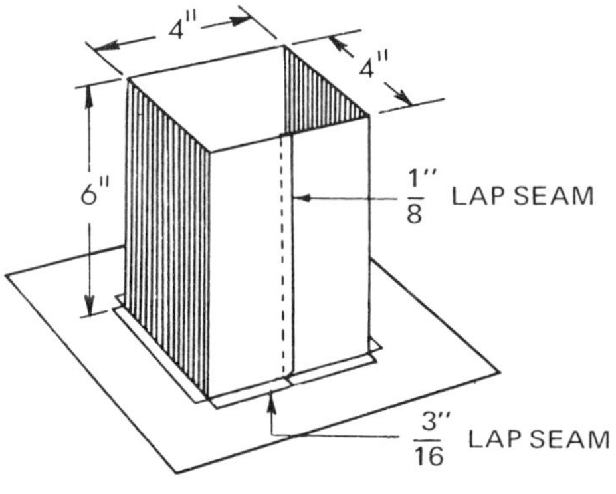

4. Compute the number of pieces of the pipe shown at the left which can be obtained from a sheet 48″ x 120″.

UNIT 79 LENGTH OF WIRE FOR EDGES OF JOBS

The length of wire (L.W.) needed for a wired edge is equal to the perimeter of the job plus an allowance for the wire. The formula for circular shapes is used for tapered jobs which are wired after they are formed. If the wiring is done before the job is formed, use πd.

SQUARE JOB

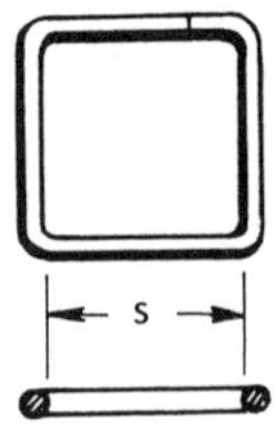

Formula: L.W. = 4s + 2D

s = length of side of job in inches

D = diameter of wire in inches

Example: If s = 18″ and D = 1/4″, find L.W.

1. Formula L.W. = 4s + 2D
2. Substitution L.W. = 4 x 18 + 2 x 1/4
3. Evaluation L.W. = 72 + 1/2, or 72 1/2″ Ans.

RECTANGULAR JOB

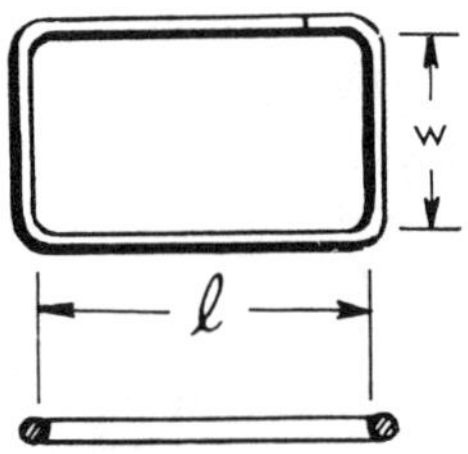

Formula: L.W. = 2ℓ + 2w + 2D

ℓ = length of job in inches

w = width of job in inches

D = diameter of wire in inches

Example: If ℓ = 10″, w = 5″, and D = 1/8″, find L.W.

1. Formula L.W. = 2ℓ + 2w + 2D
2. Substitution L.W. = 2 x 10 + 2 x 5 + 2 x 1/8
3. Evaluation L.W. = 20 + 10 + 1/4, or 30 1/4″ Ans.

CIRCULAR JOB (Tapered)

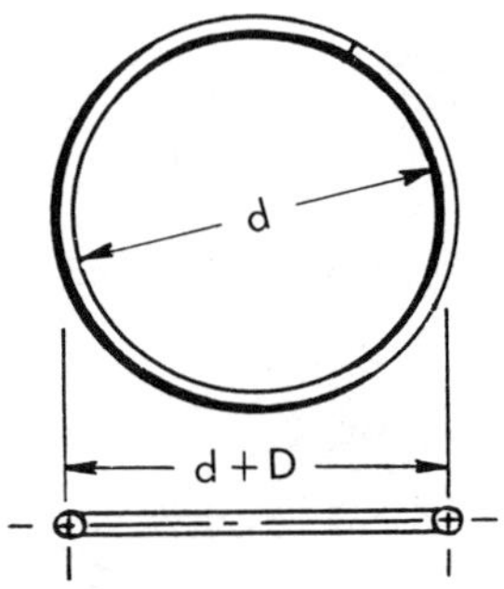

Formula: L.W. = π (d + D)

d = diameter of job in inches

D = diameter of wire in inches

Example: If d = 19 3/4″, D = 1/4″, find L.W. (32nds)

1. Formula L.W. = π (d + D)
2. Substitution L.W. = 3.14 (19 3/4 + 1/4)
3. Evaluation L.W. = 3.14 x 20

L.W. = 62.832″, or 62 27/32″ Ans.

ASSIGNMENT

A. Square Job

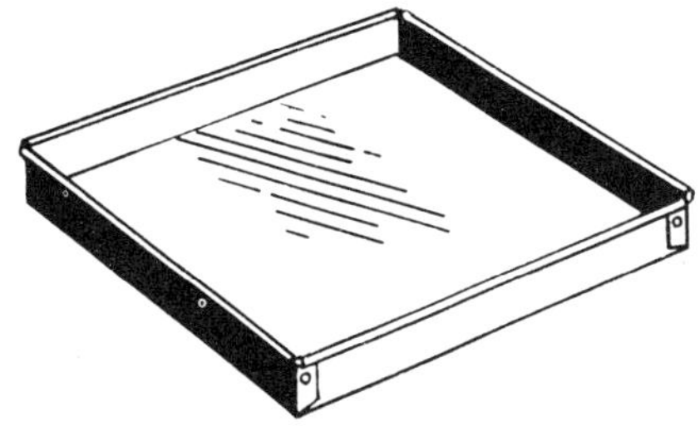

1. side = 9″, diameter of wire = 1/8″
2. side = 10 1/2″, diameter of wire = 3/32″
3. side = 2′ 6″, diameter of wire = 7/32″

B. Rectangular Job

1. length = 6″, width = 4″, diameter of wire = 1/8″
2. length = 8 1/2″, width = 5 1/4″, diameter of wire = 5/32″
3. length = 2′ 9 3/4″, width = 1′ 7 1/4″, diameter of wire = 1/4″

C. Circular Jobs

Find the length of wire to the closest 32nd of an in.

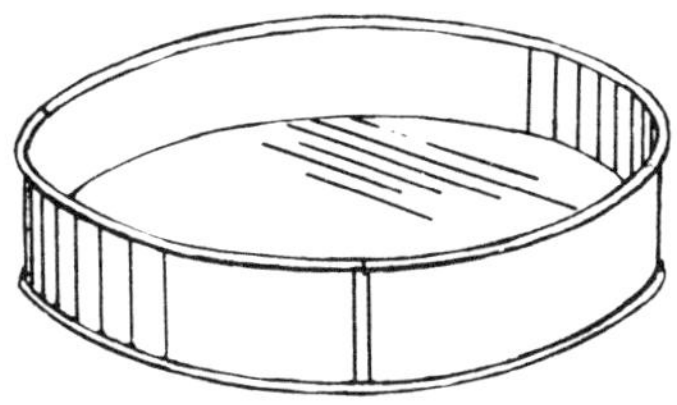

1. Jobs wired before being formed
   a. diameter of job = 10″, diameter of wire = 1/8″
   b. diameter of job = 30″, diameter of wire = 1/4″
   c. diameter of job = 25 1/2″, diameter of wire = 3/16″

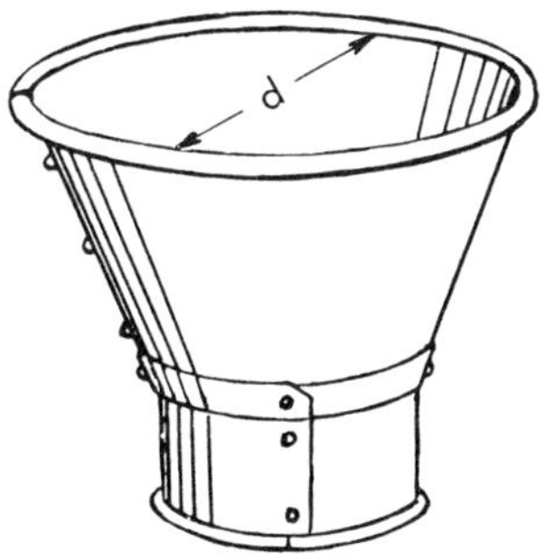

2. Jobs wired after being formed
   a. d = 9 7/8″ diameter of wire = 1/8″
   b. d = 14 13/16″, diameter of wire = 3/16″
   c. d = 20″, diameter of wire = 1/4″

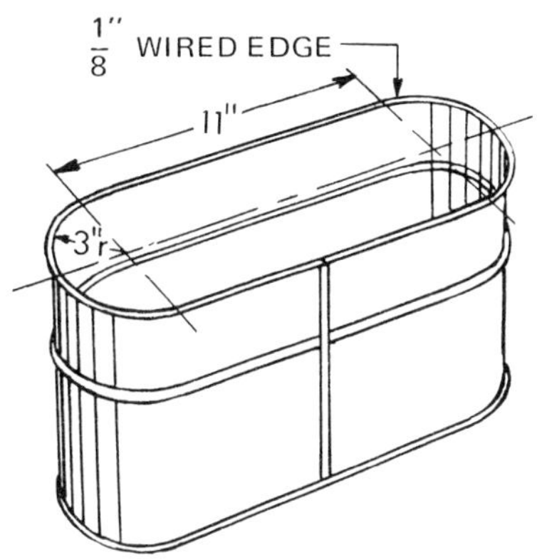

D. Additional Problems

1. Compute the length (closest 32nd of an in.) of wire needed for the edge of the tank.

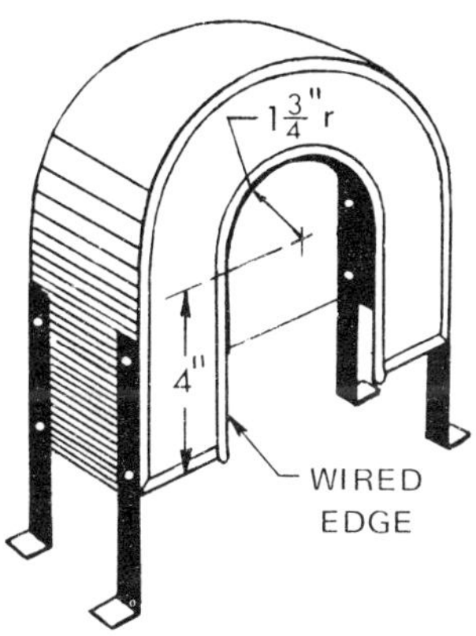

2. Compute, to the closest 32nd of an in., the length of wire needed for the inner edge of the guard. The diameter of the wire is 1/8 in.

UNIT 80 LENGTH OF ARCS OF CIRCLES

To find the length of the stretchout of the throat or the heel of a square elbow, it is necessary to know how to compute the length of an arc of a circle. A square elbow is an elbow with the heel and the throat at 90 deg. to the sides (cheeks).

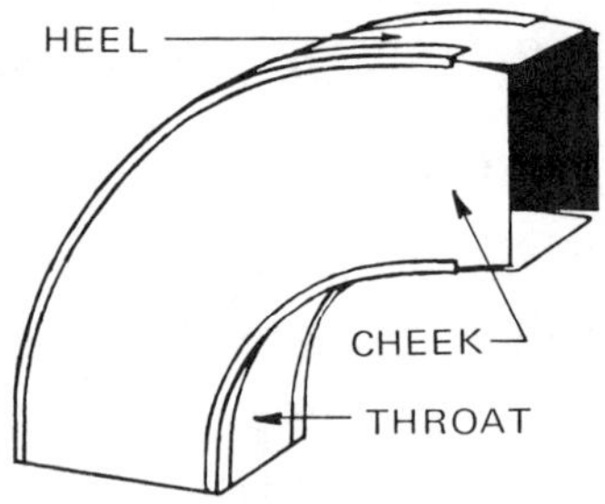

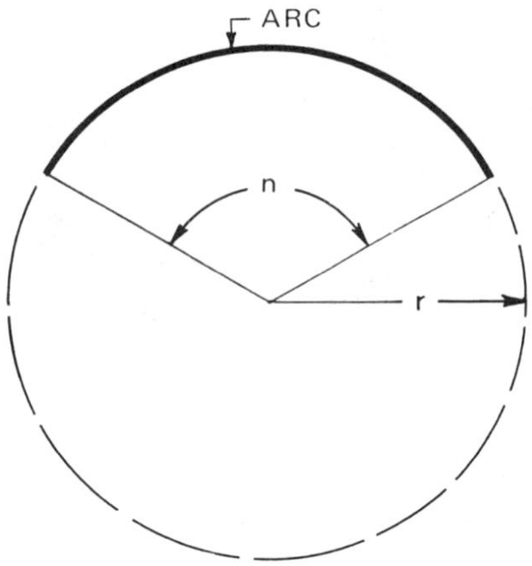

PRINCIPLE: The length of an arc of a circle is equal to the ratio of its central angle to 360 deg. times the circumference.

Formula: $\ell = \frac{n}{360} \times \pi d$, or $\frac{n\pi d}{360}$

Symbols: ℓ = length of arc

n = no. of deg. in central angle

π = 3.14

d = 2 x radius of arc, or 2r

Example: If the central angle of a circle is 45 deg., the radius of the circle is 3 in., find the length (closest 32nd of an in.) of the intercepted arc.

1. Formula $\ell = \frac{n\pi d}{360}$

2. Substitution $\ell = \frac{45 \times 3.14 \times 6}{360}$

3. Evaluation $\ell = \frac{\overset{1}{\cancel{45}} \times 3.14 \times \overset{3}{\cancel{6}}}{\underset{4}{\cancel{\underset{8}{\cancel{360}}}}}$

$\ell = \frac{3.14 \times 3}{4}$

$\ell = \frac{9.42}{4}$, or 2.355

ℓ = 2.355", or 2 11/32" Ans.

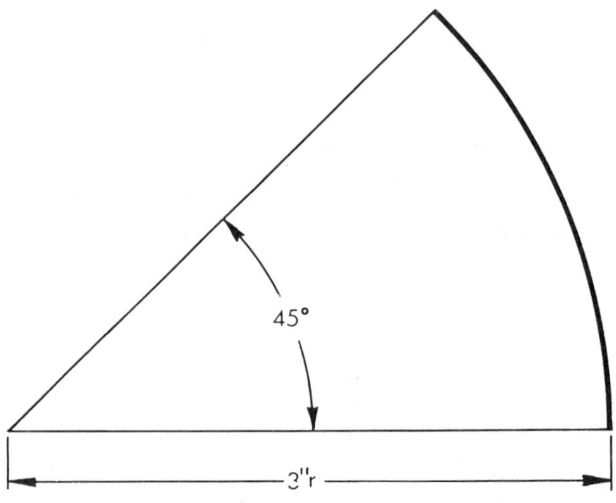

Note: This answer can be checked by measuring the length of the arc in the drawing above with a flexible rule, or by the step-off method. (See *Measurement and Layout,* page 80.)

ASSIGNMENT

A. Length of Arcs of Circles

Compute the length of the arcs in the following problems. Draw each angle and each arc, and check the computations by measurement.

1. central angle = 90 deg., r = 4″
2. central angle = 60 deg., r = 6″
3. central angle = 80 deg., r = 8″
4. central angle = 72 deg., r = 3 1/2″
5. central angle = 30 deg., r = 5″
6. central angle = 165 deg., r = 6 1/4″

B. Square Elbow (No Seams)

Compute the length of the stretchout of the throat and the heel of the elbows; add width of cheek, w, to the radius of throat to obtain the radius of heel.

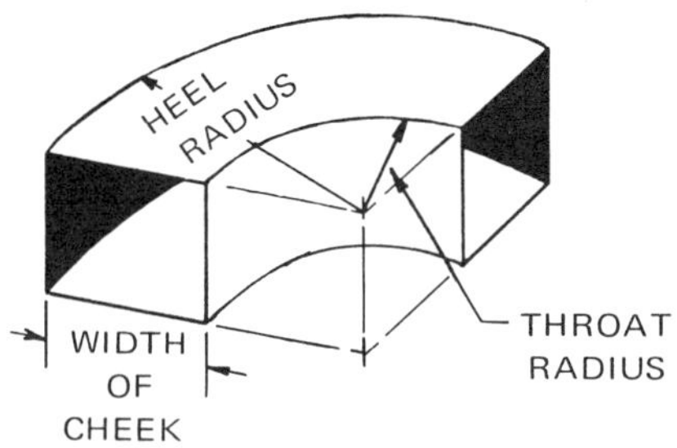

1. angle = 90 deg., throat radius = 8″, w = 8″
2. angle = 60 deg., throat radius = 6″, w = 6″
3. angle = 30 deg., throat radius = 9″, w = 5″
4. angle = 45 deg., throat radius = 9″, w = 9″
5. angle = 72 deg., throat radius = 30″, w = 20″

C. Square Elbow (Pittsburgh Lock)

Compute the size of the stretchout for the throat and the heel of each elbow listed below. **Hint:** Add 1 in. to the length of stretchout for the end seam. Add the allowance for the pocket of the Pittsburgh lock to both sides of the stretchout to obtain the width of the stretchout.

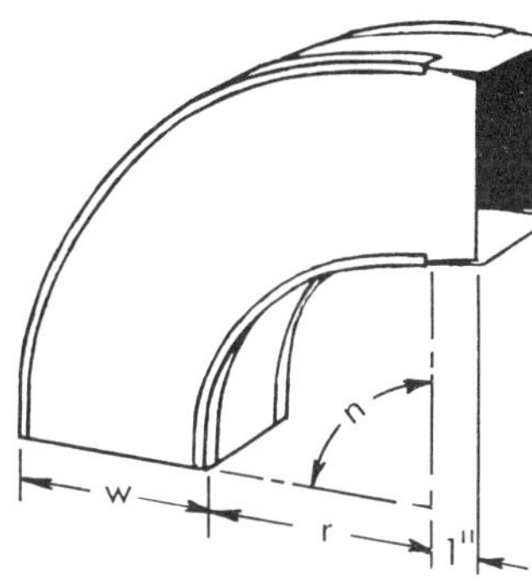

1. n = 90°, r = 10″, w = 10″, Pittsburgh lock = 1/4″
2. n = 80°, r = 8″, w = 8″, Pittsburgh lock = 1/4″
3. n = 60°, r = 20″, w = 10″, Pittsburgh lock = 1/2″

D. Square Elbow (Double Seam)

Compute the size of the stretchout of the throat and the heel of each square elbow listed below.

1. n = 60°, r = 8″, w = 6″, double seam = 1/8″
2. n = 120°, r = 9″, w = 9″, double seam = 1/8″
3. n = 72°, r = 25″, w = 20″, double seam = 1/4″

UNIT 81 DESCRIPTION OF TAPERED SOLIDS

Tapered solids refer to solids, the sides of which are tapered and meet at a point. Combinations of tapered solids and straight-sided solids mentioned occur in many sheet metal jobs.

TAPERED SOLIDS

Tapered solids may be pyramid-shaped or cone-shaped. A *pyramid* is a tapered solid whose base is a polygon whose faces meet at a common point called the vertex or the apex. A *cone* is a tapered solid whose base is a curved figure and whose lateral surface tapers uniformly to the vertex or the apex. (See drawing.)

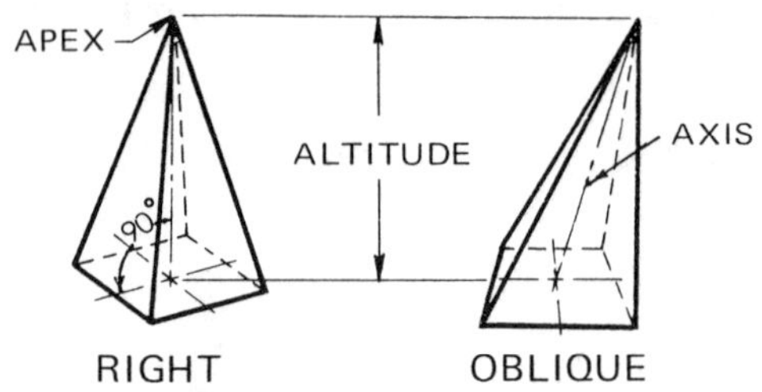

PYRAMIDS

The axis of a pyramid or a cone is the line from the apex to the center of the base. The altitude, or height, is the perpendicular distance from the plane of the base to the apex of the pyramid or the cone.

Pyramids and cones are classified as right or oblique, depending upon whether the axes are at right angles or oblique to the base. (See the drawing at right.) The shape of the base also determines the kind of pyramid or cone. Examples follow:

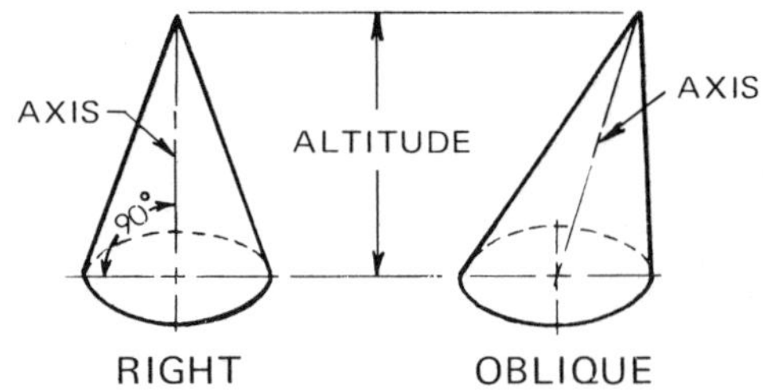

CONES

Right Square Pyramid

A right square pyramid is a right pyramid whose base is a square. The altitude, or height, is designated by *a*, or *h*, the side of the base by *s*. The slant height is a straight line from the apex at right angles to one edge of the base and divides each face into two equal right triangles. It is the shortest line that can be drawn from the apex to the edge of the base. Covers, tops, roofs, etc. are jobs of this shape.

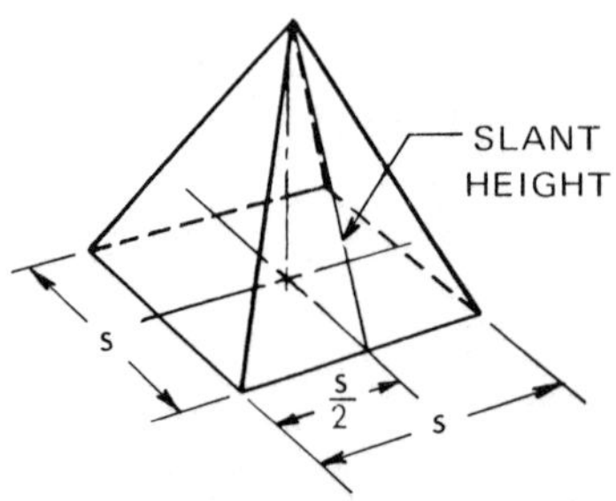

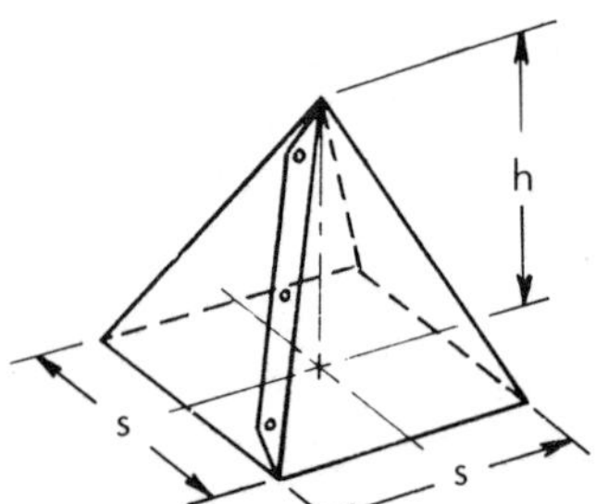

Right Rectangular Pyramid

A right rectangular pyramid is a right pyramid with a rectangular-shaped base. The altitude or the height is designated by *a*, or *h*, the edges of the base by ℓ and *w*. The slant heights of the end faces and of the side faces differ in length.

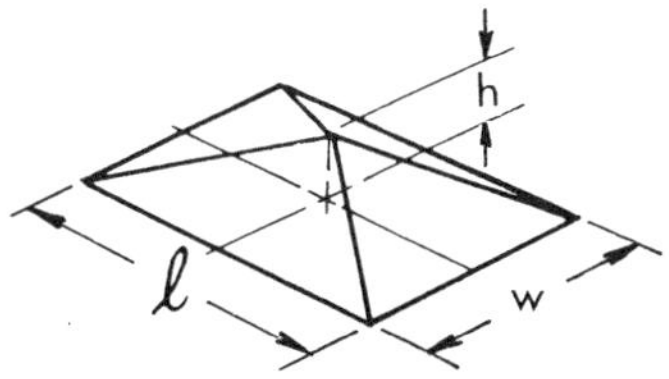

Right Circular Cone

A right circular cone is a right cone whose base is a circle. Right circular cones are usually identified by the word *cone*. The slant height of a right circular cone is a straight line drawn from the apex to the edge of the base. Tops, covers, etc. are jobs of this kind.

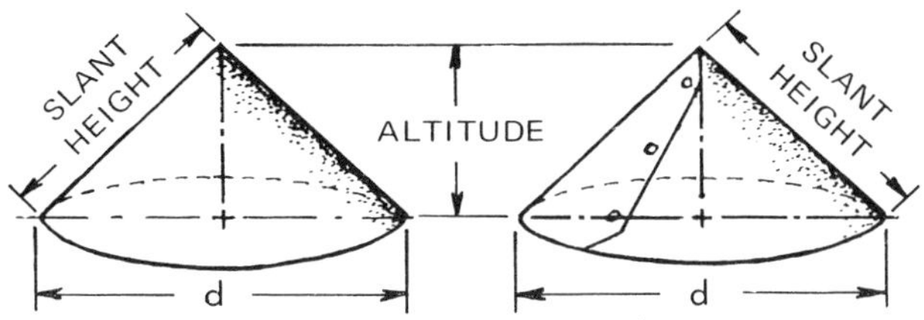

Frustums of Pyramids and Cones

The remaining part of a cone or a pyramid after the top is cut off by a plane parallel to the base is called the frustum of a pyramid or a cone. A truncated pyramid or cone is formed by a plane oblique to the base.

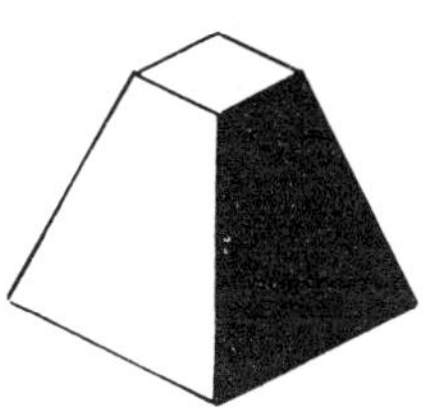

FRUSTUM OF PYRAMID

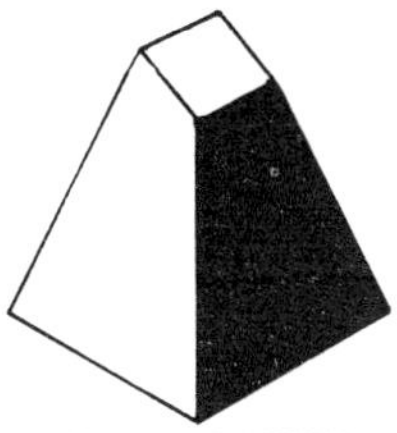

TRUNCATED PYRAMID

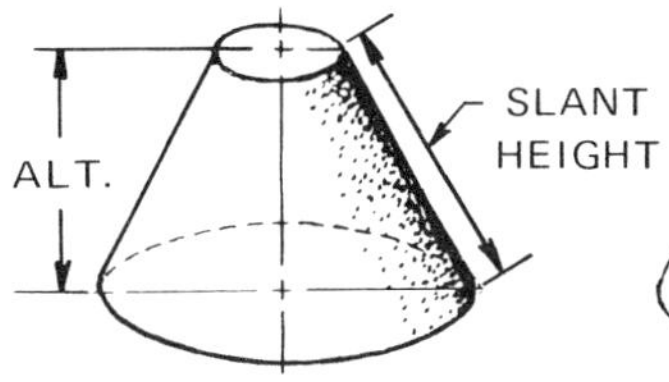

FRUSTUM OF CONE

TRUNCATED CONE

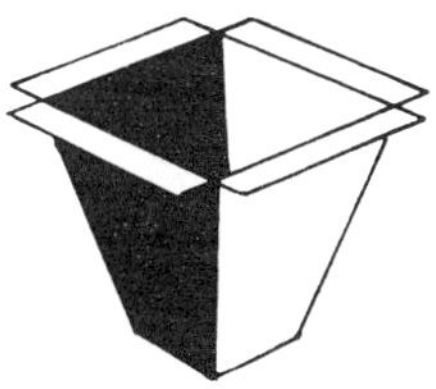

HOPPER

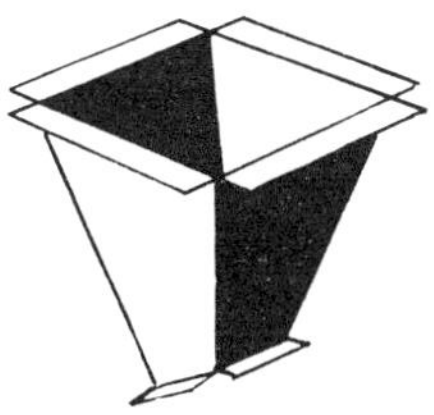

HOPPER

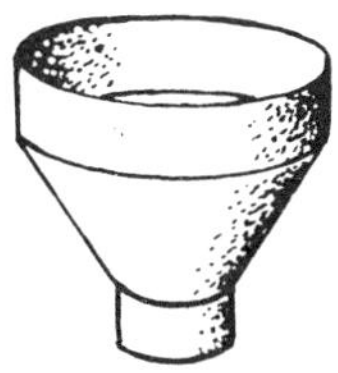

REDUCER

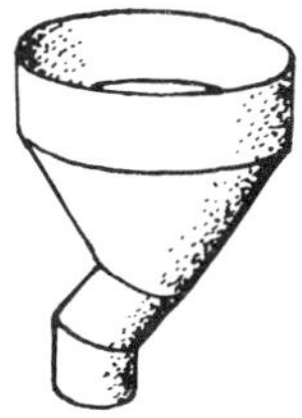

GRAIN SPOUT

Combination of Solids (Examples are shown below.)

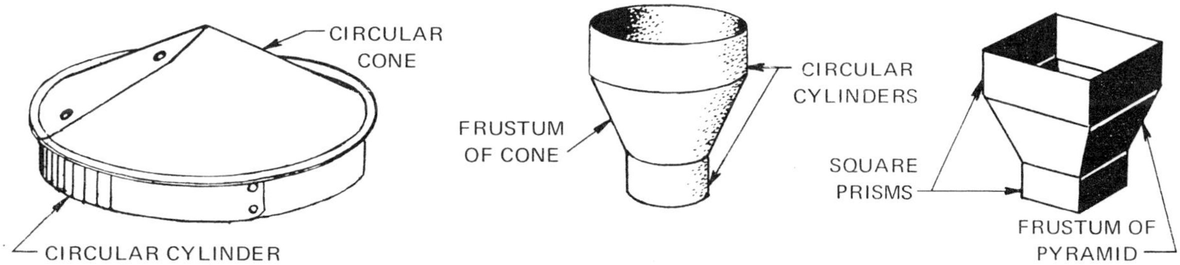

ASSIGNMENT

A. Identification

Identify with the proper mathematical term each of the following jobs or parts of jobs:

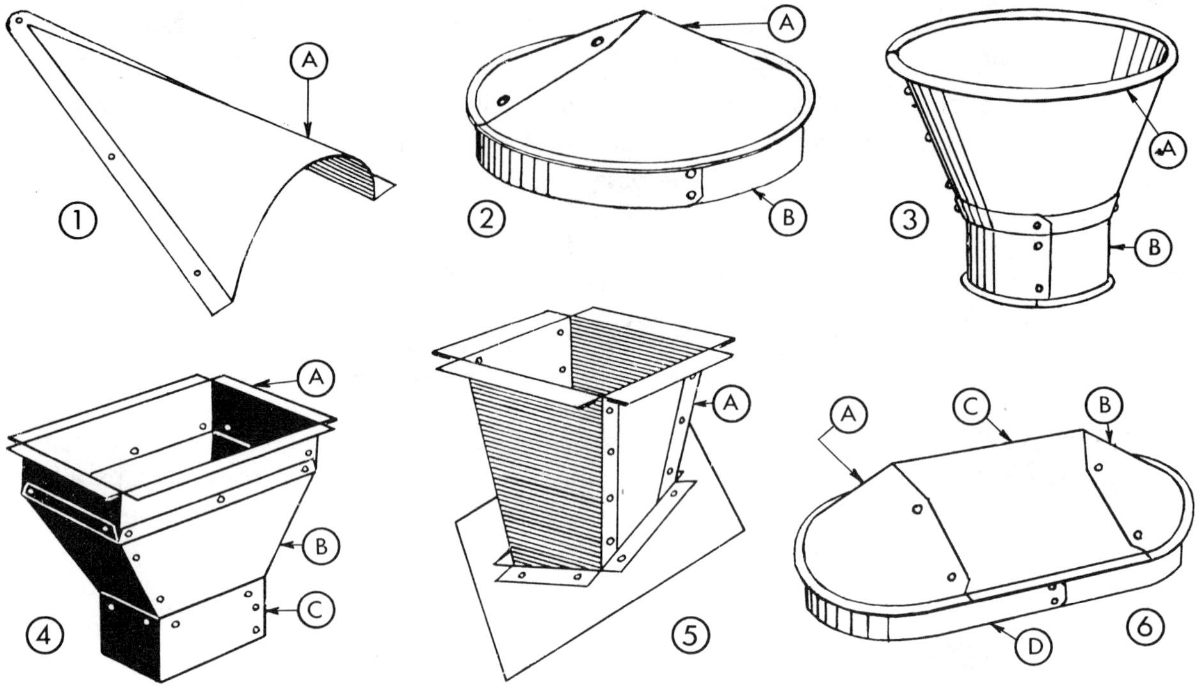

B. Problems

Answer the following questions:

1. How do a right pyramid and an oblique pyramid differ in shape?
2. What is the difference in length between the axis of a right circular cone and the altitude of a right circular cone?
3. What is the difference in shape between a right square pyramid and a right rectangular pyramid?
4. What is the relationship of the slant heights in the adjacent faces of a right rectangular pyramid?
5. Which line is longer in a right circular cone, the altitude or the slant height? Why?
6. Which line would be longer in an oblique square pyramid, the axis or the altitude? Why?
7. What is the difference between a truncated cone and a frustum of a cone?
8. Name four jobs which you have made in shop which are pyramid-shaped or cone-shaped, and designate the type of each job with the proper mathematical term.

UNIT 82 LAW OF RIGHT TRIANGLES

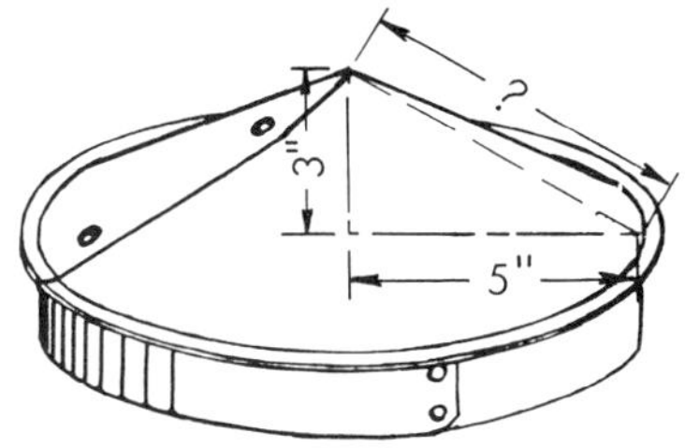

By means of the formulas based upon the law of right triangles, the sheet metal worker can quickly determine the exact measurements of many jobs. The length of the slant height of the cone shown can be computed by this method. In order to understand the formulas used, it is helpful to learn how to find the square root.

SQUARE ROOT

The *square root* of a number is that number which, when multiplied by itself, is equal to the given number. The determination of the square root of a number can be considered the opposite operation of squaring a number. A perfect square is a number which has an exact square root. For example, 64 is a perfect square having 8 as its square root. If the number is a perfect square, the square root can be found by inspection. *The approximate square root* of a number, when multiplied by itself, does not quite equal the number of which it is supposed to be a square root. For example, when finding the square root of a number, it may be necessary to take it to a number of decimal places to work it out completely. The practical method is to give the answer to two decimal places to arrive at the approximate square root.

Instead of computing the square root of a number, tables of square root may be used. Tables in the appendix may be used to calculate square and square root. The instructor will demonstrate the use of these tables and the method of keeping the decimal point in its proper place.

Examples: The square root of 9 is 3 because 3 x 3 = 9.

The square root of 16 is 4 because 4 x 4 = 16.

The square root of 25 is 5 because 5 x 5 = 25.

Square root is indicated by the symbol $\sqrt{}$. Example: $\sqrt{36} = 6$ Ans.

Finding Square Root by Computation

The determination of square roots of numbers by computation is explained by finding the square root of 123,904.

1. Divide into groups of two from the decimal point each way. Assume a decimal behind the last number if none is indicated.

   $\overline{)12_\wedge 3{,}9_\wedge 04}$

2. Find the greatest square that is contained in the first group (3), subtract 9 from 12 = 3, bring down the next group of figures (39).

$$\begin{array}{r} 3\phantom{3{,}9_\wedge 04} \\ 3\overline{)12_\wedge 3{,}9_\wedge 04} \\ 9\phantom{3{,}9_\wedge 04} \\ \hline 3\;39 \end{array}$$

3. Double the answer (3). Put this number (6) to the left of 339. Mentally add a zero to the 6. See how many times 60 goes into 339; it goes 5 times. Put the 5 in the answer above the 39, replace the zero with the 5, and multiply 65 by 5. Subtract 325 from 339, which equals 14. Bring down the next group of figures (04).

```
       3 5
  3)12,3,9,04
     9
 65)3 3 9
    3 2 5
      1 4 0 4
```

4. Double the answer (35). Put this number (70) to the left of 1404. Mentally add a zero to the 70. See how many times 700 goes into 1404; it goes 2 times. Put the 2 in the answer above the 04; replace the zero which you added mentally by the 2, and multiply 702 by 2 which equals 1404.
   There is no remainder and there are no more groups to be brought down.

   Check: 352 x 352 = 123,904

```
       3 5  2
  3)12,3,9,04
     9
 65)3 3 9
    3 2 5
702) 1 4 0 4
     1 4 0 4
352 Ans.
```

LAW OF RIGHT TRIANGLES: TO FIND THE HYPOTENUSE OF A RIGHT TRIANGLE

The Law of Right Triangles

The square of the hypotenuse of a right triangle is equal to the square of the base plus the square of the altitude. The formula is $c^2 = a^2 + b^2$.

RULE: To find the hypotenuse of a right triangle, square the altitude and square the base, add these results, and then find the square root of this sum.

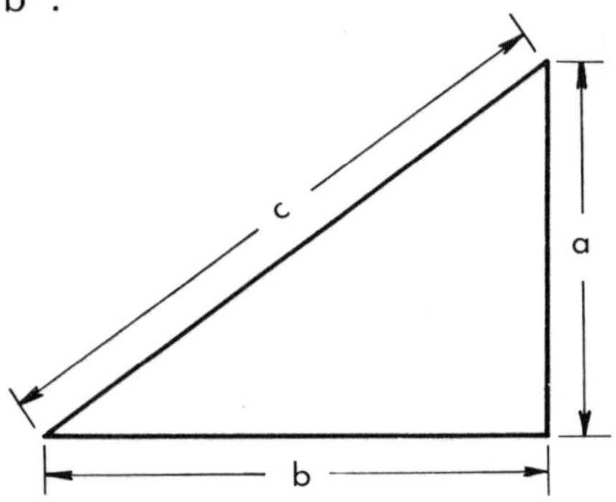

Formula: $c = \sqrt{a^2 + b^2}$

Symbols: c = length of hypotenuse
a = length of altitude
b = length of base

Example: A right triangle has an altitude of 1.5 in. and a base of 2 in. Find the length of the hypotenuse to two decimal places.

1. Formula $c = \sqrt{a^2 + b^2}$

2. Substitution $c = \sqrt{1.5^2 + 2^2}$

3. Evaluation $c = \sqrt{1.5 \times 1.5 + 2 \times 2}$

$c = \sqrt{2.25 + 4}$

$c = \sqrt{6.25}$

c = 2.5″, or 2 1/2″ Ans.

```
            2. 5
          √6.25
 2 x 2 =    4
            2 25
45 x 5 =    2 25
            0 0 0
```

Note: To check answer, measure length *c* on the drawing above. The drawing is actual size.

LAW OF RIGHT TRIANGLES TO FIND EITHER LEG OF A RIGHT TRIANGLE

RULE: To find either leg (base or altitude) of a right triangle, square the hypotenuse and the known leg, subtract the square of the known leg from the square of the hypotenuse, and then find the square root of the result.

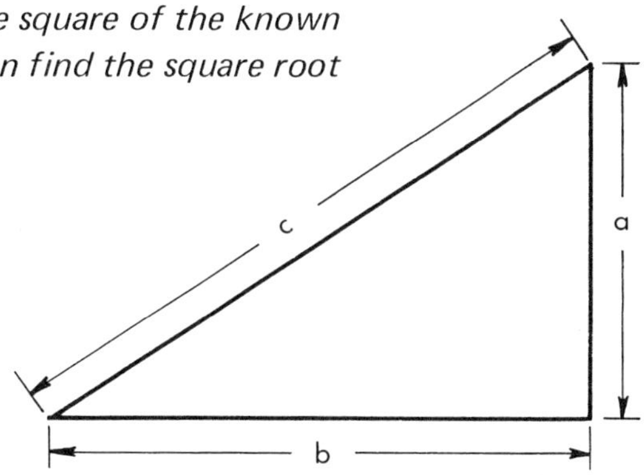

Formulas: $a = \sqrt{c^2 - b^2}$

$b = \sqrt{c^2 - a^2}$

Symbols: a = length of altitude

b = length of base

c = length of hypotenuse

Example: If the hypotenuse of a right triangle is 3 in. and the base is 2.5 in., find the length of the altitude to the closest 32nd of an inch.

1. Formula $a = \sqrt{c^2 - b^2}$
2. Substitution $a = \sqrt{3^2 - 2.5^2}$
3. Evaluation $a = \sqrt{3 \times 3 - 2.5 \times 2.5}$

$a = \sqrt{9 - 6.25}$

$a = \sqrt{2.75}$ $a = 1\frac{21''}{32}$ Ans.

```
              1. 6  5  8
            √ 2.75 00 00
   1 x 1 =    1
              1 75
  26 x 6 =    1 56
                19 00
 325 x 5 =      16 25
                 2 75 00
3308 x 8 =       2 64 64
                   10 36
```

Note: Measure the length of *a* in the drawing on the preceding page to check the correctness of the answer.

CONSTRUCTION OF A RIGHT ANGLE 3-4-5 METHOD

A right angle can be constructed by drawing a right triangle of 3 in., 4 in., and 5 in., or a multiple, or a fraction of each as follows:

1. Using the 4-in. side as a baseline and the ends of that line as centers, swing the arcs with radii of 3 in. and 5 in.

2. Connect the intersecting point with the baseline.

Note: Measure ∠ C to check if it is a right angle.

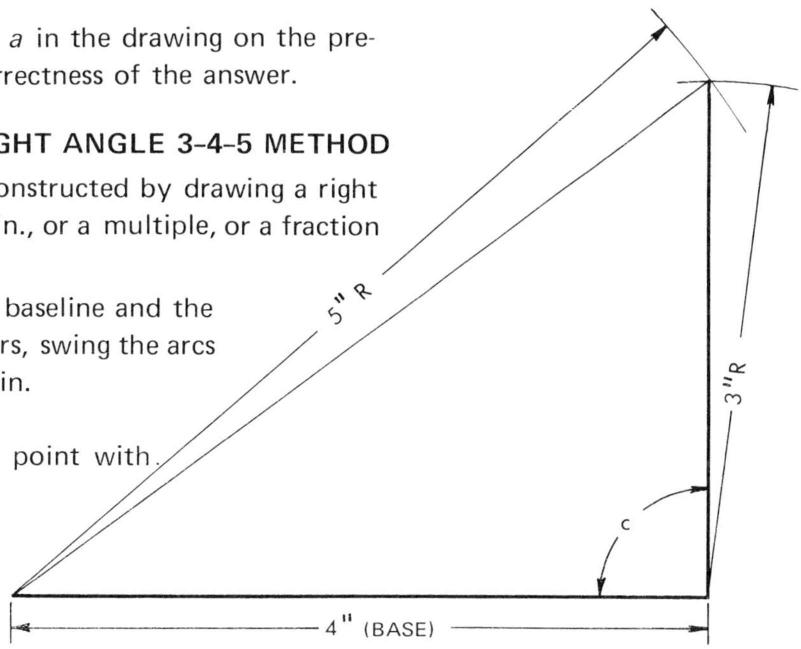

Explanation of Principle

1. Since $5^2 = 3^2 + 4^2$ (law of right triangles)
2. Then $25 = 9 + 16$
3. And $25 = 25$
4. ∴ The triangle is a right triangle.

Note: This method is also used to lay out a right angle for large work. In this case, lengths of 6 feet, 8 feet, and 10 feet can be used for the legs and the hypotenuse.

ASSIGNMENT

A. Square Root

Find each square root to three decimal places.

1. $\sqrt{4}$
2. $\sqrt{36}$
3. $\sqrt{49}$
4. $\sqrt{144}$
5. $\sqrt{121}$
6. $\sqrt{81}$
7. $\sqrt{169}$
8. $\sqrt{64}$
9. $\sqrt{3}$
10. $\sqrt{6}$
11. $\sqrt{7}$
12. $\sqrt{5}$
13. $\sqrt{10}$
14. $\sqrt{4096}$
15. $\sqrt{2500}$
16. $\sqrt{6.25}$
17. $\sqrt{81.5625}$
18. $\sqrt{15.25}$
19. $\sqrt{5.4962}$
20. $\sqrt{8.9741}$
21. $\sqrt{10.428}$
22. $\sqrt{210.29}$
23. $\sqrt{452.6188}$
24. $\sqrt{2141.21}$

B. Law of Right Triangles – To Find the Hypotenuse

Find the missing dimension of each problem to the closest 32nd of an inch. Lay out the right triangles of problems 1 to 3, and check the answers by actual measurement.

1. a = 2″, b = 3″
2. a = 3″, b = 4″
3. a = 4″, b = 5″
4. a = 7½″, b = 2″
5. a = 10″, b = 10½″
6. a = 60″, b = 11¼″
7. a = 29¾″, b = 15″
8. a = 6.22″, b = 4.56″

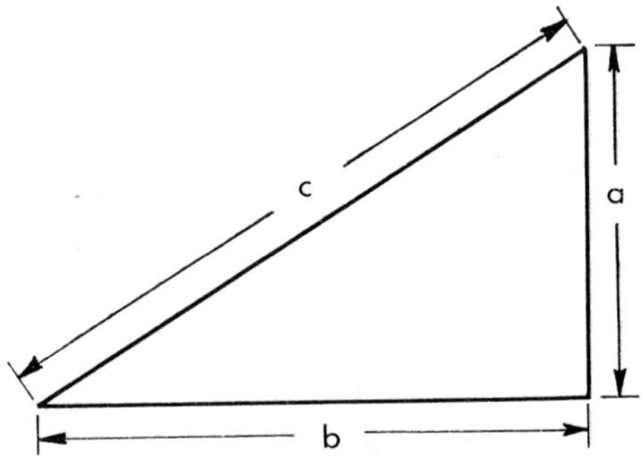

9.

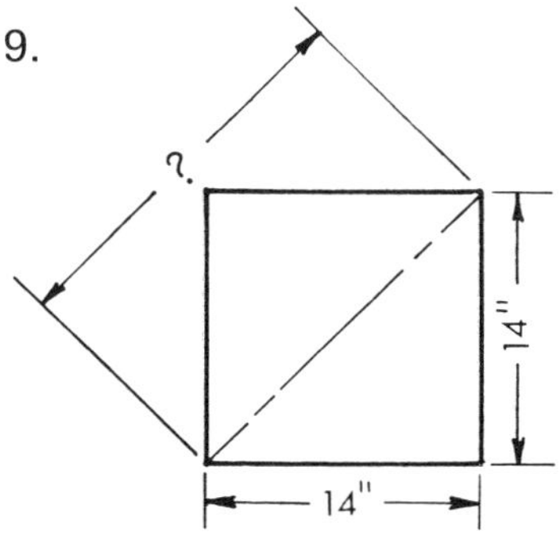

10.

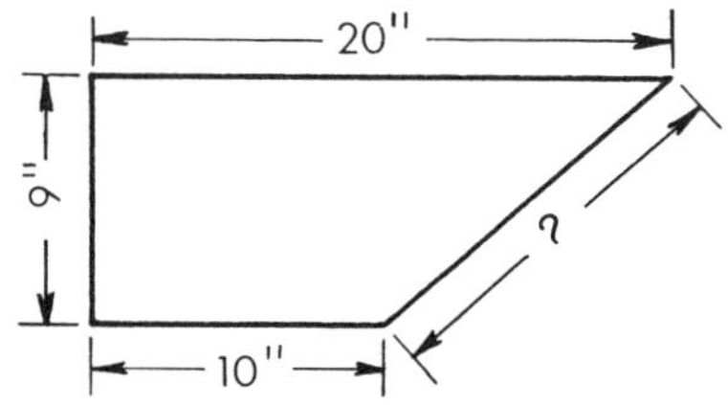

11.

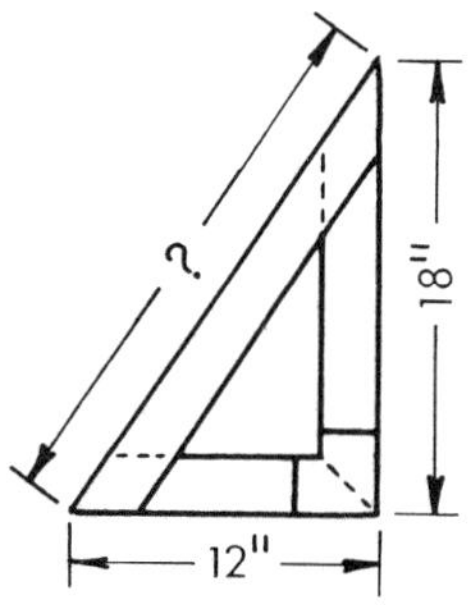

12.

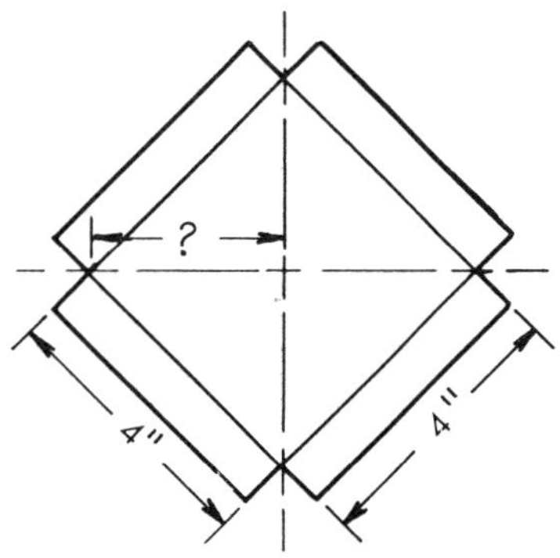

C. Law of Right Triangles - To Find Either Leg

Find the missing dimension of each problem to the closest 32nd of an inch. Lay out the right triangles of problems 1 to 3, and check the answers by actual measurement.

1. b = 2", c = 4"
2. b = 4", c = 5"
3. b = 3", c = 4"
4. a = 7½", c = 9"
5. a = 2¼", c = 4"
6. b = 5", c = 6¼"
7. a = 2.52", c = 8.28"
8. b = 7.26", c = 10.32"

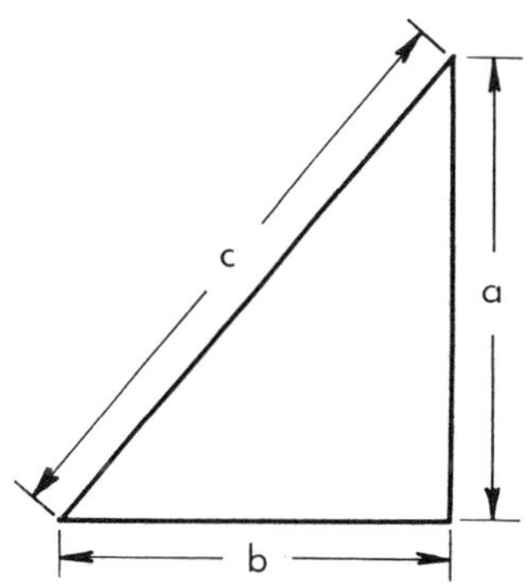

9.

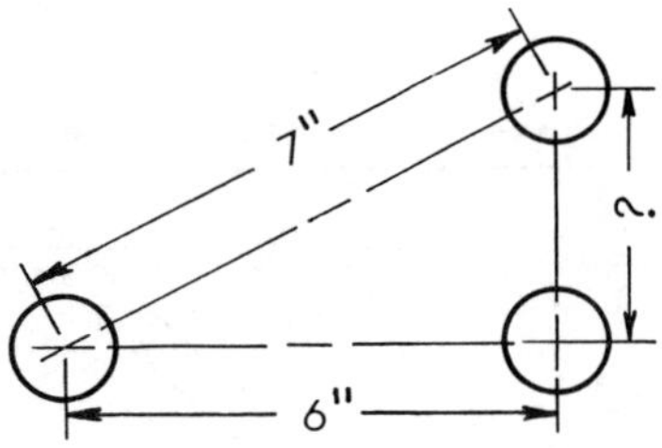

10.

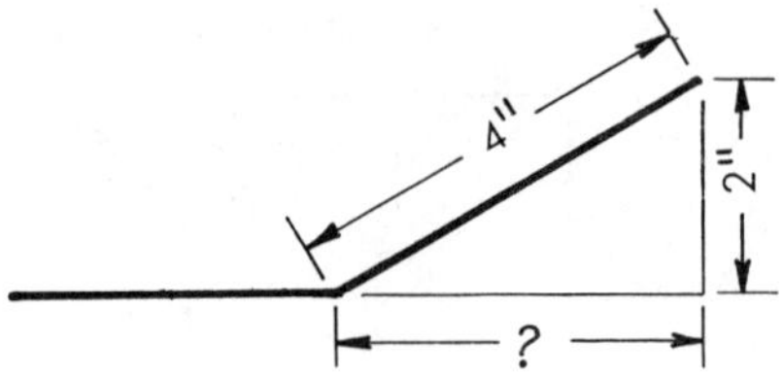

D. Construction of a Right Angle

Draw right angles using the following dimensions.

1. 4½", 6", 7½"

2. 2¼", 3", 3¾"

UNIT 83 STRETCHOUTS OF CONES

The radius and the angle of the stretchout of a right cone can be computed by means of formulas or found by the layout method. The radius of the stretchout is equal to the slant height of the cone. In certain cases, it may be advantageous to compute these measurements.

SLANT HEIGHT

The length of the slant height (H) is obtained from the law of right triangles, and is equal to the square root of the sum of the altitude (a) squared and the base (d/2) squared of one right triangle of the cone.

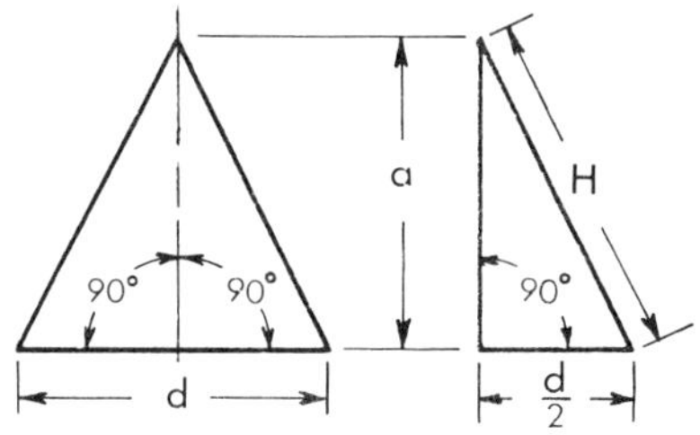

Formula: $H = \sqrt{a^2 + (d/2)^2}$

Note: The cross section of a cone can be separated into two equal right triangles. Therefore, the base of one triangle equals one-half the diameter of the cone.

Example: The altitude of a cone is 3¾ in.; the dia. of the base is 4 in. Find the slant height to the nearest 32nd in.

1. Formula $H = \sqrt{a^2 + (d/2)^2}$
2. Substitution $H = \sqrt{(3.75)^2 + (4/2)^2}$
3. Evaluation $H = \sqrt{14.0625 + 4}$

   $H = \sqrt{18.0625}$

   $H = 4.25''$, or 4¼″ Ans.

ANGLE OF STRETCHOUT

The angle of stretchout (n) is equal to the ratio of the circumference of the base (πd) of the cone to the circumference of the stretchout circle (2πH), times 360 deg.

Formula: $n = \frac{\pi d}{2\pi H} \times 360$, or $n = \frac{180d}{H}$

Note: $\frac{\overset{1}{\cancel{\pi}}d}{\underset{1}{\cancel{2}}\underset{1}{\cancel{\pi}}H} \times \overset{180}{\cancel{360}} = \frac{180d}{H}$

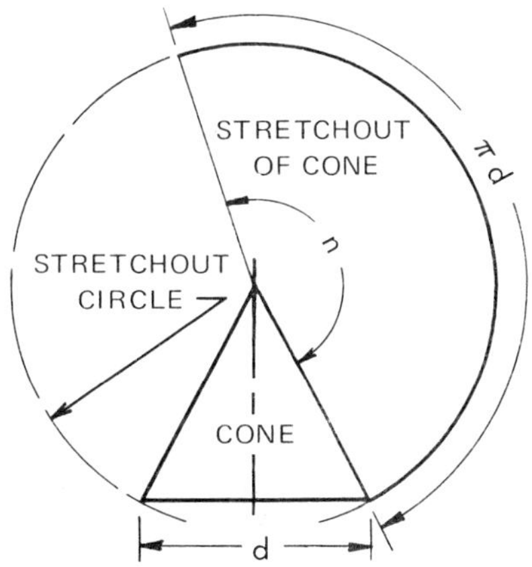

Example: Find the angle (closest 1/2 deg.) of the stretchout in the above example.

1. Formula $n = \frac{180d}{H}$
2. Substitution $n = \frac{180 \times 4}{4.25}$
3. Evaluation $n = \frac{720}{4.25}$

   $n = 169.4$, or 169½° Ans.

ASSIGNMENT

A. Slant Height of a Right Cone

Compute the slant height of each cone to the nearest 32nd of an inch.

1. dia. = 20″, alt. = 10″
2. dia. = 10″, alt. = 6″
3. dia. = 5½″, alt. = 9″
4. dia. = 10′, alt. = 2′ 1″

B. Radius and Angle of Stretchout of a Right Cone

Compute the radius of the stretchout to the nearest 32nd of an inch and the angle of the stretchout to the nearest half-degree of each cone. Lay out problems 1 to 3 to actual size, and check the computed angle by measuring the layout.

1. dia. = 4″, alt. = 3″
2. dia. = 3″, alt. = 2″
3. dia. = 5″, alt. = 3″
4. dia. = 12″, alt. = 25″
5. dia. = 10½″, alt. = 4″
6. dia. = 10′, alt. = 3′ 6″

C. Additional Problems

1. Compute the radius and the angle of stretchout of the top part of the cone-shaped cover.

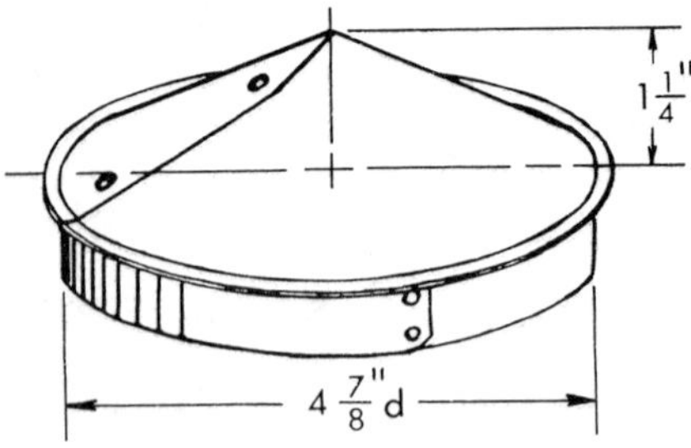

2. What is the radius and the angle of stretchout of the air scoop?

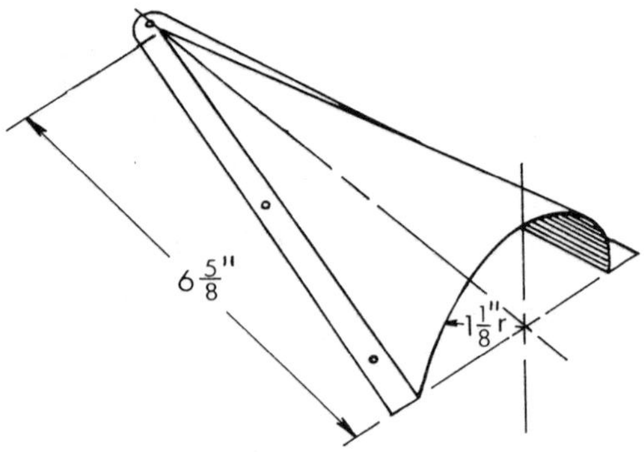

3. Compute the radius and the angle of stretchout of the half cones in the top of the cover.

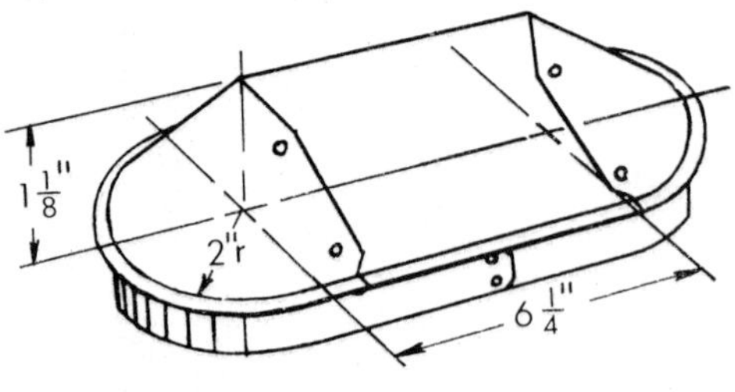

4. What is the length of the slant height of the square pyramid?

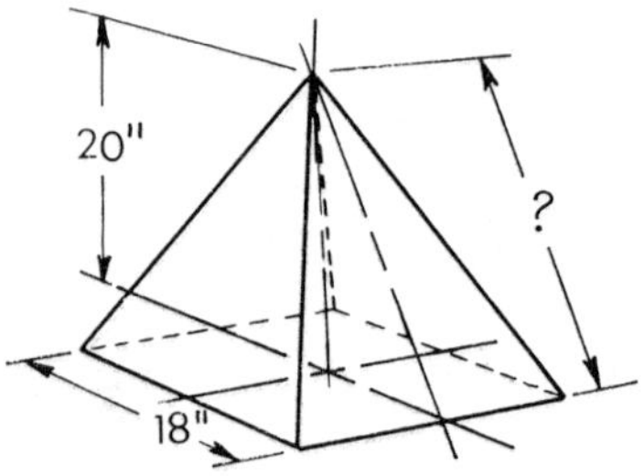

UNIT 84 ROOF PITCHES

Since sheet metal is widely used as roofing material, the contractor or the workman must be able to compute the pitch (slope) and other dimensions of a roof. Pitch is expressed as the ratio of the rise to the span. For example, if the span of a roof is 22 ft. and the rise is 11 ft., the pitch is 11/22, or 1/2. Some of the pitches in common use are 1/4, 1/3, 1/2, 2/3, 5/12, and 7/12.

RISE OF ROOF

The rise of a roof is equal to the pitch multiplied by the span.

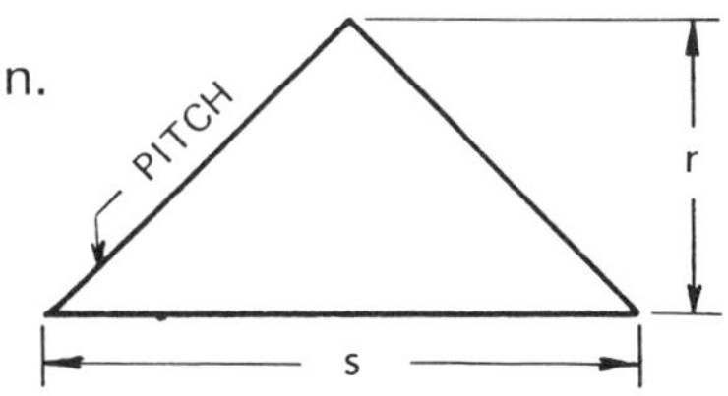

Formula: $r = p \times s$

Symbols: p = pitch
r = rise; s = span

Example: A roof has a 22-ft. span and a 1/2 pitch. Find the rise.

1. Formula $r = p \times s$
2. Substitution $r = 1/2 \times 22$
3. Evaluation $r = 11$ ft. Ans.

LENGTH OF ROOF SLOPE

The length of the slope is obtained from the law of right triangles and is equal to the square root of the sum of the rise (r) squared and one-half the span (s/2) squared.

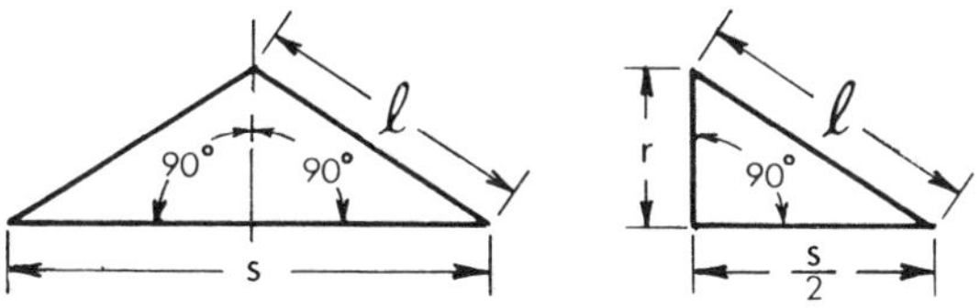

Formula: $\ell = \sqrt{r^2 + (s/2)^2}$

Symbols: ℓ = length of roof slope
r = rise
s = span; $s/2$ = run

Example: The span of a roof is 24 ft., and the pitch is 1/3. Compute the length of the roof slope in ft. and in. (closest 16th.)

Rise:
1. $r = p \times s$
2. $r = 1/3 \times 24$
3. $r = 8'$ Ans.

Slope:
1. $\ell = \sqrt{r^2 + (s/2)^2}$
2. $\ell = \sqrt{8^2 + (24/2)^2}$
3. $\ell = \sqrt{64 + 144}$
4. $\ell = \sqrt{208}$, or $14.422'$
5. $\ell = 14'\ 5.064''$, or $14'\ 5\ 1/16''$ Ans.

ASSIGNMENT

A. Pitch and Rise of Roofs

Compute the pitch or the rise of the roof in each of the following problems.

1. rise = 13′, span = 26′
2. rise = 10′, span = 30′
3. rise = 10′, span = 24′
4. rise = 9′ 6″, span = 28′ 6″
5. pitch = 1/2, span = 28′
6. pitch = 5/12, span = 24′
7. pitch = 7/12, span = 30′
8. pitch = 2/3, span = 33′

B. Length of Roof Slopes

Compute the length of the roof slopes in each of the following problems.

1. pitch = 1/3, span = 30′
2. pitch = 1/2, span = 32′
3. pitch = 2/3, span = 24′
4. pitch = 7/12, span = 24′
5. pitch = 3/4, span = 22′
6. pitch = 1 1/2, span = 10′

C. Additional Problems

Compute the length of the sloping line in each of the drawings below. Double the run to find the span. **Hint:** Double the run to find the span.

1.

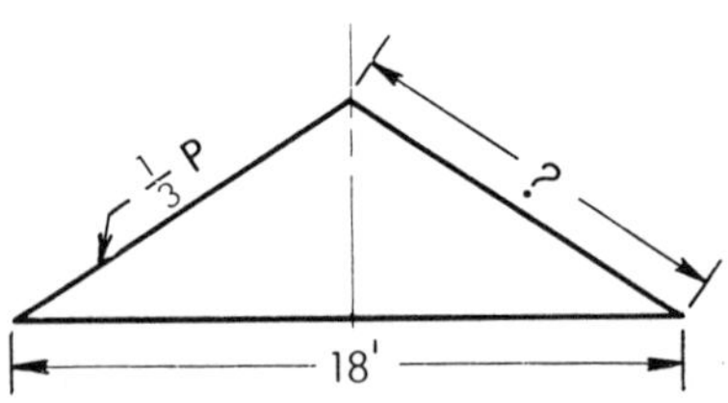

2.

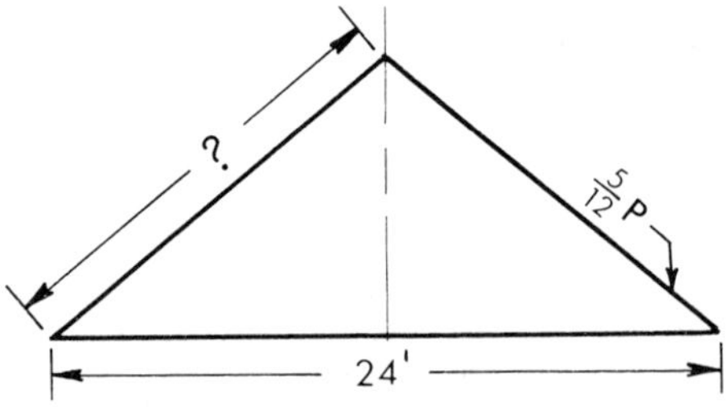

3.

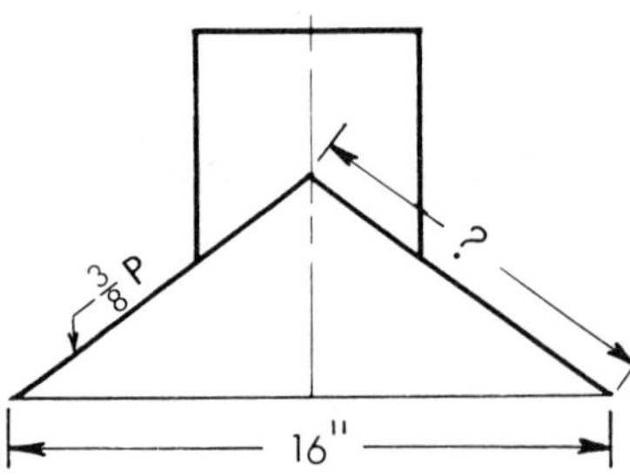

4.

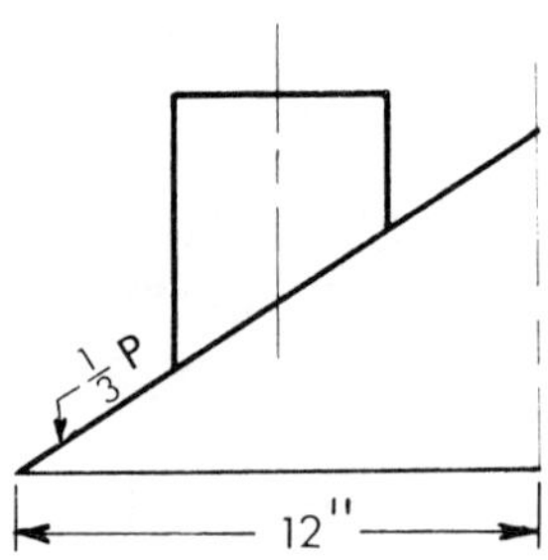

UNIT 85 BEND ALLOWANCES

Bend allowance is the length of material which must be added to the straight parts of the job to allow for forming. There are two formulas which are used for finding the bend allowance: the approximate formula for sharp bends of 90 deg., and the exact formula for bends with an inside raduis.

APPROXIMATE ALLOWANCES FOR SHARP BENDS

Add 1/2 the metal thickness to the inside dimensions of the job for each 90 deg. bend. Subtract 1/2 the thickness from the outside dimensions for each 90 deg. bend.

Formula: B.A. = $\ell + L \pm t/2$

Symbols: B.A. = bend allowance (in.)
ℓ, L = straight parts (in.)
t = metal thickness (in.)

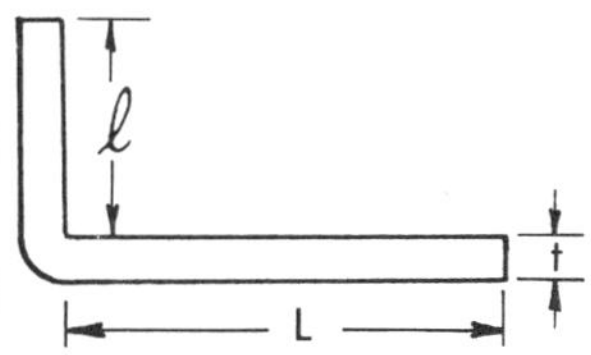

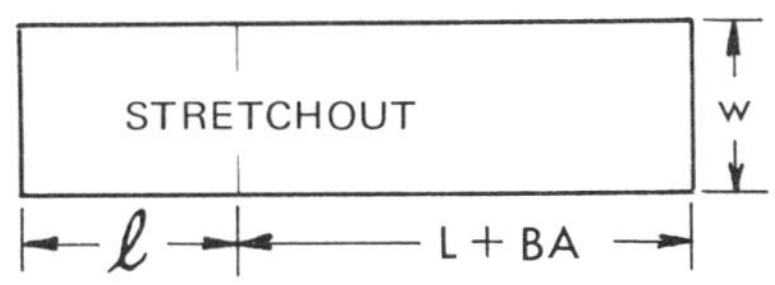

Example: Find the length of material needed if a piece of 1/8-in. band iron is to be formed to the following dimensions: inside lengths of straight parts, 5/8 in. and 5 1/4 in.; angle of bend 90 deg.

1. L.S. = $\ell + L + t/2$
2. L.S. = 5/8 + 5 1/4 + 1/2 x 1/8, or 5 15/16″ Ans.

EXACT ALLOWANCES FOR RADIUS BENDS

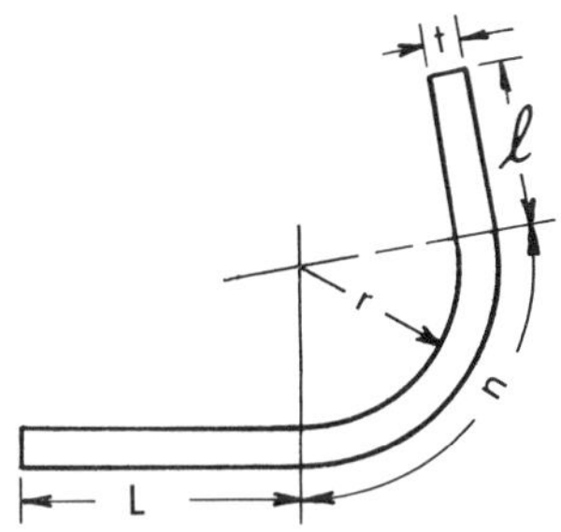

Formula: B.A. = n (.01743r + .0078t)

Symbols: B.A. = bend allowance in inches
n = number of deg. in bend
r = inside radius of bend in inches
t = thickness of metal in inches

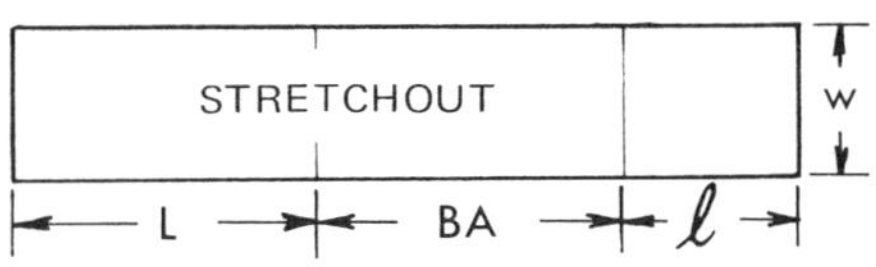

Example: Find the bend allowance (64th of an in.) for metal, thickness .0125 in., formed to a 100-deg. angle and a .5-in. radius.

1. Formula	B.A.	=	n (.01743r + .0078t)
2. Substitution	B.A.	=	100 (.01743 x .5 + .0078 x .0125)
3. Evaluation	B.A.	=	100 (.008715 + .00009750)
	B.A.	=	100 x .0088125, or .88125
	B.A.	=	.881 in., or 7/8 in. Ans.

ASSIGNMENT

A. Approximate Allowances for Sharp Bends

Find the length of the stretchout for each job.

1.

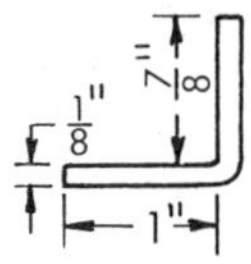

2.

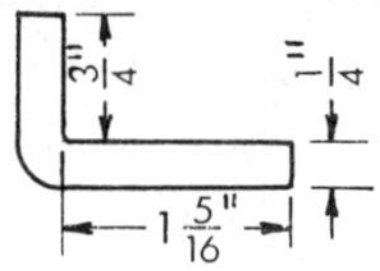

3.

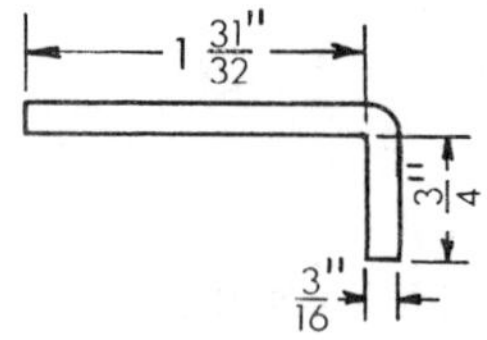

4.

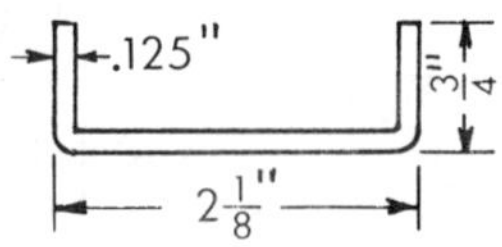

5.

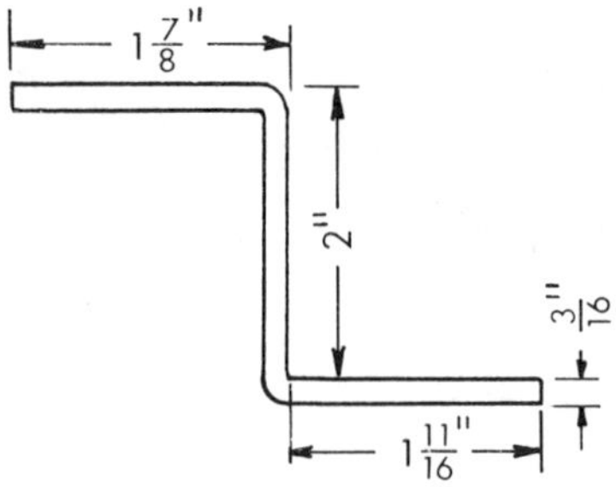

6.

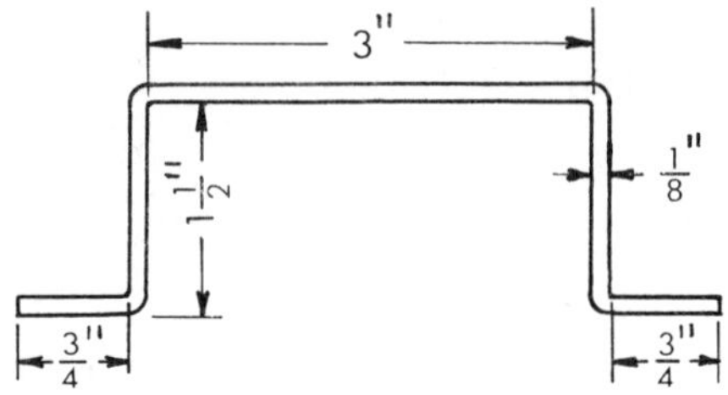

B. Exact Allowances for Radius Bends

Find the length of metal needed for each job to the closest 64th of an inch. Draw the stretchouts of jobs 1 to 6 (make the width of each job 2 in.)

1. central angle = 50°, inside radius = 1″, thickness = .040″
2. central angle = 80°, inside radius = 1¼″, thickness = .0598″
3. central angle = 62°, inside radius = .375″, thickness = .080″

4.

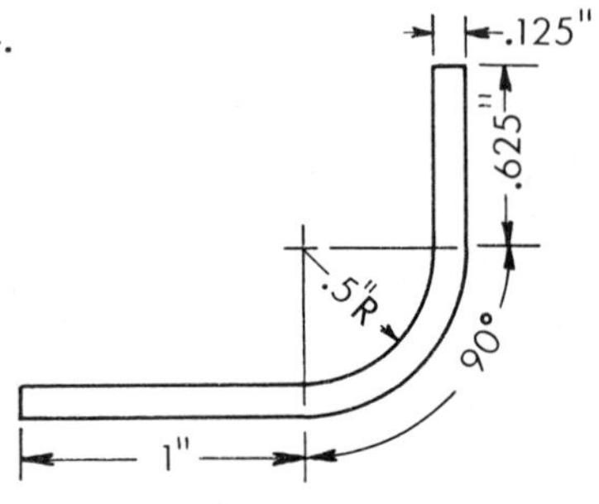

5.

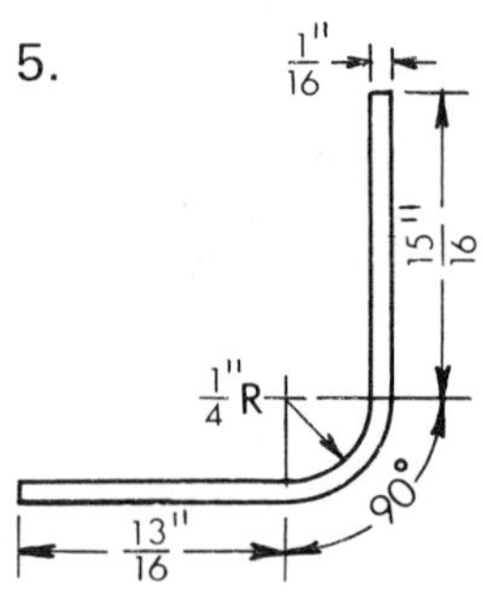

6.

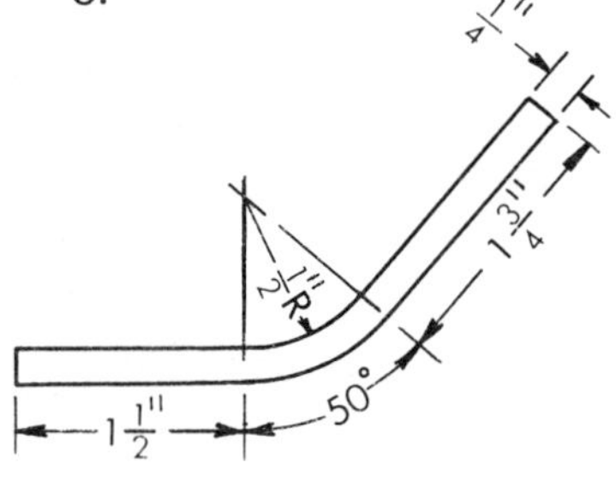

7.

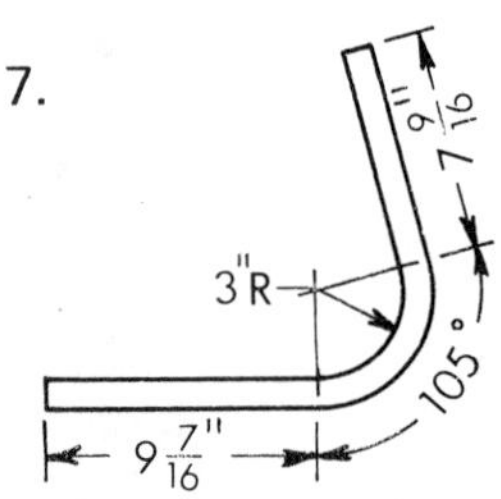

8.

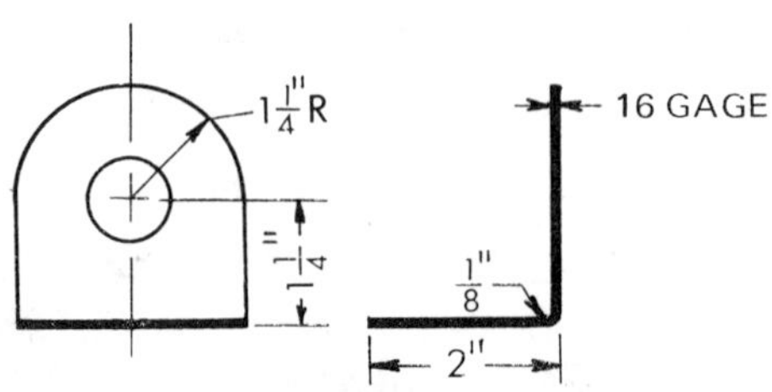

9.

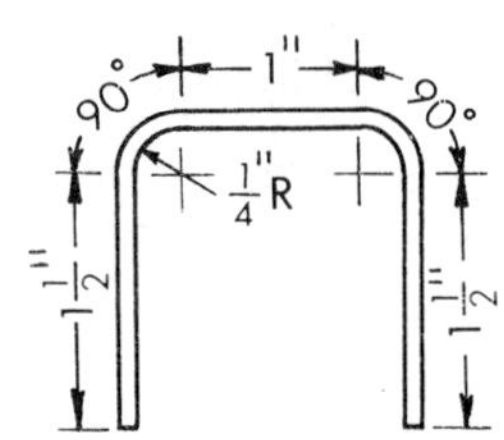

UNIT 86 SURFACE SPEEDS OF ROTATING CYLINDERS

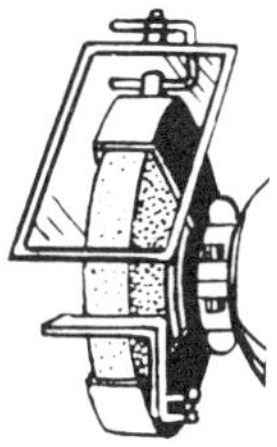

Surface speed refers to the linear distance a point on a circle or a cylinder moves when the circle turns about its center or axis in a certain time. The correct cutting (surface) speeds of drills and grinding wheels, and the correct lineal (surface) speeds of fans, belts, and gears must be used for the efficient and safe operation of such tools and machines.

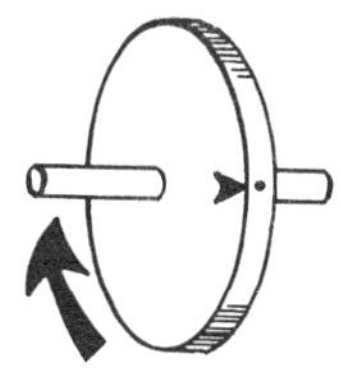

SURFACE SPEEDS IN F.P.M.

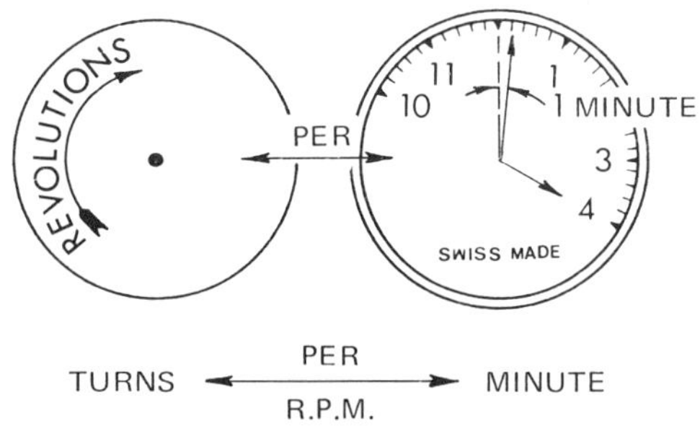

Surface speed is expressed in feet per minute (f.p.m.) and is obtained by using the revolutions per minute and the circumference of the circle.

One revolution or one complete turn of a circle in one minute around its center, is called one revolution per minute. Revolutions per minute (r.p.m.) means the number of turns per minute.

RULE: Surface speed in feet per minute is equal to the circumference of the circle multiplied by the number of turns in a minute: feet per minute = circumference x revolutions per min.

Formula: $S = \frac{\pi dn}{12}$

Symbols: S = surface speed in f.p.m.
d = dia. of circle in in.
n = r.p.m.; π = 3.14
c = circumference in feet

Explanation of Formula

1. See rule and symbols. $S = c \times n$
2. Change πd to feet. $c = \frac{\pi d}{12}$
3. Substitute $\frac{\pi d}{12}$ for c. $\therefore \quad S = \frac{\pi dn}{12}$

Example: A 10-in. grinding wheel rotates at 1000 r.p.m. Find the surface speed to the closest foot per minute.

1. Formula $S = \frac{\pi dn}{12}$
2. Substitution $S = \frac{3.14 \times 10 \times 1000}{12}$
3. Evaluation $S = \frac{31400}{12}$, or 2616.6, or 2627 f.p.m. Ans.

ASSIGNMENT

1. The diameter of a wheel is designated as *d*, and the revolutions per minute are designated as r.p.m. Compute the surface speed to the closest foot per minute in each of the following problems:

a. d = 14″, r.p.m. = 1000	d. d = 20″, r.p.m. = 1000
b. d = 8″, r.p.m. = 2000	e. d = 13″, r.p.m. = 1250
c. d = 9″, r.p.m. = 1500	f. d = 2′ 6″, r.p.m. = 550

2. If a 1/4-in. drill rotates at 800 r.p.m., compute the cutting speed. Since the cutting speed of a drill is the speed of the outer edge of the lip, use d = 1/4″.

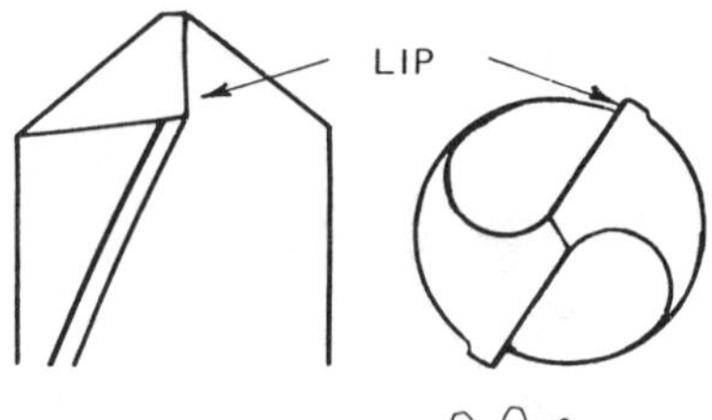

3. A 1/8-in. drill rotates at 1500 r.p.m. If the correct cutting speed for this drill and for the material used is 80 f.p.m. to 100 f.p.m., is the drill rotating at the correct speed?

4. What is the lineal speed of the tip of a tooth on a gear the outer dia. of which is 13 in., r.p.m. 950?

5. What is the lineal speed of the tip of a tooth on a gear the outer dia. of which is 9 in., r.p.m. 1475?

6. Compute the peripheral (lineal) speed of the tip of a 40-in. fan blade, r.p.m. 350. If the maximum peripheral speed allowable for this fan is 6000 f.p.m., is it turning too fast?

7. Compute the surface speeds of the grinding wheel shown below turning at 2000 r.p.m., before and after it is worn.

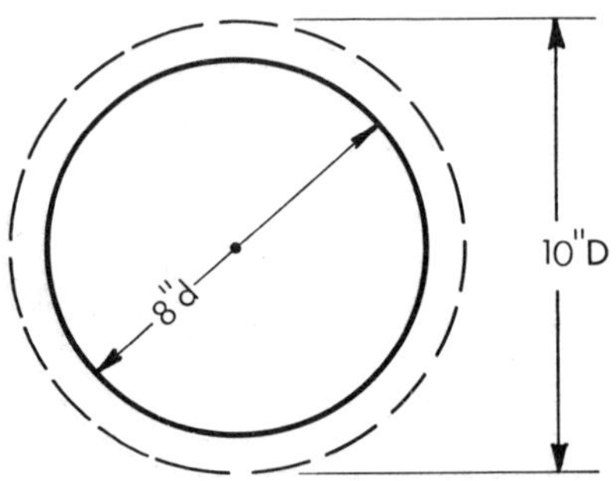

8. A 12-in. grinding wheel turning at 1600 r.p.m. wears 2 in. on each side of the wheel. If the correct surface speed for this type of wheel and for the material used is 5000 to 6000 f.p.m., is the surface speed satisfactory after the wheel is worn?

9. If a 12-in. wheel is used on a grinder designed for a 10-in. wheel with an r.p.m. of 2000, would the surface speed of the 12-in. wheel be dangerous? (The maximum limit of surface speed for that wheel is 6000 f.p.m.)

10. Find the lineal speed of a belt running on a 30-in. pulley which revolves at 250 r.p.m. Assume that there is no slippage between the pulley and belt.

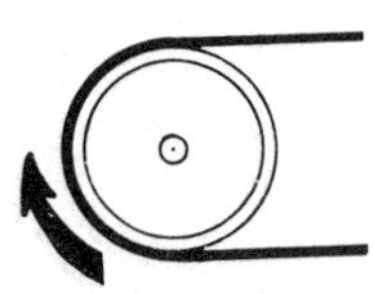

SECTION 11 — TRIGONOMETRY

UNIT 87 INTRODUCTION TO TRIGONOMETRY

Trigonometry means triangle measure. Using trigonometric formulas, we can calculate the size of the angles of a right triangle, or the length of its sides.

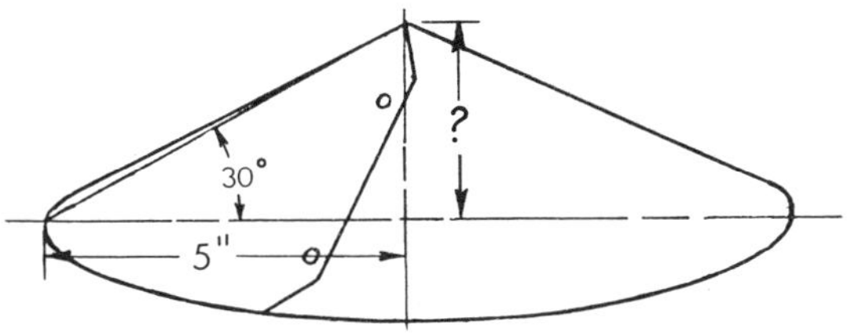

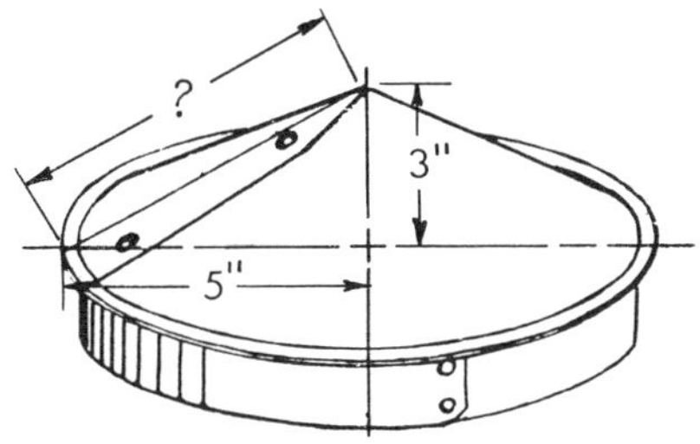

In this section, the use of three of the six basic formulas which apply to right triangles will be studied: the tangent (tan), the sine (sin), and the cosine (cos.)

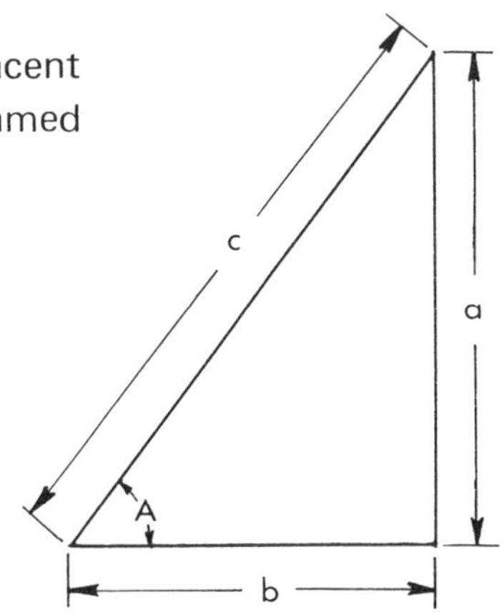

Formulas: If the opposite side *a* is divided by the adjacent side *b*, the ratio *a/b* is obtained, which is named the tangent of the angle A.

$$\tan A = \frac{\text{opp.}}{\text{adjac.}}, \text{ or } \frac{a}{b}$$

$$\text{Similarly, } \sin A = \frac{\text{opp.}}{\text{hypot.}}, \text{ or } \frac{a}{c};$$

$$\cos A = \frac{\text{adjac.}}{\text{hypot.}}, \text{ or } \frac{b}{c}$$

Symbols: a = altitude, or opposite side (opp.) to $\angle$ A
b = base, or adjacent side (adjac.) to $\angle$ A
c = hypotenuse (hypot.)

READING THE TABLE OF NATURAL TRIGONOMETRIC FUNCTIONS

These three ratios are called *trigonometric functions.* Their values have been computed for angles from 0° to 90° and are listed in tabular form on the next page. To learn the use of this table, substitute the lengths of the sides (a = 2'', b = 1.5'', c = 2.5'') of the triangle above, in the three formulas and divide (5 decimal places).

$\tan A = \frac{a}{b}$, or $\frac{2}{1.5}$, or 1.33333 $\therefore \angle A = 53°$ (Use the tangent column to find the angle closest to 1.3333.)

$\sin A = \frac{a}{c}$, or $\frac{2}{2.5}$, or .80000 $\therefore \angle A = 53^\circ$ (Find the angle closest to .8000 in the sine column.)

$\cos A = \frac{b}{c}$, or $\frac{1.5}{2.5}$, or .60000 $\therefore \angle A = 53^\circ$ (Find the angle closest to .6000 in the cosine column.)

Note: If the angle A in the triangle on page 229 is measured with a protractor, it will be found to be 53°.

TABLES OF NATURAL TRIGONOMETRIC FUNCTIONS

ANGLE	SINE	COSINE	TANGENT	ANGLE	SINE	COSINE	TANGENT
1°	.0175	.9998	.0175	46°	.7193	.6947	1.0355
2°	.0349	.9994	.0349	47°	.7314	.6820	1.0724
3°	.0523	.9986	.0524	48°	.7431	.6691	1.1106
4°	.0698	.9976	.0699	49°	.7547	.6561	1.1504
5°	.0872	.9962	.0875	50°	.7660	.6428	1.1918
6°	.1045	.9945	.1051	51°	.7771	.6293	1.2349
7°	.1219	.9925	.1228	52°	.7880	.6157	1.2799
8°	.1392	.9903	.1405	53°	.7986	.6018	1.3270
9°	.1564	.9877	.1584	54°	.8090	.5878	1.3764
10°	.1736	.9848	.1763	55°	.8192	.5736	1.4281
11°	.1908	.9816	.1944	56°	.8290	.5592	1.4826
12°	.2079	.9781	.2126	57°	.8387	.5446	1.5399
13°	.2250	.9744	.2309	58°	.8480	.5299	1.6003
14°	.2419	.9703	.2493	59°	.8572	.5150	1.6643
15°	.2588	.9659	.2679	60°	.8660	.5000	1.7321
16°	.2756	.9613	.2867	61°	.8746	.4848	1.8040
17°	.2924	.9563	.3057	62°	.8829	.4695	1.8807
18°	.3090	.9511	.3249	63°	.8910	.4540	1.9626
19°	.3256	.9455	.3443	64°	.8988	.4384	2.0503
20°	.3420	.9397	.3640	65°	.9063	.4226	2.1445
21°	.3584	.9336	.3839	66°	.9135	.4067	2.2460
22°	.3746	.9272	.4040	67°	.9205	.3907	2.3559
23°	.3907	.9205	.4245	68°	.9272	.3746	2.4751
24°	.4067	.9135	.4452	69°	.9336	.3584	2.6051
25°	.4226	.9063	.4663	70°	.9397	.3420	2.7475
26°	.4384	.8988	.4877	71°	.9455	.3256	2.9042
27°	.4540	.8910	.5095	72°	.9511	.3090	3.0777
28°	.4695	.8829	.5317	73°	.9563	.2924	3.2709
29°	.4848	.8746	.5543	74°	.9613	.2756	3.4874
30°	.5000	.8660	.5774	75°	.9659	.2588	3.7321
31°	.5150	.8572	.6009	76°	.9703	.2419	4.0108
32°	.5299	.8480	.6249	77°	.9744	.2250	4.3315
33°	.5446	.8387	.6494	78°	.9781	.2079	4.7046
34°	.5592	.8290	.6745	79°	.9816	.1908	5.1446
35°	.5736	.8192	.7002	80°	.9848	.1736	5.6713
36°	.5878	.8090	.7265	81°	.9877	.1564	6.3138
37°	.6018	.7986	.7536	82°	.9903	.1392	7.1154
38°	.6157	.7880	.7813	83°	.9925	.1219	8.1443
39°	.6293	.7771	.8098	84°	.9945	.1045	9.5144
40°	.6428	.7660	.8391	85°	.9962	.0872	11.4301
41°	.6561	.7547	.8693	86°	.9976	.0698	14.3007
42°	.6691	.7431	.9004	87°	.9986	.0523	19.0811
43°	.6820	.7314	.9325	88°	.9994	.0349	28.6363
44°	.6947	.7193	.9657	89°	.9998	.0175	57.2900
45°	.7071	.7071	1.0000	90°	1.0000	.0000	

ASSIGNMENT

Use the table of trigonometric functions on the preceding page to solve the following problems.

1. Find the number (ratio) corresponding to each of the following angles:

Tangent	Sine	Cosine
a. 10°	g. 10°	m. 10°
b. 20°	h. 20°	n. 20°
c. 89°	i. 25°	o. 39°
d. 1°	j. 29°	p. 59°
e. 35°	k. 14°	q. 81°
f.	l.	r.

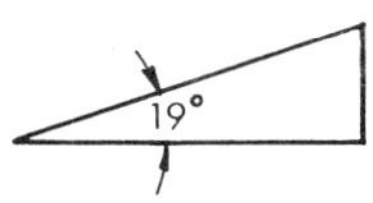

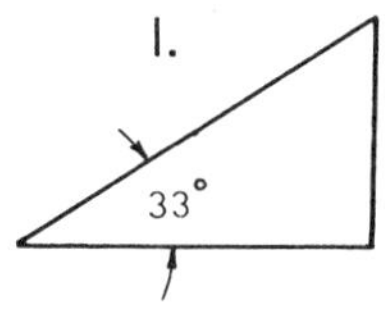

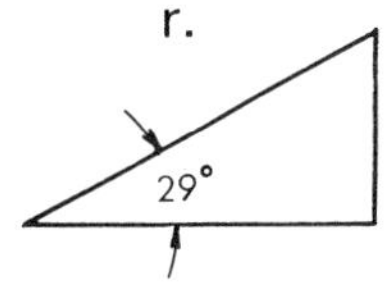

2. Find the value of the angle corresponding to each of the following numbers (ratios):

Tangent	Sine	Cosine
a. .5774	h. .2588	o. .6428
b. .7265	i. .5736	p. .3420
c. 11.4301	j. .9848	q. .9998
d. 1.800	k. .9972	r. .9830
e. 1.501	l. .4300	s. .4100
f. 1/2	m. 1/4	t. 7/8
g. 2/4	n. 3/4	u. 5/8

UNIT 88 USING THE TANGENT FORMULA

The *tangent* formula is used to find the acute angle of a right triangle if the opposite side and the adjacent side are known, or to find the side of a right triangle if the acute angle opposite this side and the adjacent side are known.

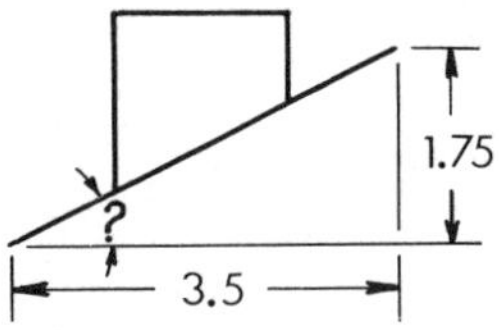

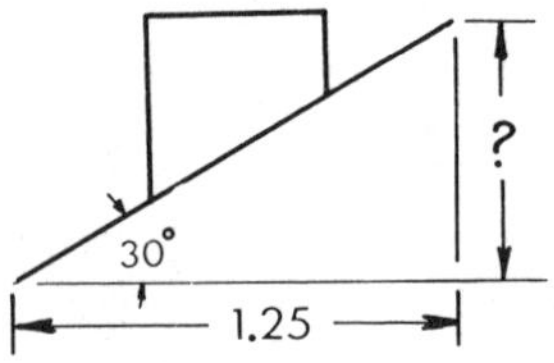

COMPUTATION OF ANGLE

Example: a = 2¼″, b = 1¾″

1. Formula $\tan A = \frac{\text{opp.}}{\text{adjac.}}, \frac{a}{b}$

2. Substitution $\tan A = \frac{2\frac{1}{4}}{1\frac{3}{4}}$

3. Evaluation $\tan A = \frac{2.25}{1.75}$, or 1.28571

$\tan A = 1.2857$

$\therefore$ Angle A = 52° Ans.

Note: Check this answer by measuring angle A of the triangle.

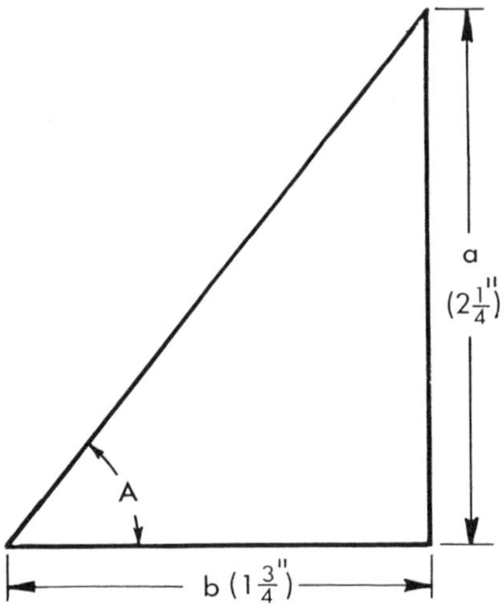

COMPUTATION OF OPPOSITE SIDE

Example: Angle A = 40°, b = 2″

1. Formula $\tan A = \frac{\text{opp.}}{\text{adjac.}}$, or $\frac{a}{b}$

2. Substitution $\tan 40° = \frac{a}{2}$

3. Evaluation $\frac{a}{2} = .8391$
(tan 40° = .8391)

$\frac{2 \times a}{2} = .8391 \times 2$

$a = 1.6782''$

$a = 1.678''$, or $1\frac{43''}{64}$ Ans.

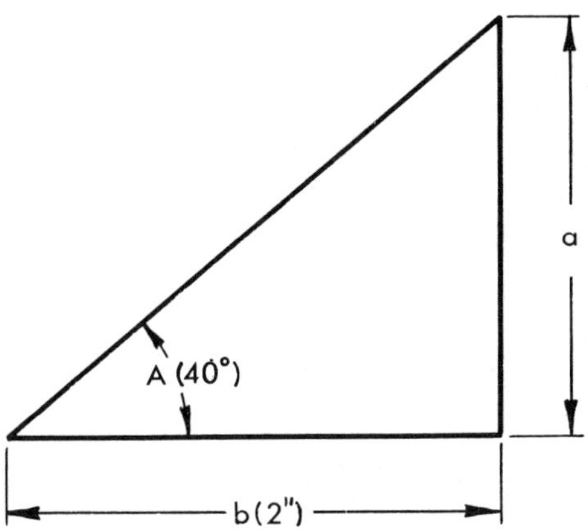

Note: Check this answer by measuring altitude *a* of the triangle.

ASSIGNMENT

A. Computation of Angle

Compute the value of angle A in each of the following right triangles. Check each answer by constructing the triangle and measuring the angle A.

1. a = 3″, b = 4″
2. a = 4″, b = 2½″
3. a = 4.5″, b = 1.5″
4. a = 5″, b = 3.25″

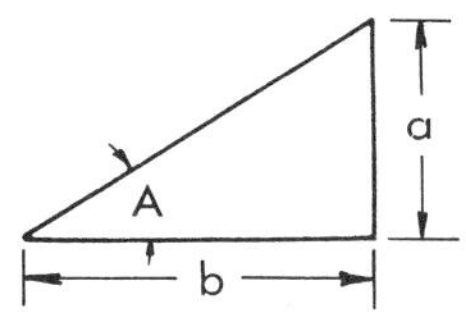

B. Computation of Opposite Side

Compute the length of side *a* of each triangle to the accuracy indicated. Check problems 1 to 4 by constructing each triangle and measuring its altitude.

1. ∠A = 20°, b = 5″, (32nds)
2. ∠A = 30°, b = 4″, (32nds)
3. ∠A = 35°, b = 3″, (32nds)
4. ∠A = 49°, b = 3″, (64ths)
5. ∠A = 50°, b = 19″, (16ths)
6. ∠A = 42°, b = 25½″ (16ths)
7. ∠A = 9°, b = 2.58″, (100ths)
8. ∠A = 81°, b = 8′ 6″, (16ths)

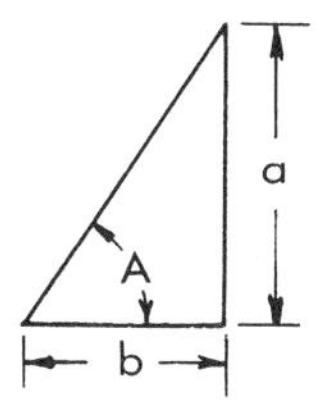

C. Trade Problems

1. Compute the value of angle A.

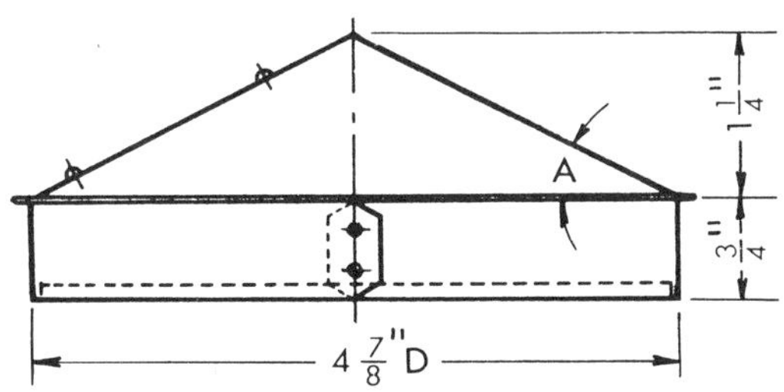

2. Compute length of *a* (32nds).

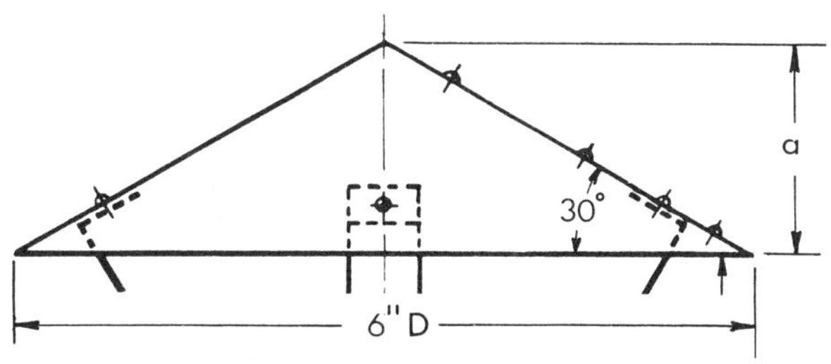

3. Compute the value of angle A.

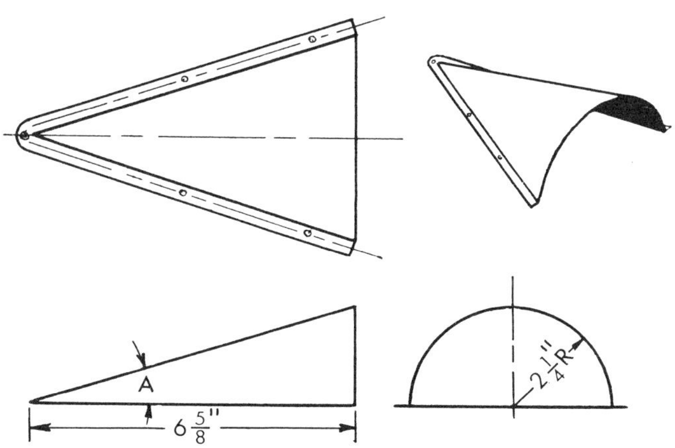

4. Compute length of *y* (32nds). Find the length of the base of the right triangle by adding 3.5 and 2.5 ÷ 2.

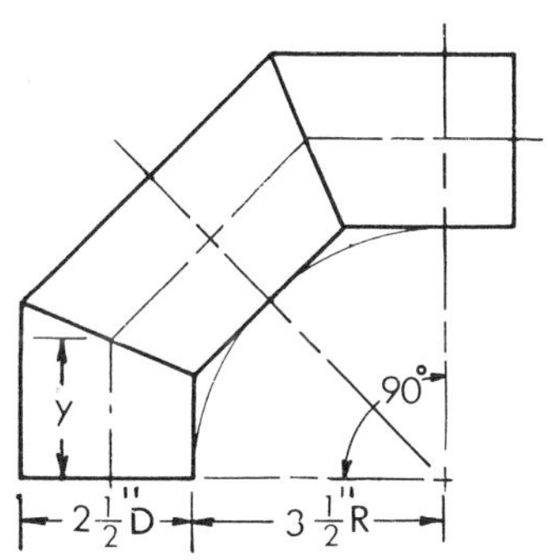

UNIT 89 USING THE SINE FORMULA

The *sine* formula is used to find the acute angle of a right triangle if the opposite side and the hypotenuse are known, or to find the side if the acute angle opposite this side and the hypotenuse are known.

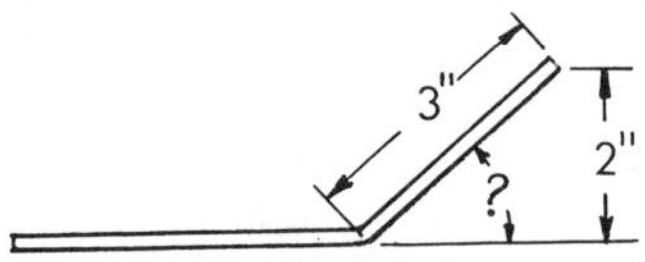

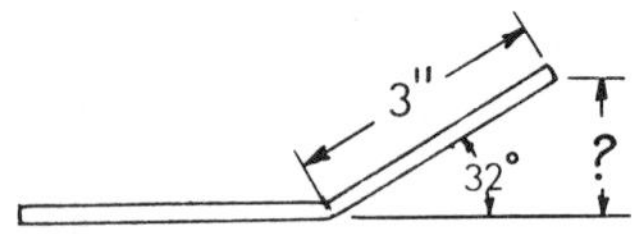

COMPUTATION OF ANGLE

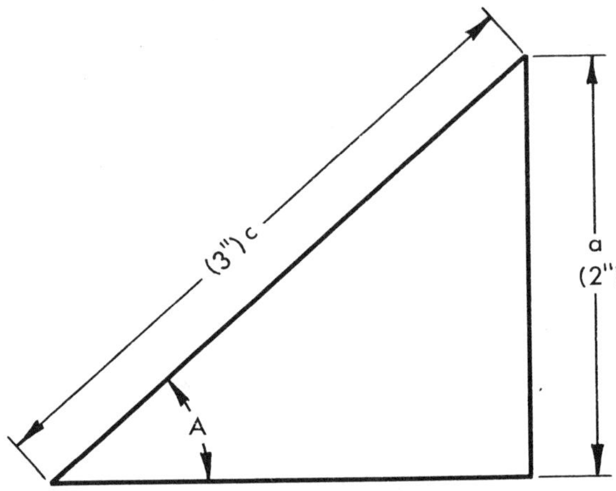

Example: A = 2″, c = 3″

1. Formula $\sin A = \frac{\text{opp.}}{\text{hypot.}}$, or $\frac{a}{c}$
2. Substitution $\sin A = \frac{2}{3}$
3. Evaluation $\sin A = \frac{2}{3}$, .66666

$\sin A = .6667$

$\therefore$ Angle A = 42° Ans.

Note: Check this answer by measuring angle A of the triangle.

COMPUTATION OF OPPOSITE SIDE

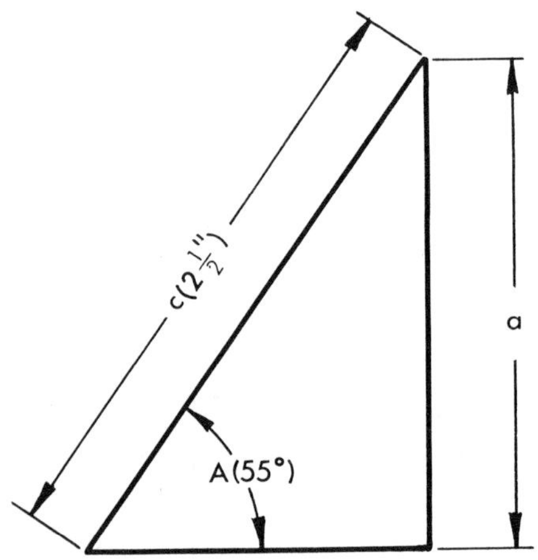

Example: Angle A = 55°, c = 2½″

1. Formula $\sin A = \frac{\text{opp.}}{\text{hypot.}}$, or $\frac{a}{c}$
2. Substitution $\sin 55° = \frac{a}{2\frac{1}{2}}$
3. Evaluation $.8192 = \frac{a}{2.5}$

$$\frac{a}{2.5} = .8192$$

$$\frac{2.5 \times a}{2.5} = .8192 \times 2.5$$

$$a = 2.04800''$$

$$a = 2.048'', \text{ or } 2\frac{3''}{64} \text{ Ans.}$$

Note: Check this answer by measuring altitude *a* of the triangle.

ASSIGNMENT

A. Computation of Angle

Compute the value of angle A in the following right triangles. Check the answers of problems 1 to 3 by constructing each triangle and measuring the angle A.

1. a = 1.5″, c = 2.5″
2. a = 3.0″, c = 5.0″
3. a = 2¼″, c = 3½″
4. a = 20.2″, c = 45.8″
5. a = 3′ 9″, c = 5′ 6″
6. a = 4′ 8″, c = 9′ 7″

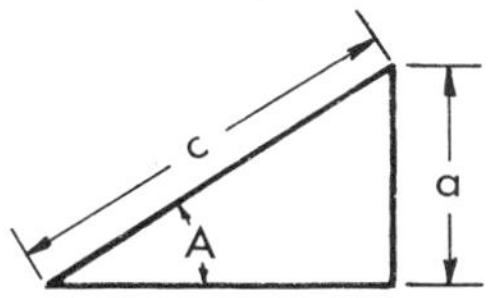

B. Computation of Opposite Side

Compute the length of side *a* of each triangle to the accuracy indicated (see triangle above for meaning of letters). Check problems 1 to 4 by constructing each triangle and measuring the length of the altitude.

1. ∠A = 60°, c = 3″, (32nds)
2. ∠A = 60°, c = 4″, (32nds)
3. ∠A = 30°, c = 3″, (32nds)
4. ∠A = 21°, c = 5″, (64ths)
5. ∠A = 69°, c = 12½″, (16ths)
6. ∠A = 75°, c = 16½″, (16ths)
7. ∠A = 54°, c = 3.02″, (100ths)
8. ∠A = 38°, c = 2′ 9″, (16ths)

C. Trade Problems

1. Compute height *a* (64ths).

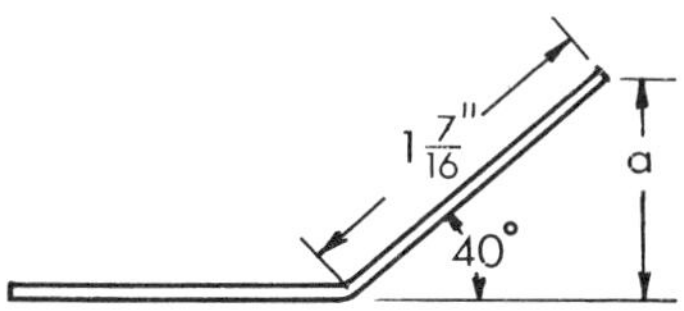

2. Compute the value of angle A.

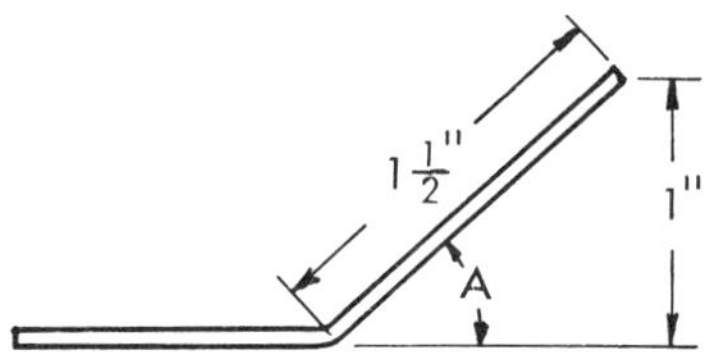

3. Compute length of *a* (16ths).

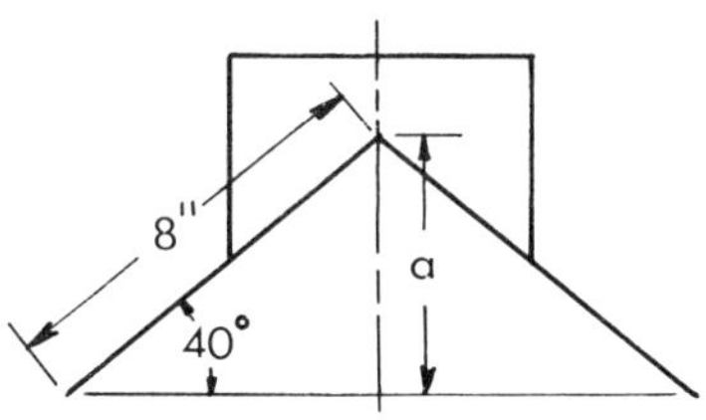

4. Compute distance *a* (16ths)

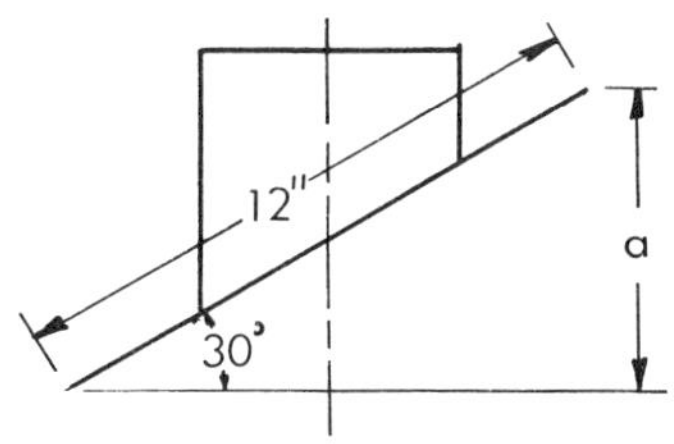

5. Compute the value of angle A.

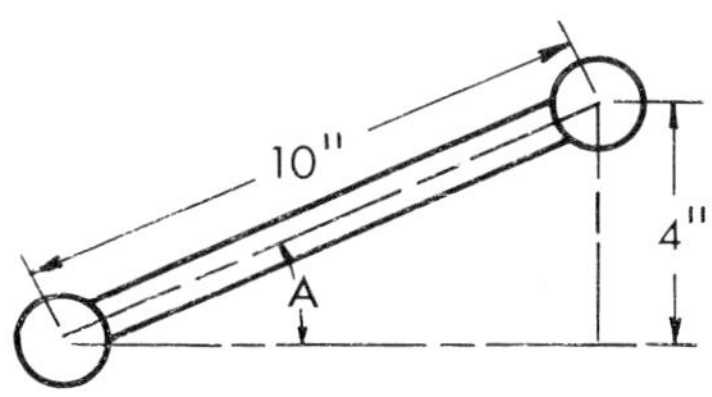

6. Compute the value of angle A.

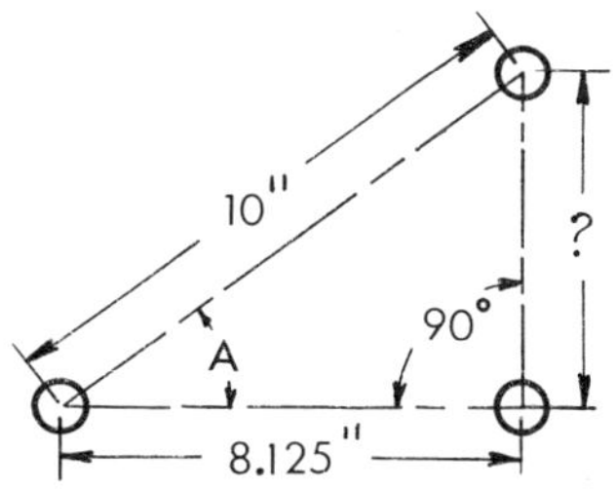

UNIT 90 USING THE COSINE FORMULA

The *cosine* formula is used to find the acute angle of a right triangle if the adjacent side and hypotenuse are known, or to find the side if the acute angle adjacent to this side and the hypotenuse are known.

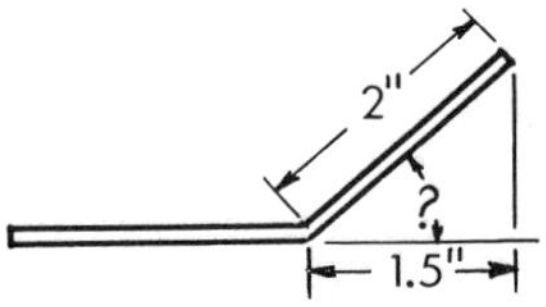

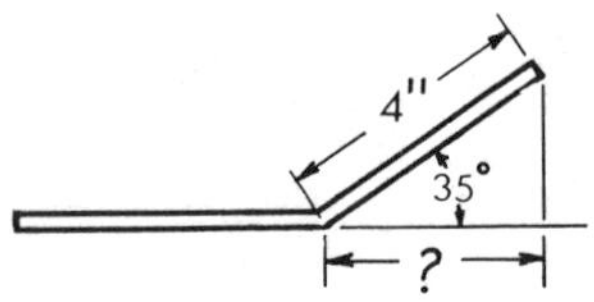

COMPUTATION OF ANGLE

Example: b = 2¼", c = 3"

1. Formula $\cos A = \dfrac{\text{adjac.}}{\text{hypot.}}, \text{ or } \dfrac{b}{c}$

2. Substitution $\cos A = \dfrac{2\frac{1}{4}}{3}$

3. Evaluation $\cos A = \dfrac{2.25}{3}$

$\cos A = .7500$

$\therefore$ Angle A $= 41°$ Ans.

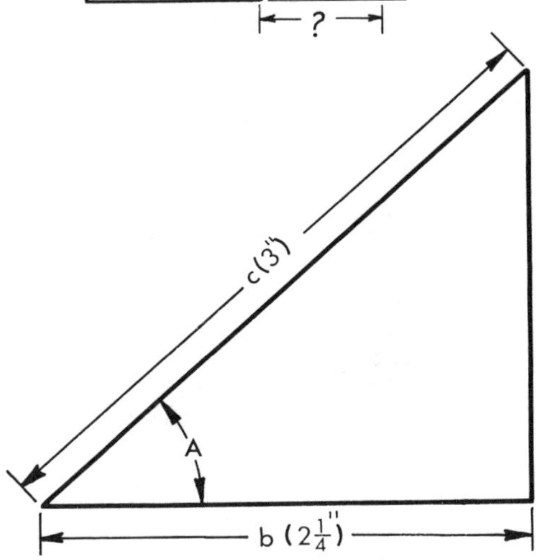

Note: Check this answer by measuring angle A of the triangle.

COMPUTATION OF ADJACENT SIDE

Example: Angle A = 58°, c = 3¼"

1. Formula $\cos A = \dfrac{\text{adjac.}}{\text{hypot.}}, \text{ or } \dfrac{b}{c}$

2. Substitution $\cos 58° = \dfrac{b}{3\frac{1}{4}}$

3. Evaluation (cos 58° = .5299) $.5299 = \dfrac{b}{3.25}$

$\dfrac{b}{3.25} = .5299$

$\dfrac{3.25 \times b}{3.25} = .5299 \times 3.25$

$b = 1.72217''$

$b = 1.722''$, or $1\dfrac{23''}{32}$

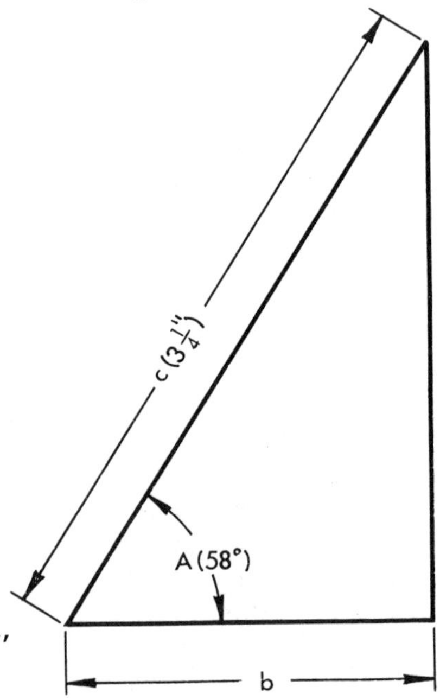

Note: Check this answer by measuring base *b* of the triangle.

ASSIGNMENT

A. Computation of Angle

Compute the value of angle A in the following right triangles. Check the answers of the first three problems by constructing the triangle and measuring the angle A.

1. c = 4″, b = 3″
2. c = 3″, b = 2.5″
3. c = 3 ½″, b = 2 ¾″

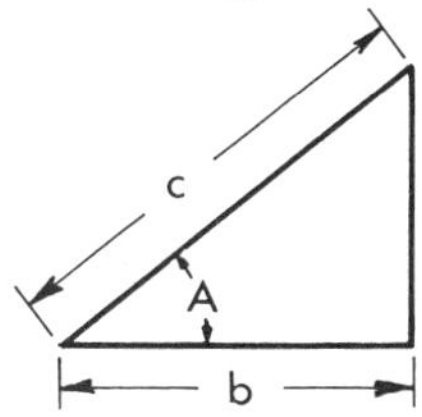

4. c = 50 ¼″, b = 39 ¾″
5. c = 2.02″, b = 1.53″
6. c = 10′ 3″, b = 5′ 4″

B. Computation of Adjacent Side

Compute length of side *b* of each triangle to the accuracy indicated (see triangle above for meaning of letters). Check the first four problems by constructing each triangle and measuring the length of the base.

1. ∠A = 60°, c = 3″ (32nds)
2. ∠A = 11°, c = 4″, (32nds)
3. ∠A = 40°, c = 4 ¼″, (32nds)
4. ∠A = 45°, c = 3 ½″, (64ths)
5. ∠A = 27°, c = 15 ¾″, (16ths)
6. ∠A = 44°, c = 10.22″, (50ths)
7. ∠A = 20°, c = 28′ 3″, (16ths)
8. ∠A = 31°, c = 14′ 8″, (16ths)

C. Trade Problems

1. Compute the value of angle A.

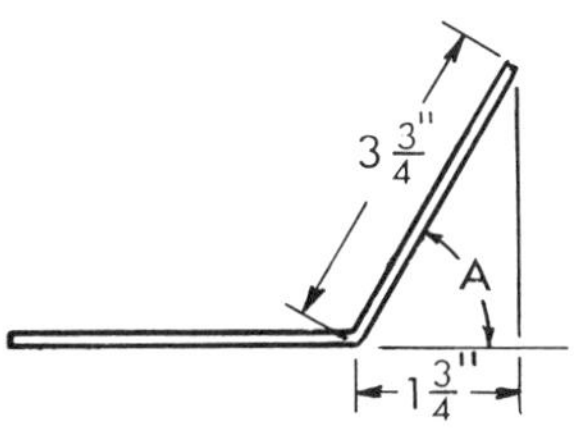

2. Compute length of *b* (64ths).

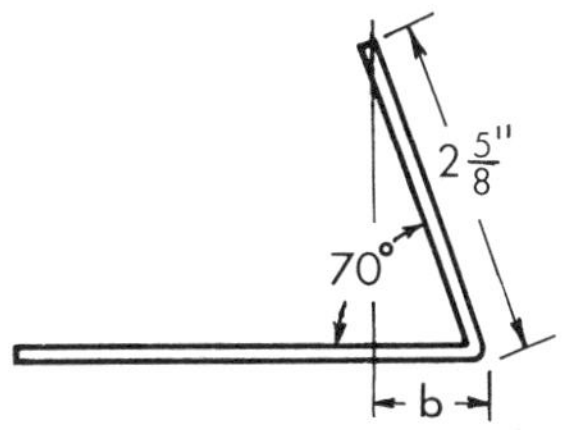

3. Compute length of *b* (16ths).

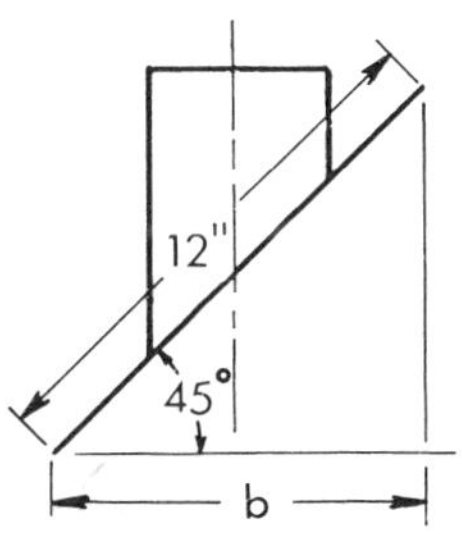

4. Compute length of *b* (16ths). Find the length of the base of the right triangle; double this result to obtain *b*.

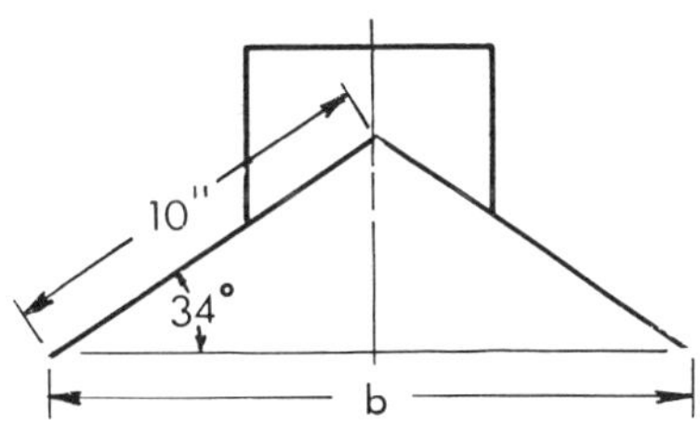

UNIT 91 SELECTION OF FORMULAS

In order to select the proper trigonometric formula for a problem, it is necessary to label a right triangle correctly. The 90-deg. angle is always labeled capital C; the hypotenuse is labeled with a small letter *c*. In a similar manner, one of the acute angles is labeled A and the side opposite *a*; the other angle B and the side opposite *b*.

Formulas:

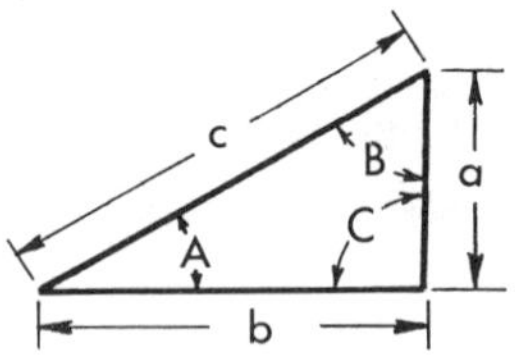

$$\tan A = \frac{a}{b}$$

$$\sin A = \frac{a}{c}$$

$$\cos A = \frac{b}{c}$$

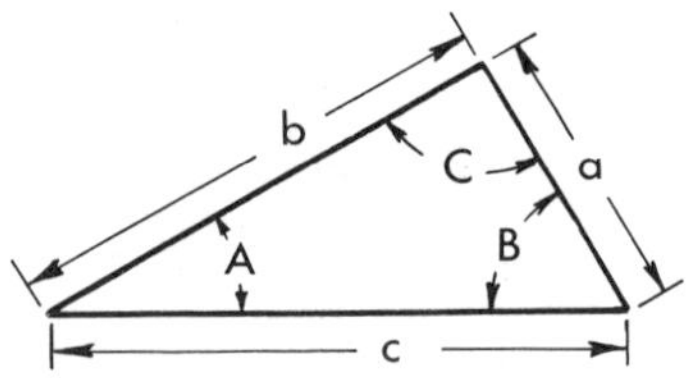

COMPUTATION OF ACUTE ANGLES

Known Data: Two sides of a right triangle, or one side and the hypotenuse

1. Draw the right triangle for the problem and label it.
2. Write the names of the two known quantities, in fractional form. In this problem, ∠ A is wanted; the adjacent side (2 in.) and the hypotenuse (5 in.) are known.
3. Select the formula which applies. If necessary, invert the fraction in order to use the tangent, sine, or cosine formula. Use the cosine formula for this problem.

   Note: To find ∠ B, subtract ∠ A from 90°.

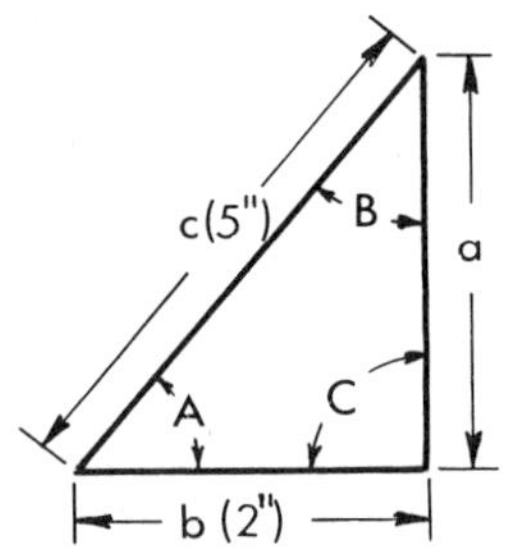

$$\frac{\text{adjac.}}{\text{hypot.}} = \frac{b}{c}$$

$$\cos A = \frac{b}{c}, \text{ or } \frac{2}{5}$$

COMPUTATION OF SIDES

Known Data: One side and one acute angle of a right triangle, or the hypotenuse and one acute angle

1. Draw the right triangle and label it.
2. Write the name of the side to be found as a numerator and the name of the known side as a denominator. In this problem, side *a* is wanted; the hypotenuse (3 in.) and ∠ A (40 deg.) are known.
3. Select the proper formula. Use the sine formula for this problem.

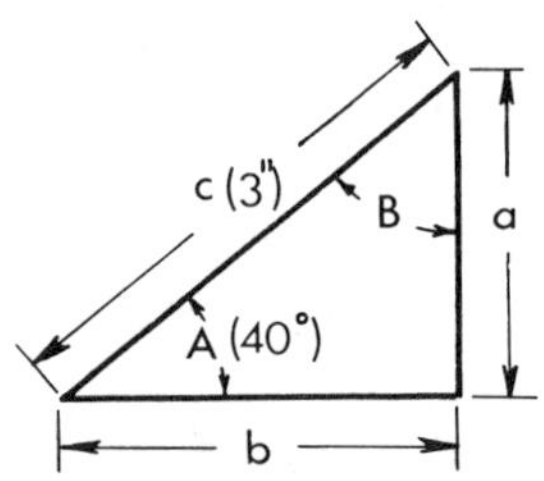

$$\frac{\text{opp.}}{\text{hypot.}} = \frac{a}{c}$$

$$\sin A = \frac{a}{c}, \text{ or } \sin 40^\circ = \frac{a}{3}$$

ASSIGNMENT

Using the right triangles shown, solve the following problems to the accuracy indicated.

1. Find the distances *x* and *y* (64ths) of the fitting shown below.

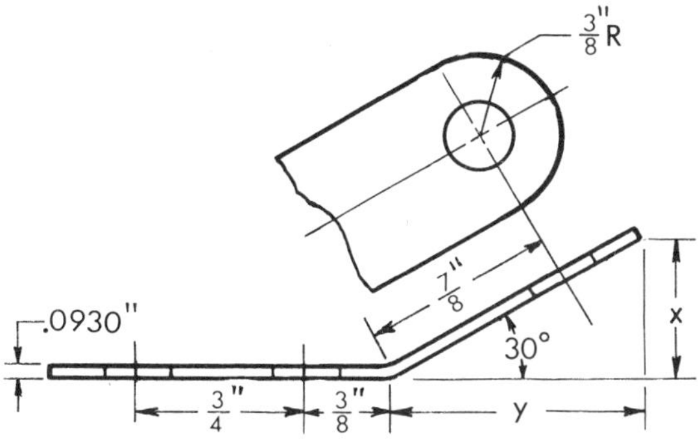

2. Compute the value of angle A of the cover.

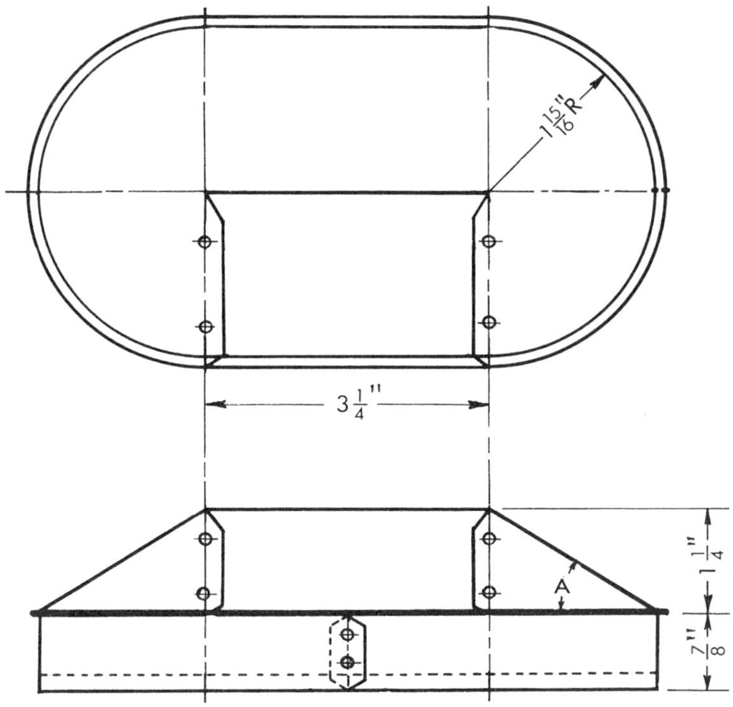

3. Compute the heights *m* and *n* (32nds) of the elbow. (Find angle A and then side *a*.)

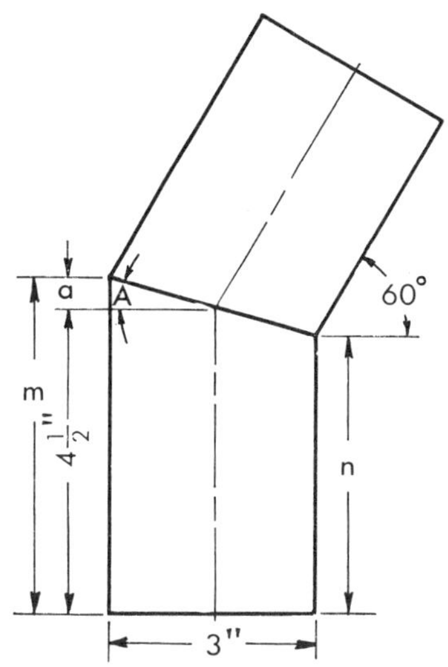

4. The span of a roof is 24 ft. and the pitch is 1/3. Compute the angle of slope of the roof. (Find the rise of the roof before proceeding with the problem.)

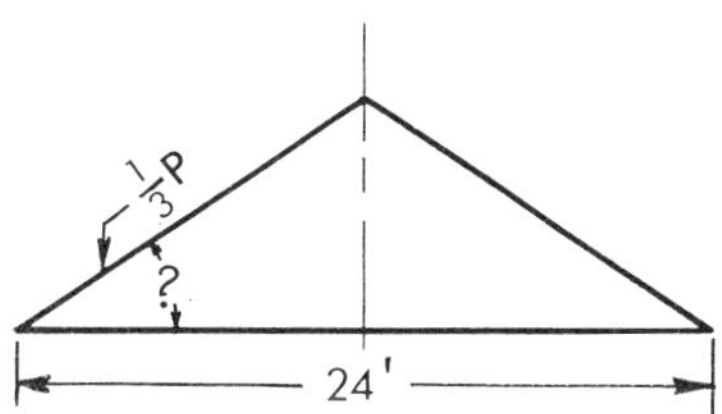

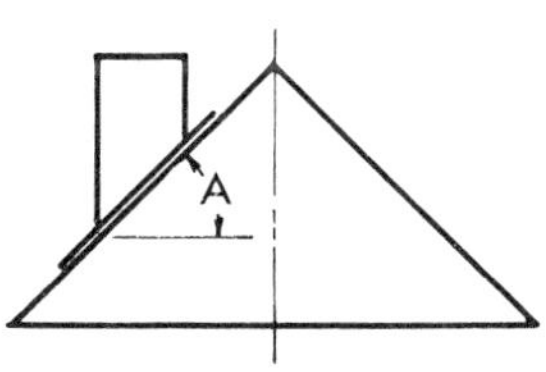

5. The span of the roof is 22 ft.; the pitch is 1/2. Compute the angle A of the roof flange at the left.

6. Compute distances *m* and *n* (32nds). (Find ∠ A and side *a*.)

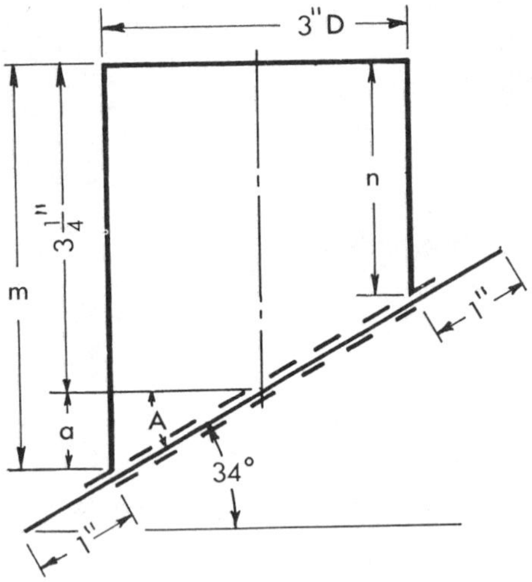

7. Compute the values of the angles A and B. $b = \dfrac{D - d}{2}$

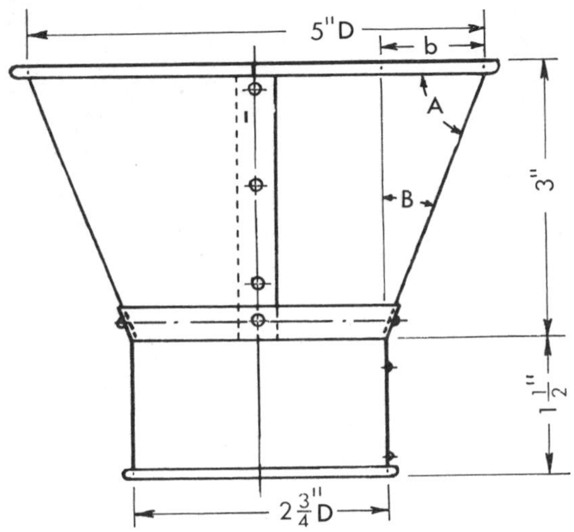

8. Compute distance *x* (32nds). (Find ∠ A and side *a*.)

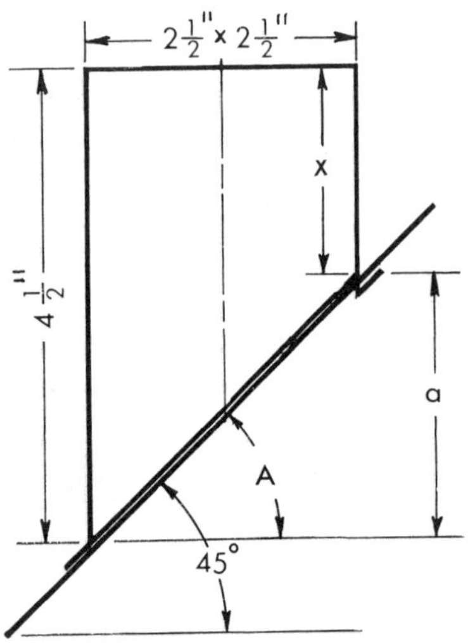

9. Compute distance *y* (32nds). (Find ∠ A and then side *a*.)

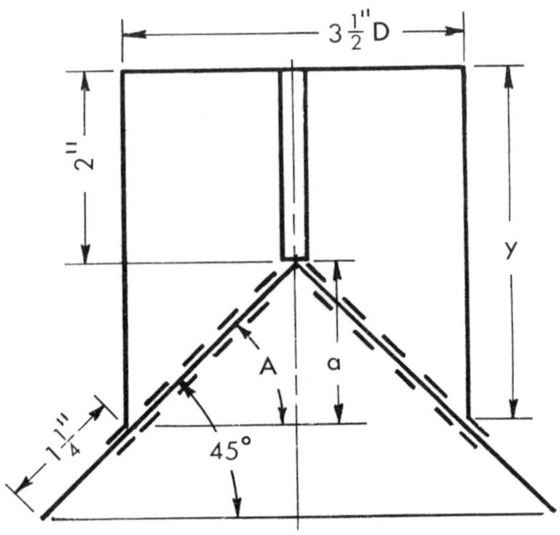

10. Compute the hole spacing *a* (64ths) of the plate shown at the right. The diameter of the large circle is 10 in. (Find ∠ A and then a/2.)

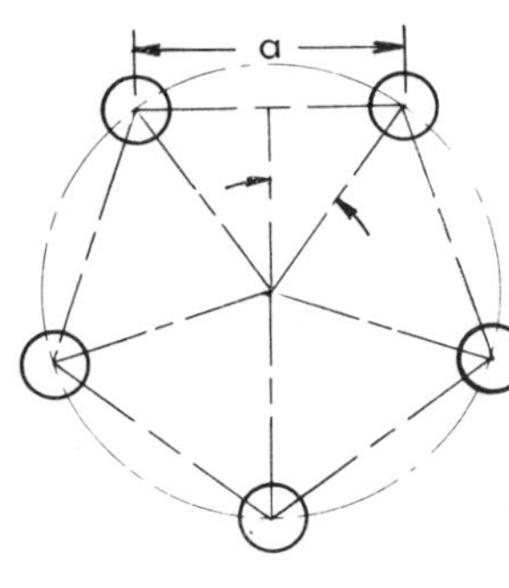

SECTION 12 GRAPHS

UNIT 92 BAR GRAPHS

Graphs are used to show statistical data pictorially. They often show relationships, especially increases and decreases, more readily than numbers. The dates and quantities used in the following graphs and problems are for illustrative purposes only.

There are several types of graphs, such as picture graphs, bar graphs, circle graphs, and line graphs. Each type has its particular advantage. The picture graph is perhaps the simplest graph to read, but the symbols are usually difficult to draw. Furthermore, the picture graph can be only approximately accurate. When constructing graphs, the use of squared paper or graph paper will make counting and measuring of distances more convenient and accurate.

The use of bar graphs permits the more accurate representation of data than the picture graph. In the bar graph the lengths of the bars represent the numbers (data).

Example: Using the following data, make a bar graph showing the aircraft production of the ABC Corporation:

Year		*Aircraft*	*Year*		*Aircraft*
1970	–	453	1973	–	1903
1971	–	1507	1974	–	1850
1972	–	1800	1975	–	1750

PROCEDURE

1. Arrange the data according to some sequence: according to years, from smallest to largest, or from largest to smallest.
2. Choose a vertical scale so that the largest number will not make the graph too large and yet the smallest number can be represented. In this example, let one small square represent fifty aircraft. Since 1903/50 equals 38, thirty-eight small squares will represent 1903 aircraft. Similarly, 453/50 equals 9 small squares.

3. Print the vertical scale at one side of the graph, numbering the large squares only. Since 5 small squares (50 aircraft) equal one large square, then each large square will represent 250 aircraft.

4. Choose the proper distance (horizontal) between bars so that the graph will look well balanced, and print the number or name of each bar under it. (Usually the distance between the bars should be equal to the width of the bars.)

5. Using the original data, compute the length of all the bars and draw the bars the proper lengths. The number of small squares can be changed to large squares by dividing each number by 5, that is, 9 small squares equal 1 large square plus 4 small squares.

453/50 = 9 small squares	1903/50 = 38 small squares
1507/50 = 30 small squares	1850/50 = 37 small squares
1800/50 = 36 small squares	1750/50 = 35 small squares

6. Print the title of the graph in a suitable place.

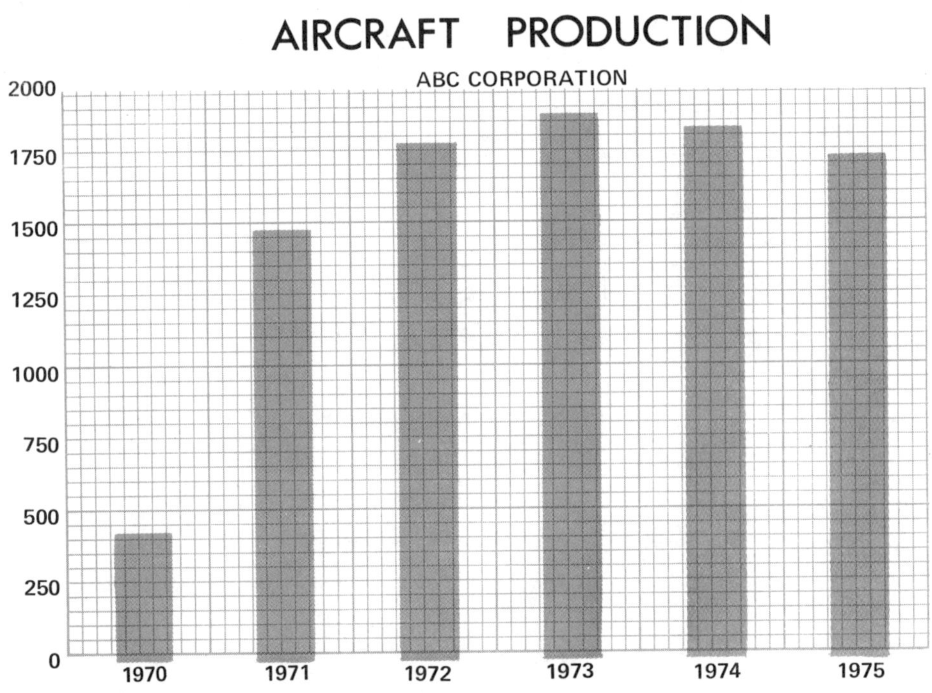

ANALYSIS OF BAR GRAPH -- AIRCRAFT PRODUCTION

1. Which year shows the greatest increase in production? Answer: 1971

2. Which year shows the maximum (greatest) production? Answer: 1973

3. Which year shows the minimum (least) production? Answer: 1970

4. What is the approximate ratio of the maximum and the minimum production? Answer: 4 to 1

ASSIGNMENT

1. Using the following data, make a bar graph showing the production of an industrial firm:

Year		*Units*	*Year*		*Units*
1970	–	100	1973	–	740
1971	–	400	1974	–	820
1972	–	580	1975	–	900

2. Make a bar graph showing the following accidents incurred with hand tools. (Use the letters A, B, etc. to represent the different types of accidents.)

A.	Unsafe or improper use of equipment	59%
B.	Overloading, poor arranging.	19%
C.	Unnecessary exposure to danger	17%
D.	Nonuse of personal protective equipment . .	2%
E.	All others.	1%
F.	No unsafe acts	2%

3. Using the graph below, answer the following questions:
   a. Which year showed the greatest production of welding wire?
   b. Which year showed the least production of welding wire?
   c. What is the approximate ratio of the maximum amount of welding wire produced to the minimum amount produced?
   d. Make a list of the pounds of welding wire produced each year.

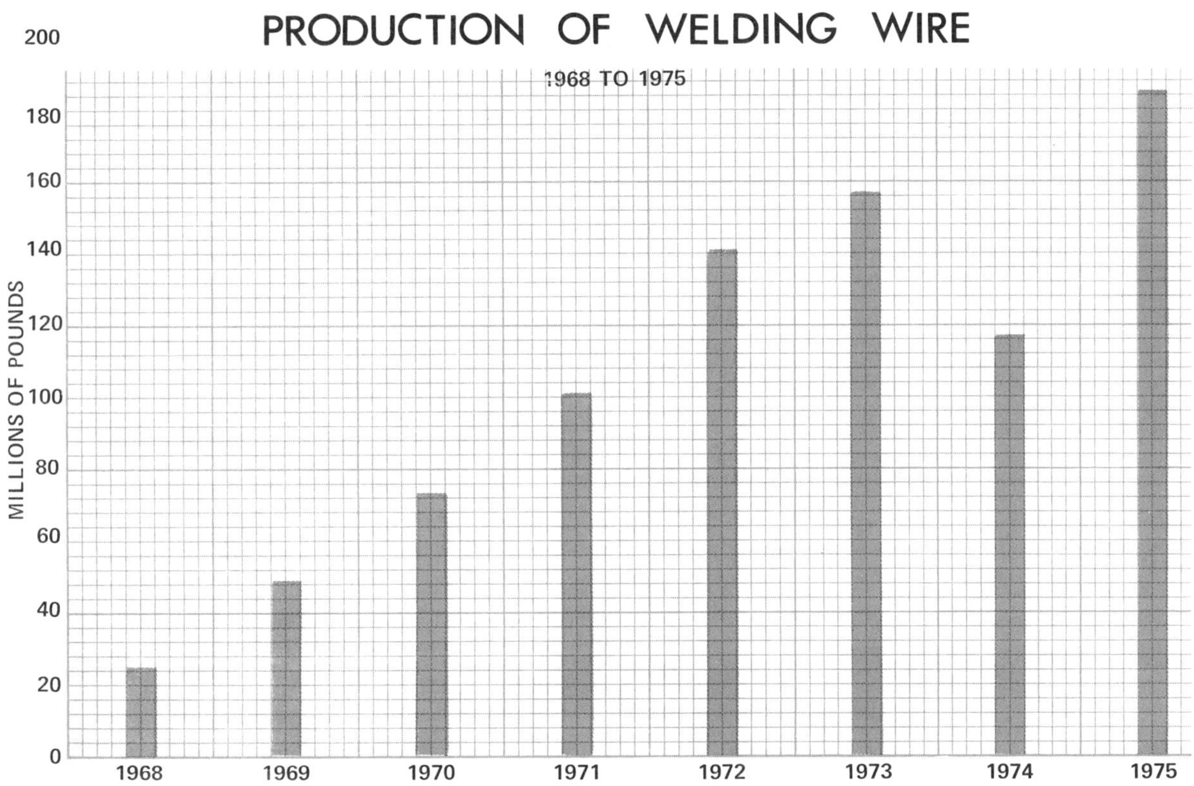

UNIT 93 CIRCLE GRAPHS

Circle graphs are useful in showing plainly the relationship of certain items to each other, and especially in showing the relationship of certain parts or items in respect to the whole.

Example: Using the following data, construct a circle graph showing the relative numbers of the deaths from the various kinds of accidents in 1972: Home – 6,000; Motor Vehicle – 16,500; Occupational – 18,500; Public (except motor vehicle) – 6,500.

PROCEDURE

1. Find the sum of the items.
2. Find the decimal part, and the percent that each item is of the whole. The sum of the percents should equal 100.0%.

   Note: The item, .3894, was made equal to .390 so the sum would equal 100%.
3. Multiply the decimal equivalent of each item by 360 deg. to find its angular equivalent. The sum of the angles should equal 360 deg.

   Note: The angle, 140.4 deg., was made 141 deg. so that the sum would equal 360 deg.
4. Draw a circle of convenient size and lay out in this circle the central angle corresponding to each item.
5. Label each angular part with its name and percent. Give the graph a title.

Total accidents = 47,500

$$\frac{6,000}{47,500} = .1263, \text{ or } 12.6\%$$
$$\frac{16,500}{47,500} = .3473, \text{ or } 34.7\%$$
$$\frac{18,500}{47,500} = .3894, \text{ or } 39.0\%$$
$$\frac{6,500}{47,500} = .1368, \text{ or } \underline{13.7\%}$$
$$100.0\%$$

$$.126 \times 360^\circ = 45^\circ$$
$$.347 \times 360^\circ = 125^\circ$$
$$.390 \times 360^\circ = 141^\circ$$
$$.137 \times 360^\circ = \underline{49^\circ}$$
$$360^\circ$$

DEATHS DUE TO ACCIDENTS

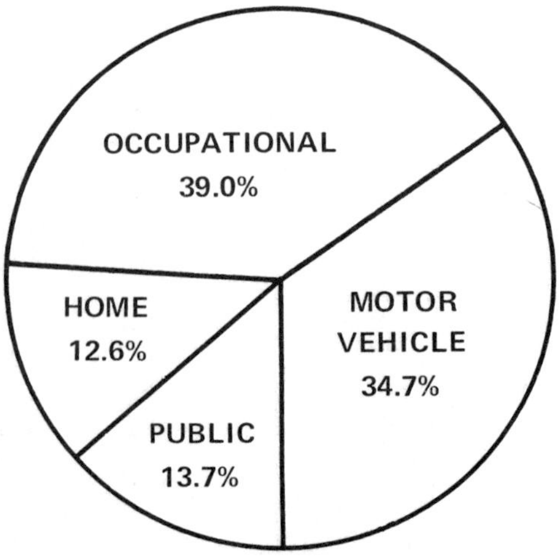

1972

ANALYSIS OF CIRCLE GRAPH

1. What is the approximate relationship of the number of deaths due to occupational accidents and those due to motor vehicle accidents? Answer: approximately equal
2. What is the approximate fractional relationship of the deaths caused by motor vehicles to the total number of deaths due to accidents? Answer: 1/3

ASSIGNMENT

1. During the school year, 1969 – 1970, 5703 boys were registered in the vocational schools of a certain city, and 5298 boys were registered in the academic schools of the same city. Construct a circle graph to represent the above data. Using this graph, answer the following questions:
   a. What approximate fractional part of the total boys attend a vocational school?
   b. What approximate fractional part of the total boys attend an academic school?
   c. What is the approximate ratio of the boys who attend a vocational school to those who attend an academic school?
2. Deaths due to various types of accidents in 1974 are as follows:

Pedestrian	13,550
Two-vehicle	12,500
Railroad	1,840
Others	12,080

   Construct a circle graph to represent this data and then answer the questions below:
   a. What is the approximate ratio of the pedestrian deaths and the two-vehicle deaths?
   b. What is the approximate ratio of the pedestrian deaths to the total deaths, the two-vehicle to the total?
3. Draw a circle graph to represent the steel production of various countries in 1975 and then answer the questions.

Russia	13%
France	6%
Great Britain	10%
Germany	19%
United States	35%
All others	17%

   a. What was the approximate ratio of the steel production of the United States to that of Germany?
   b. What was the approximate ratio of the steel production of the United States to that of Russia?
   c. What was the approximate ratio of the steel production of the United States to the total production?

UNIT 94 LINE GRAPHS

Three types of line graphs and their general use are shown below:

BROKEN LINE	STRAIGHT LINE	CURVED LINE
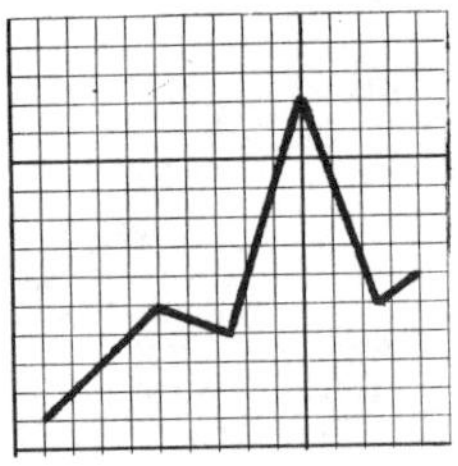	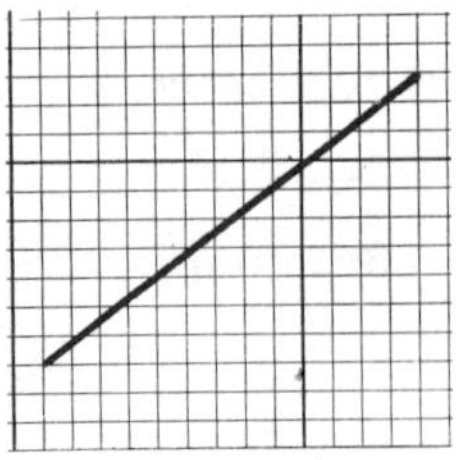	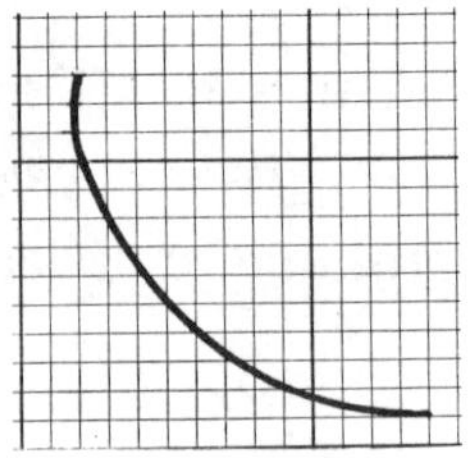
Unrelated data, such as temperature, production, etc.	Related data, such as the surface speeds of grinding wheels	Related data, such as cross-section areas of square pipe.

Broken-line graphs are often used for representing the same kind of data as bar graphs, but they show rates of change better than bar graphs. They are made in a somewhat similar manner except that the points are connected with straight lines as shown below.

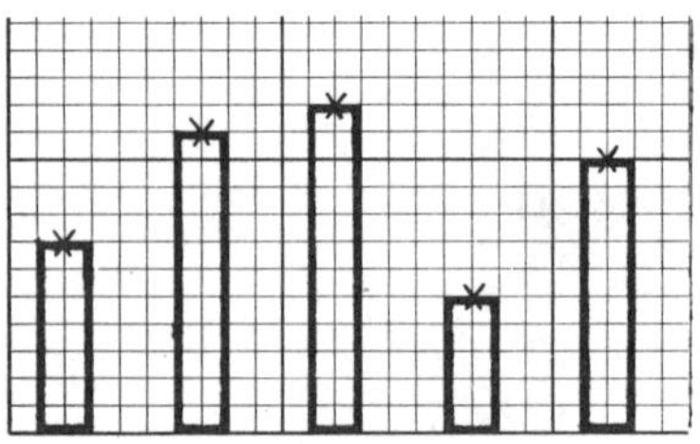

BAR GRAPH

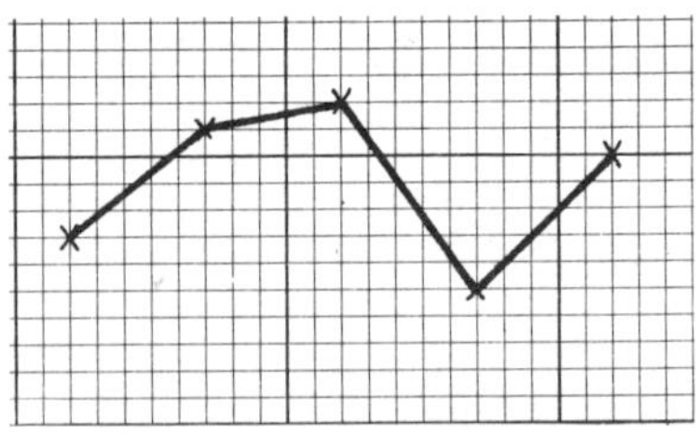

BROKEN–LINE GRAPH

Straight-line and curved-line graphs are used to picture data which is obtained from formulas, or other related sources. After such a graph has been made, other values can be obtained from the graph without additional computations.

Example: Make a graph showing the surface speeds of a grinding wheel turning at 2000 r.p.m. Use wheels varying from 12 in. dia. to 6 in. dia. with 1 in. intervals.

PROCEDURE

1. Using the formula $S = \pi dn/12$, compute values of the surface speeds for 12″, 11″, 10″, 9″, 8″, 7″, 6″ diameters to the closest r.p.m.
2. List these values in tabular form and round off each surface speed to the nearest 50 r.p.m. (See the table on the next page.)

Dia. in (″)	f.p.m.	Dia. in (″)	f.p.m.
12	6280 = 6300	8	4187 = 4200
11	5757 = 5750	7	3667 = 3650
10	5233 = 5250	6	3140 = 3150
9	4710 = 4700		

3. Choose suitable vertical and horizontal scales so that the graph will be approximately square-shaped. Label the graph with these scales. In this graph, 10 small squares vertically equals 1 in. in diameter and 10 small squares horizontally equals 500 f.p.m.
4. Plot (locate) the successive points as indicated by the pairs of numbers listed above. The first point is d = 12″, s = 6300 f.p.m. Connect the points with a continuous line (straight or curved).

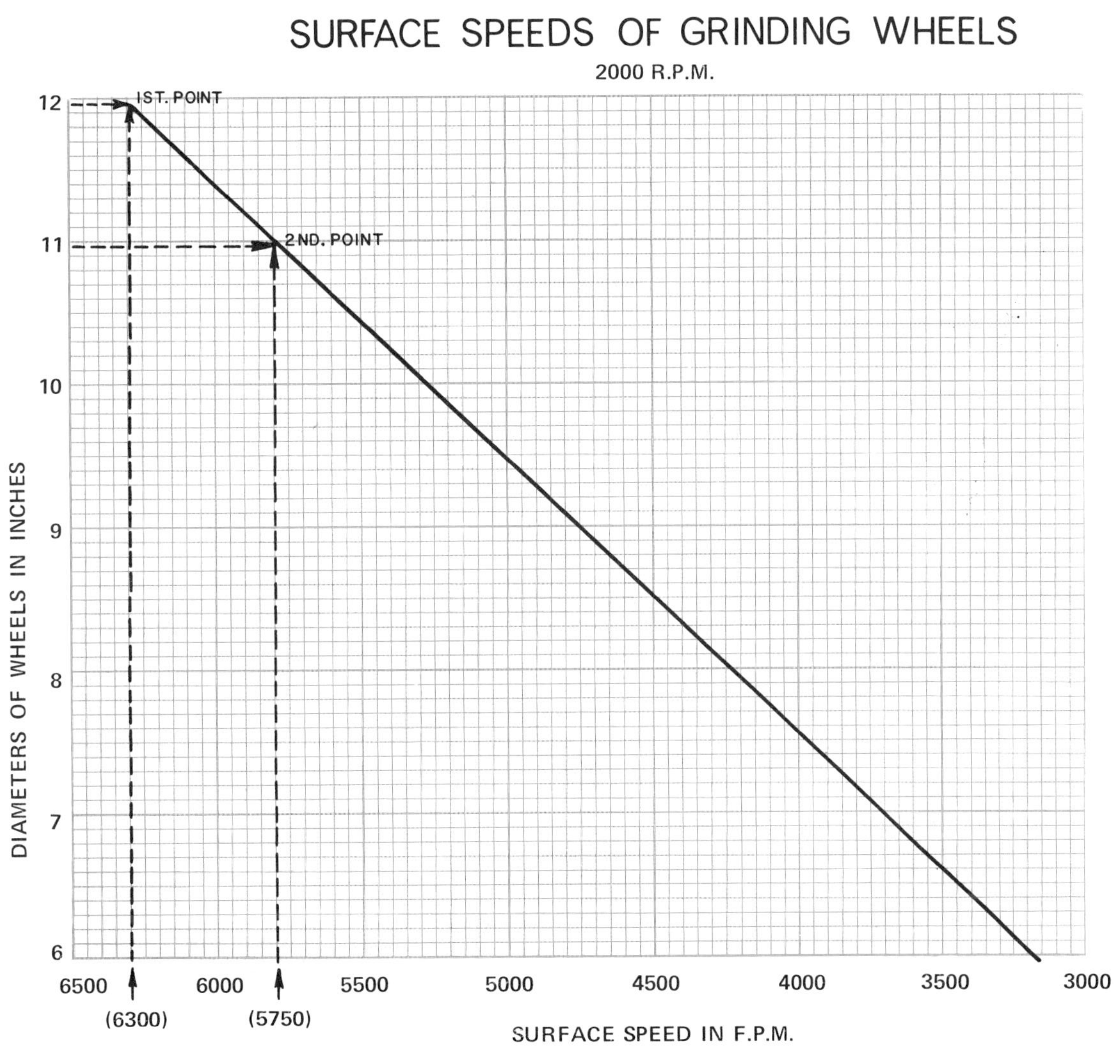

READING A LINE GRAPH

1. Line graphs may be used to find approximate values without computation.
   a. To obtain the surface speed for a 10.5-in. wheel rotating at 2000 r.p.m., read the points A and B as shown in the graph below. The surface speed for a 10.5-in. wheel rotating at 2000 r.p.m. is 5500 f.p.m. Check this result by computation.
   b. To obtain the diameter for a surface speed of 4000 f.p.m. if the wheel rotates at 2000 r.p.m., read the points C and D as shown in the graph below. The diameter is 7.6 in. for a wheel rotating at 2000 r.p.m. Check this result by computation.

SURFACE SPEEDS OF GRINDING WHEELS

2000 R. P. M.

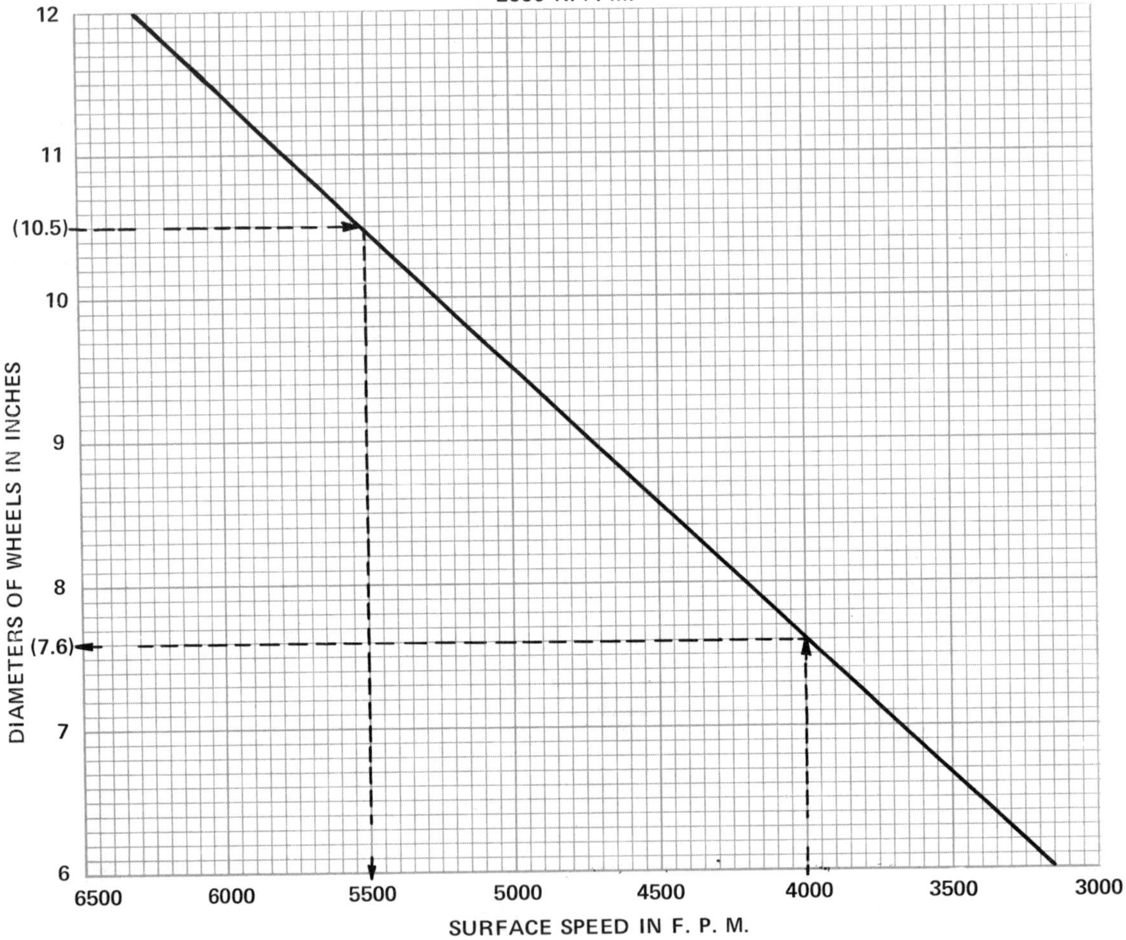

2. A line graph also shows variations or changes. For instance, the graph shows clearly that the surface speed decreases as the diameter of the wheel decreases.

ASSIGNMENT

1. Make a line graph showing a company's annual production of trucks.

Year		*Trucks*		*Year*		*Trucks*
1975	–	5500	–	1972	–	4815
1974	–	6325	–	1971	–	4205
1973	–	5250	–	1970	–	5200

   a. Which type of line graph is this?

   b. Was there an increase in production each year? Which year had the greatest increase in production?

2. Make a line graph showing the lengths of sides of square pipes in relation to their cross-section areas. Use the following lengths of sides: 1″, 2″, 3″, 4″, 5″, 6″, 7″, 8″, 9″.

   a. Find the areas of the following pipes from the graph: s = 2.5″; s = 6.5″; s = 8.7″.

   b. Check these values by computation. Is this graph accurate enough for your work?

   c. What type of line graph is this?

3. Make a line graph of the diameters of round pipes in relation to their cross-section areas. Use the following diameters: 1″, 2″, 3″, 4″, 5″, 6″, 7″, 8″, 9″.

   a. Use the graph to find the cross-section areas of pipes having the following diameters: 1.5″, 4.6″, 5.25″.

   b. Check these values by computation. Is this graph accurate enough for your work?

4. Make a line graph showing the gage number of black iron in relation to the weight per square foot. Use the following gages: 10, 12, 14, 16, 18, 20, 22, 24, 26, 28, 30, 32.

   a. How does the weight per square foot increase compared to the gage numbers?

   b. Between which gages does the weight per square foot increase most rapidly; the least rapidly?

5. Make a graph for galvanized iron similar to the one in problem 4. Use the same piece of graph paper and the same scales but draw the curve with a dash line instead of a solid line.

6. a. Make up a graph showing the amount of air moved through pipes of 3 in., 5 in., 7 in., and 9 in. diameter at a pressure of a 5-in. water column. (Information to be obtained from other text or a library)

   b. From this graph, how much air can be moved through a 6-in. pipe? 5¾-in. pipe?

7. a. Make a graph showing the amount of air moved through a 5-in. pipe at the following pressures: 3-in. water column, 5-in. water column, 7-in. water column.

   b. From the graph, what is the reading for pressure of 4¼-in. water column? 3½-in. water column? 5¾-in. water column?

SECTION 13 PRACTICAL PROBLEMS

UNIT 95 ON–THE–JOB APPLICATIONS

Note: Insert local costs and wages where indicated.

1. Aluminum sheets .032 in. thick are used to hook up a roof-mounted air-conditioning system. This metal weighs .45/lb./sq. ft. and costs $105.23/cwt. How much would the material cost to make up 385′ of 22″ circular duct? (1½-in. seam allowance; allow 3% for joining section.)
2. How much would it cost to make up a rectangular duct of the same length, and the same cross-sectional area as the duct in question 1 with a 2½-in. seam allowance (allow 2% for joining sections) and one 20-in. side?
3. A chemical plant needs a vent for toxic and corrosive gases. The most practical size is 12″ x 6″. The plastic costs __________ per 4′ x 8′ sheet. It takes a journeyman __________ hours to fabricate the 32′ needed for the job. At a rate of __________ per hour, what will be the cost of the duct after addition of 30% for overhead and profit?
4. A sheet metal shop orders the following goods just prior to a cost increase:

   $1,400.00 for 26-gage galvanized sheet metal

   $ 375.00 for 20-gage galvanized sheet metal

   $ 900.00 for .013 aluminum sheets

   $ 187.75 for copper sheets

   In order to take advantage of this, the shop owner borrows money from the bank at the rate of 10½% per year. In five months time he receives payment from a contract and pays off the bank loan in a lump sum. How much is this lump sum going to be?
5. A shop bids and gets a contract for $9,235.00. The terms are:

   25% at beginning of job

   25% at approximately halfway on job

   25% on job completion

   25% six months after job completion

   The outstanding amount at the end of the job is to collect interest at 1% per month. What will be the amount of the check for the final payment? **Note:** This interest rate is not compounded. Do not pay interest on the interest.
6. A research building needs several stainless steel trays, each 28″ x 60″ x 6″ deep with a lip or rim 4″ wide. They require a thickness of __________ and stainless steel costs __________ /cwt. To make up these trays requires four (4) hours labor each at __________ per hour. What is the shop cost? If you owned the shop, what would you bid, using 40% for overhead and profit?

7. You own a sheet metal shop employing 12 journeymen at ________ per hour, three third-year apprentices at ________ per hour, five second-year apprentices at ________ per hour, and two first-year apprentices at ________ per hour. They get a wage increase of 7½% when their new contract is signed. How much more are you going to have to pay in wages per 40-hour week?
8. You own a sheet metal shop and have to purchase a new van for delivering purposes. The quoted list price is ________ ; however, the salesman, hoping to get more business offers you a discount of 15% off list price. What saving does this represent?
9. You are approached with the following problem. A man has built a boat and wants to fabricate two fuel tanks to mount on the transom. The size of each tank is 8 in. deep at top, 6 in. on bottom, 22 in. wide, and 18 in. high. How many gallons will he be able to carry with full tanks?

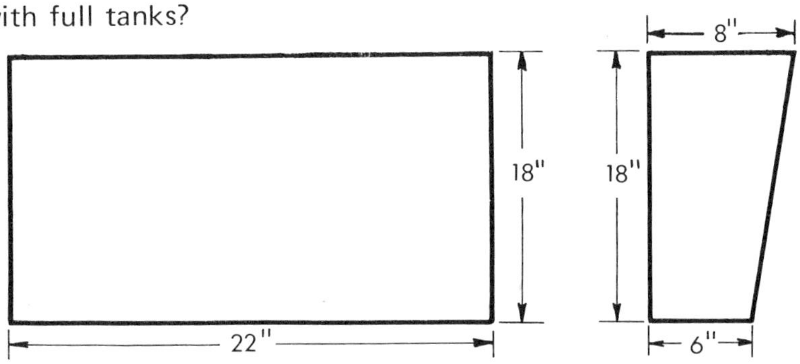

10. You construct a conical bottom for a tank which is 6 ft. in diameter and the depth of the cone is 4 ft. How many cubic feet will it hold?
11. An inverted pyramid hopper bottom has a top measuring 4′ x 5′ 10″; the vertical height is 4′ 5″. If the contents weigh 94¼ lb. per cubic foot, how much will the total contents of this hopper weigh?
12. The tank above the hopper bottom measures 4′ x 5′ 10″ x 18′ 1″. What will the combined weight of the contents of tank and hopper bottom be?
13. The above tank and bottom are made of aluminum of .1144 thickness, weighing 1.605 lb./sq. ft. How much will the empty tank weigh?
14. What is the combined weight of tank and material?
15. Compute the cost to make up a circular duct 28 in. in diameter, 185 ft. long of 26-gage aluminum, weighing .253 lb. per sq. ft. The seam and joint allowance is 7½%. Cost of material is ________ per cwt., and ________ hours of labor at ________ per hour is required for fabrication.
16. What is the bill for fabricating a rectangular duct measuring 18″ x 24″ x 123′ long? Allowance for seams and locks is 3½%. The material is 24-gage galvanized iron weighing 1.156 lb./sq. ft. costing ________ per hundred pounds. Labor for forming locks, etc., is ________ hours at ________ per hour.

17. You are to make a cylindrical tank with a conical bottom; this unit is going to be used as a dry chemical mixer.

    Dimensions:

    Diameter 60″
    Overall height 9′
    Depth of conical bottom 42″

    In the center of this tank is a tube 6 in. in diameter into which a 5½-in. auger will be inserted.

    What is the volume in cubic feet if the tank is filled within 8 in. of the top? What will be the volume if filled to within 3 ft. of the top?

    (Consider the 6-in. tube as being empty and include the volume of the conical bottom and cylindrical tank.)

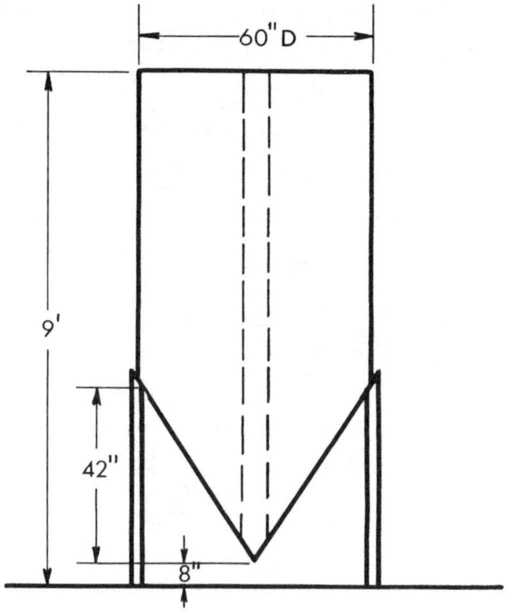

18. What will the weight of the tank be if it is made out of 18-gage galvanized sheet metal? (The weight of frame and 5½-in. auger is 185 lb.) This weight includes the 6-in. tube.
19. What will the weight of the tank and contents be if the material to fill it weighs 48 lb. per cubic foot? (Consider the tank to be filled to within 8 in. of the top and the 6-in. tube to be empty.)
20. How many square feet of material is needed to make up a 5-in. diameter pipe for a run of 290 ft., allowing 17% waste, seams, and joints?

APPENDIX

This appendix can be used by the student and by the instructor to find quickly any information on terms, abbreviations, symbols, units of measure, and constants which has been explained and used in a previous unit. In addition, an index of the mathematical tables and the tables of trade data used in the course is included. The same abbreviation is used for the singular and plural form of a term.

The formulas which have been developed and used in this course are also listed in the appendix. This list will provide a ready reference for subsequent problems, or for review.

TERMS, ABBREVIATIONS AND SYMBOLS

Term	Abbreviation/Symbol	Meaning
acute angle		angle less than 90 deg. and greater than 0 deg.
adjacent	adjac.	next to, neighboring
alternate		every other one
altitude	alt.	perpendicular distance between bases, or base to vertex
angle	∠	space bounded by two intersecting lines
arc		portion of curved line of circle
average	av.	mid, or most likely value
axis	Ax.	straight line around which object may turn
bar	—	line over a term, means treat as a whole
barrel	bbl.	liquid measure, 31.5 gallons (U.S.), 45 (Imperial)
bend allowance	B.A.	length of material needed to allow for forming
bisect		divide into two equal parts
black iron	blk. iron	uncoated iron or steel plate
braces	{ }	means treat as a whole
brackets	[ ]	means treat as a whole
bushel	bu.	dry measure, 2,150.42 cu. in., or 1.24446 cu. ft.
capacity	cap.	contents
cent	c. (¢)	1/100 of a dollar
central angle		angle with vertex at center of circle
cheek		side of a square elbow
chord		straight line between the ends of an arc
circle	cir. ⊙	plane curve, all points being equidistant from center
circumference		linear distance around a circle
coefficient	coeff.	any symbol placed before another as a multiplier
complementary angles		two angles whose sum equals 90 deg.
concentric		having the same center
cone		tapered solid whose base is a curved (circular) figure
constant		numerical quantity expressing a relation that retains same value provided conditions remain unchanged
cosine	cos.	ratio of adjacent side to hypotenuse (right triangle)
cross section		figure formed by cutting plane at right angles to axis
cube		solid with six equal square faces
cubic foot	cu. ft., $ft.^3$	measure of volume, 1728 cu. in.
cubic inch	cu. in., $in.^3$	measure of volume
cubic yard	cu. yd., $yd.^3$	measure of volume, 27 cu. ft.
cylinder	cyl.	solid with equal, parallel, curved (circular) bases
day	da.	time measure, 24 hours
degree	deg. (°)	angular measure
diagonal		slant line connecting corners of a polygon
diameter	dia. (d,D)	straight line through center of circle ending at curve
division	div. (–, ÷, /)	divided by, part of

Term	Abbreviation	Meaning
dollar	dol. ($)	measure of money, 100 cents
dozen	doz.	measure of counting, 12 units
equals	(=)	is equal to
equation	eq.	an expression of equality
equiangular		having equal angles
equidistant		equal distance apart at all points
equilateral		having equal sides
equivalent	equiv.	equal in measure but not always in shape
evaluation		to do the computations indicated
exponent		number of times a quantity is to be taken as a factor
exterior	ext.	outside of
factor		one number or part of a product
figure	fig.	number, an object having shape or form
foot	ft. (′)	linear measure, 12 inches
frustum		part of solid left after cutting off top parallel to base
gage	ga.	number corresponding to a size
gallon	gal.	liquid measure, 231 cu. in. (7.48 gal. equal 1 cu. ft.)
galvanized iron	galv. iron	sheets of iron or steel coated with zinc
gill	gi.	liquid measure, 1/4 pint
gross	gr.	measure of counting, 144 units
heel		outside piece of a square elbow
horizontal	hor.	parallel to the horizon
hour	hr.	measure of time, 60 minutes
hundredweight	cwt.	measure of weight, 100 pounds
hypotenuse	hypot.	slanting side of a right triangle
inch	in. (″)	linear measure (12 in. = 1 ft.)
interior	int.	inside of
inverse		opposite in order
isosceles triangle		triangle having two equal sides
lateral		side
linear (lineal) foot		a foot measured in a straight line
maximum	max.	largest, greatest, highest point
mile	mi.	linear measure, 5280 feet
minimum	min.	smallest, least, lowest point
minute	min. (′)	angular measure (60 min. = 1 deg.), time measure
month	mo.	measure of time (12 mo. = 1 yr.)
multiply	×	times
oblique	obl.	slanting, inclined
obtuse angle		angle greater than 90 deg. and less than 180 deg.
opposite	opp.	across from, facing
ounces	oz.	measure of weight (16 oz. = 1 lb.)
parallel	par. (‖)	equidistant at all points

parallelogram		a four-sided figure, opposite sides parallel
parentheses	par. ( )	treat as a whole
pattern	pat.	flat layout which when formed makes a job or part
peck	pk.	dry measure (8 qts. = 1 pk.)
percent	%	hundredths
perimeter		the linear distance around a polygon
peripheral		outer
perpendicular	$\perp$	at right angles, or 90 deg.
pi	π	3.1415976534 (usually 3.14, or 3.1416)
pint	pt.	liquid, or dry measure, (2 pts. = 1 qt.)
pitch		ratio of rise of a roof to the span
plane		a flat or level surface
polygon		plane figure formed by three or more straight lines
pound	lb. (#)	measure of weight (16 oz. = 1 lb.)
power	pr.	result of squaring, cubing, etc.
prism		solid whose bases are parallel, equal polygons
pyramid		tapered solid whose base is a polygon
quart	qt.	liquid, or dry measure, (2 pts. = 1 qt.)
radius	r,R	straight line from center of a circle to the curve
ratio	(:, —, /)	one number divided by another
rectangle	▭	four-sided plane figure, opposite sides equal, adjacent sides at 90 deg.
revolutions per min.	r.p.m.	the number of turns about the axis in one minute
right angle		90 deg.
right triangle		triangle with one right angle
rise		perpendicular distance from roof apex to span line
run		one-half the span of a roof
scale		a rule, or ratio of drawing dimensions to job dimensions
second	sec. (″)	angular measure (60 sec. = 1 min.) or time measure
sine	sin	ratio of opposite side to hypotenuse (right triangle)
slant height		straight line from apex of cone to edge of base, altitude of one face of a pyramid
span		length of the base of triangle formed by a roof
square	sq. (□)	four-sided plane figure, sides equal and at 90 deg.
square elbow		elbow with heel and throat at 90 deg. to cheeks
square foot	sq. ft. (ft.^{2}, □′)	area measure (144 sq. in. = 1 sq. ft.)
square inch	sq. in. (in.^{2}, □″)	area measure
square root	$\sqrt{}$	number taken twice as factor equals original number
square yard	sq. yd. (yd.^{2})	area measure (9 sq. ft. = 1 sq. yd.)
stretchout		flat layout which when formed makes a job or part
straight angle		angle whose sides are at 180 deg.
subtraction		minus or taking away

Term	Abbreviation	Definition
supplementary angles		two angles whose sum is 180 deg.
surface speed		distance a point on a circle moves in a certain time
tangent (circle)		straight line touching a circle at one point only
tangent	tan.	ratio of opposite side to adjacent side (right triangle)
tapered		becoming smaller at one end
therefore	(∴)	for this or that reason, consequently
throat		the inside piece of a square elbow
ton	t.	weight measure (2000 lb. = 1 ton)
trapezoid		a four-sided plane figure with two opposite sides parallel
triangle	△	plane figure formed by three straight lines
truncated		cut at an angle
vertex		point where sides of angle meet, point opposite base
vertical angles		opposite angles formed when two straight lines intersect
vertical	vert.	perpendicular to plane of horizon (plumb line)
week	wk.	time measure (1 wk. = 7 days)
year	yr.	time measure (365 days = 1 calendar yr.)
yard	yd.	linear measure (3 ft. = 1 yd.)

FORMULAS

Section 7

Perimeters — P = perimeter

Figure	Formula	Definitions
Square	$P = 4s$	s = length of side
Rectangle	$P = 2\ell + 2w$	ℓ = length w = width
Circle	$C = \pi d$, or $2\pi r$ $d = 2r$	d = diameter; r = radius $\pi = 3.14$ C = circumference
Semicircular-sided figure	$P = \pi d + 2w$	d = diameter of semicircles w = distance between centers of semicircles

Section 8

Area Measure — A = area

Figure	Formula	Definitions
Square	$A = s^2$	s = length of side
Rectangle	$A = \ell w$	ℓ = length w = width
Triangle	$A = \frac{ab}{2}$	a = altitude b = base

Trapezoid	$A = \frac{(B + b)\,a}{2}$	B = length of large base b = length of small base a = altitude
Circle	$A = \pi r^2$, or $\frac{\pi d^2}{4}$	π = 3.14 r = radius d = diameter
Semicircular-sided figure	$A = \pi r^2 + 2rw$	π = 3.14 r = radius of semicircles w = distance between centers of semicircles
Spherical-ended solid	$A = \pi d \ell$ or $2\pi r \ell$	A = area r = radius π = 3.14 ℓ = overall length d = diameter

Section 9

Volume Measure		V = volume
Cube	$V = s^3$	s = length of one side
Square prism	$V = s^2 h$	s = length of side of base h = height of prism
Rectangular solid	$V = \ell wh$	ℓ = length of base w = width of base h = height of solid
Cone	$V = \frac{\pi r^2 h}{3}$	V = volume r = radius π = 3.14 h = height
Cylinder (circular)	$V = \pi r^2 \ell$	π = 3.14 r = radius ℓ = length of cylinder
Right circular cone	$V = \frac{(\pi r^2)h}{3}$	V = volume r = radius h = height π = 3.14
Right square pyramid	$V = \frac{s^2 h}{3}$	V = volume s = length of base sides h = height
Right rectangular pyramid	$V = \frac{(\ell w)h}{3}$	V = volume ℓ = length of base w = width of base h = height

Equilateral triangular base pyramid	$V = (1/2\,a \times h) \times H$	V = volume a = length of side h = perpendicular height of triangle H = perpendicular height of pyramid
Semicircular-sided solid	$V = (\pi r^2 + 2\,rw)\,h$	π = 3.14 h = height of solid w = distance between centers of semicircles r = radius of semicircles
Spherical-end solids	$V = \frac{4}{3}\pi r^3 + \pi r^2 h$	V = volume r = radius h = distance between centers of semispheres

Section 10

Special Problems

Edge Allowances (one side)

Single hem	W	W = width of hem
Double hem	2W	W = width of hem
Wired edge	2½D	D = diameter of rivet

Seam Allowances

Lap (one side)	W	W = width of seam
Grooved (total	3W	W = width of seam
Riveted (one side)	2D	D = diameter of rivet
Dovetailed (one side)	W	W = width of seam
Burred bottom (one side)	W	W = width of lap or burr
Set-in bottom (one side)	Body = 2W Bottom = W	W = width of seam
Single, double (one side)	Body = W Bottom = 2W	W = width of flange
Pittsburgh lock (one side)	Flange = W Pocket = $2W + \frac{3''}{16}$	W = width of pocket

Stretchouts

The allowances for edges and seams given above must be added to formulas below.		L.S. = length of stretchout W.S. = width of stretchout
Square pipes	L.S. = $4s$ W.S. = ℓ	s = length of side or end ℓ = length of pipe
Rectangular pipes	L.S. = $2h + 2w$ W.S. = ℓ	h = height of end w = width of end ℓ = length of pipe

Circular jobs $\quad L.S. = \pi d \quad W.S. = \ell$

d = diameter
ℓ = length of job

Semicircular-sided jobs $\quad L.S. = \pi d + 2w \quad W.S. = \ell \quad \text{Roll} = \frac{\pi d}{2}$

d = diameter of semicircles
w = distance between centers of semicircles
ℓ = length of job

Boxes $\quad L.S. = \ell + 2h \quad W.S. = w + 2h$

ℓ = length
w = width
h = height

Length of Wire for Edges of Jobs

L.W. = length (in.) of wire

Square $\quad L.W. = 4s + 2D$

s = length (in.) of side of job
D = diameter (in.) of wire

Rectangle $\quad L.W. = 2\ell + 2w + 2D$

ℓ = length (in.) of job
w = width (in.) of job
D = diameter (in.) of wire

Circular $\quad L.W. = \pi d$

π = 3.14
d = diameter (in.) of job

Circular (tapered) $\quad L.W. = \pi(d + D)$

π = 3.14
d = diameter (in.) of job
D = diameter (in.) of wire

Length of Arcs of Circles

$$\ell = \frac{n\pi d}{360}$$

ℓ = length of arc
n = deg. in central angle
π = 3.14
d = 2 x radius of arc

Law of Right Triangles

$$c = \sqrt{a^2 + b^2}$$

a = altitude

$$b = \sqrt{c^2 - a^2}$$

b = base

$$a = \sqrt{c^2 - b^2}$$

c = hypotenuse

Constants for 45°

Length of hypotenuse = length of side x 1.414
Length of side = length of hypotenuse x .707

Stretchouts of Cones

Slant height $\quad H = \sqrt{a^2 + (d/2)^2}$

H = slant height of cone
a = altitude of cone
d = diameter of base

Angle of Stretchout $\quad n = \frac{180d}{H}$

n = angle of stretchout
d = diameter of base
H = slant height of cone

Roof Pitches

Pitch	$p = \frac{r}{s}$	r = rise p = pitch s = span
Rise	$r = ps$	
Length of roof slope	$\ell = \sqrt{r^2 + (s/2)^2}$	ℓ = length of roof slope r = rise s = span

Bend Allowances

Approximate formulas

Inside dimensions	$L.S. = \ell + L + \frac{t}{2}$	L.S. = length of stretchout t = thickness of metal
Outside dimensions	$L.S. = \ell + L - \frac{t}{2}$	L = length of long leg ℓ = length of short leg

Exact Formula	B.A. = n (.01743r + .0078t)	B.A. = bend allowance (in.) n = number of deg. in bend r = inside radius (in.) of bend t = thickness (in.) of metal

Surface Speeds (circular)

$$S = \frac{\pi dn}{12}$$

S = speed in feet per min.
π = 3.14
d = diameter (in.) of wheel
n = r.p.m.

Section 11

Trigonometry

Tangent	$\tan A = \frac{a}{b}$	a = side opposite ∠ A b = side adjacent ∠ A c = hypotenuse A = angle next to side **b**
Sine	$\sin A = \frac{a}{c}$	
Cosine	$\cos A = \frac{b}{c}$	

TABLES OF MEASUREMENT, WEIGHT, AND TIME

Linear Measure

12 inches = 1 foot (ft.)
3 feet = 1 yard (yd.)
5280 feet = 1 mile (mi.)

Inches are designated with a double prime (″). (3 inches = 3″)
Feet are designated with a single prime (′). (5 feet = 5′)

Metric System: Linear Measure

10 millimeters (mm.)	=	1 centimeter (cm.)
10 cm. or 100 mm.	=	1 decimeter (dm.)
10 dm. or 100 cm.	=	1 meter (m.)
10 m. or 1000 cm.	=	1 decameter (Dm.)
10 Dm. or 100 m.	=	1 Hectometer (Hm.)
10 Hm. or 1000 m.	=	1 kilometer (Km.)

Conversion Constants

1 inch = 2.54 cm., 1 meter = 39.37 in. ,
1 kilometer = .62137 mi.

Angular Measure

One circle	=	360 degrees (deg.)
One degree	=	60 minutes (min.)
One minute	=	60 seconds (sec.)

Degrees are designated by a small zero such as 2 deg. = 2°.
Minutes are designated by a single prime such as 5 min. = 5′.
Seconds are designated by a double prime such as 30 sec. = 30″.

Area Measure

144 square inches	=	1 square foot (sq. ft.)
9 square feet	=	1 square yard (sq. yd.)
100 square feet	=	1 square (roofing, etc.)

Cubic Measure

1728 cubic inches	=	1 cubic foot (cu. ft.)
27 cubic feet	=	1 cubic. yard (cu.yd.)

Avoirdupois Weight (commercial)

16 ounces (oz.)	=	1 pound (lb.)
100 pounds	=	1 hundredweight (cwt.)
2000 pounds	=	2 ton (t.)

United States Liquid Measure

Liquid Measure

4 gills	=	1 pint (pt.)
2 pints	=	1 quart (qt.)
4 quarts	=	1 gallon (gal.)

Conversion Constants

1 cu. ft.	=	7.48 U.S. gal.
231 cu. in.	=	1 U.S. gal.
31½ gal.	=	1 barrel (bbl.)

Imperial Liquid Measure

Conversion U.S. to Imperial Liquid Measure

1 Imp. gal. = 1.2 U.S. gal.
1 U.S. gal. = 5/6 Imp. gal.
1 Imp. gal. weighs 10 lb.
1 U.S. gal. weighs 8.33 lb.

Conversion Constants

1 cu. ft.	=	6.25 gal.
277 cu. in.	=	1 Imp. gal.
45 Imp. gal.	=	1 bbl.

Dry Measure

Dry Measure

2 pints	=	1 quart (qt.)
8 quarts	=	1 peck (pk.)
4 pecks	=	1 bushel (bu.)

Conversion Constants

1 bushel	=	2150.42 cu. in.
1 bushel	=	1.24446 cu. ft.

Counting Table

12 units = 1 dozen (doz.) 12 dozen = 1 gross (144 units)

Time

60 seconds (sec.)	=	1 minute (min.)	7 days	=	1 week (wk.)
60 minutes	=	1 hour (hr.)	52 weeks	=	1 year (yr.)
24 hours	=	1 day (da.)	365 days	=	1 year (calendar)

DECIMAL EQUIVALENT TABLE

Fraction	Decimal	Fraction	Decimal	Fraction	Decimal	Fraction	Decimal
1/64	.015625	17/64	.265625	33/64	.515625	49/64	.765625
1/32	.03125	9/32	.28125	17/32	.53125	25/32	.78125
3/64	.046875	19/64	.296875	35/64	.546875	51/64	.796875
1/16	.0625	5/16	.3125	9/16	.5625	13/16	.8125
5/64	.078125	21/64	.328125	37/64	.578125	53/64	.828125
3/32	.09375	11/32	.34375	19/32	.59375	27/32	.84375
7/64	.109375	23/64	.359375	39/64	.609375	55/64	.859375
1/8	.125	3/8	.375	5/8	.625	7/8	.875
9/64	.140625	25/64	.390625	41/64	.640625	57/64	.890625
5/32	.15625	13/32	.40625	21/32	.65625	29/32	.90625
11/64	.171875	27/64	.421875	43/64	.671875	59/64	.921875
3/16	.1875	7/16	.4375	11/16	.6875	15/16	.9375
13/64	.203125	29/64	.453125	45/64	.703125	61/64	.953125
7/32	.21875	15/32	.46875	23/32	.71875	31/32	.96875
15/64	.234375	31/64	.48438	47/64	.734375	63/64	.984375
1/4	.250	1/2	.500	3/4	.750	1	1.000

TABLES OF TRADE DATA

Weights and List Prices of Black Band Iron

Size	Wt. per Lineal Ft. (lb.)	Length (ft.)	List Price Per 100 lb.
1/2 in. x 12 gage	.185	12	$ 20.70
3/4 in. x 12 gage	.278	12	19.70
1/2 in. x 1/8 in.	.212	20	20.50
3/4 in. x 1/8 in.	.318	20	19.50
1 in. x 1/8 in.	.425	20	19.10
1 1/2 in. x 1/8 in.	.6375	20	19.00
3/4 in. x 3/16 in.	.478	20	19.40
1 1/2 in. x 3/16 in.	.956	20	18.90

Thicknesses and Weights of Black and Galvanized Iron

Gage U.S.	BLACK IRON Thick. In.	BLACK IRON Lb. per Sq. Ft.	GALV. IRON Lb. per Sq. Ft.	Gage U.S.	BLACK IRON Thick. in.	BLACK IRON Lb. per Sq. Ft.	GALV. IRON Lb. per Sq. Ft.
32	.0097	.40625	.56250	19	.0418	1.75	1.90625
31	.0105	.4375	.59375	18	.0478	2.	2.15625
30	.0120	.5	.65625	17	.0538	2.25	2.40625
29	.0135	.5625	.71875	16	.0598	2.5	2.65625
28	.0149	.625	.78125	15	.0673	2.8125	2.96875
27	.0164	.6875	.84375	14	.0747	3.125	3.28125
26	.0179	.75	.90625	13	.0897	3.75	3.90625
25	.0209	.875	1.03125	12	.1046	4.375	4.53125
24	.0239	1.	1.15625	11	.1196	5.	5.15625
23	.0269	1.125	1.28125	10	.1345	5.625	5.78125
22	.0299	1.25	1.40625	9	.1495	6.25	6.40625
21	.0329	1.375	1.53125	8	.1644	6.875	7.03125
20	.0359	1.5	1.65625				

Gages and Thicknesses of Sheet Steel – Black Iron

U.S. Gage	Thickness Inches	U.S. Gage	Thickness Inches	U.S. Gage	Thickness Inches
38	.0060	26	.0179	14	.0747
37	.0064	25	.0209	13	.0897
36	.0067	24	.0239	12	.1046
35	.0075	23	.0269	11	.1196
34	.0082	22	.0299	10	.1345
33	.0090	21	.0329	9	.1495
32	.0097	20	.0359	8	.1644
31	.0105	19	.0418	7	.1793
30	.0120	18	.0478	6	.1943
29	.0135	17	.0538	5	.2092
28	.0149	16	.0598	4	.2242
27	.0164	15	.0673	3	.2391

Gages and Thicknesses of Tin Plate

Base Wt. Lb.	Trade Term	Approx. U.S. Gage	Approx. Thick. In.	Lb. Per Sq. Ft.	Base Wt. Lb.	Trade Term	Approx. U.S. Gage	Approx. Thick. In.	Lb. Per Sq. Ft.
55		38	.0061	.253	155	2X	27	.0175	.712
60		37	.0068	.276	168	3XL	26	.0189	.771
65		36	.0073	.298	175	3X	26	.0197	.804
70		35	.0079	.321	180	DX	25	.0203	.827
75		34	.0084	.344	188	4XL	25	.0211	.863
80		33	.0090	.367	195	4X	25	.0220	.895
85		32	.0095	.390	208	5XL	25	.0238	.955
90		31	.0101	.413	210	D2X	25	.0241	.964
95		31	.0107	.436	215	5X	25	.0246	.987
100	1CL	30½	.0113	.459	228	6XL	24	.0261	1.047
107	1C	30	.0120	.491	235	6X	24	.0269	1.079
118		29	.0133	.542	240	D3X	24	.0275	1.102
123		29	.0138	.564	248	7XL	23	.0284	1.139
128	1XL	28	.0144	.588	255	7X	23	.0292	1.171
135	1X	28	.0152	.620	268	8XL	23	.0307	1.231
139	DC	28	.0156	.638	270	D4X	23	.0310	1.240
148	2XL	27	.0167	.680	275	8X	22	.0315	1.263
150		27	.0169	.688					

Notes: The preceding table have two valuable uses:

a. To find the thickness of metal if the gage is given.

b. To find the gage of metal if the thickness is given.

Coated Tinners' Rivets

Size	Diameter In.	Length In.	Size	Diameter In.	Length In.
8 oz.	.089	5/32	3½ lb.	.165	21/64
10 oz.	.095	11/64	4 lb.	.176	11/32
12 oz.	.105	3/16	5 lb.	.186	3/8
14 oz.	.109	3/16	6 lb.	.203	25/64
1 lb.	.112	13/64	7 lb.	.220	13/32
1¼ lb.	.120	7/32	8 lb.	.224	7/16
1½ lb.	.130	15/64	9 lb.	.230	29/64
1¾ lb.	.134	1/4	10 lb.	.238	15/32
2 lb.	.144	17/64	12 lb.	.259	1/2
2½ lb.	.148	9/32	14 lb.	.284	33/64
3 lb.	.160	5/16	16 lb.	.300	17/32

POWERS AND ROOTS OF NUMBERS (1 through 100)

Number	Powers		Roots		Number	Powers		Roots	
	Square	Cube	Square	Cube		Square	Cube	Square	Cube
1	1	1	1.000	1.000	51	2,601	132,651	7.141	3.708
2	4	8	1.414	1.260	52	2,704	140,608	7.211	3.733
3	9	27	1.732	1.442	53	2,809	148,877	7.280	3.756
4	16	64	2.000	1.587	54	2,916	157,464	7.348	3.780
5	25	125	2.236	1.710	55	3,025	166,375	7.416	3.803
6	36	216	2.449	1.817	56	3,136	175,616	7.483	3.826
7	49	343	2.646	1.913	57	3,249	185,193	7.550	3.849
8	64	512	2.828	2.000	58	3,364	195,112	7.616	3.871
9	81	729	3.000	2.080	59	3,481	205,379	7.681	3.893
10	100	1,000	3.162	2.154	60	3,600	216,000	7.746	3.915
11	121	1,331	3.317	2.224	61	3,721	226,981	7.810	3.936
12	144	1,728	3.464	2.289	62	3,844	238,328	7.874	3.958
13	169	2,197	3.606	2.351	63	3,969	250,047	7.937	3.979
14	196	2,744	3.742	2.410	64	4,096	262,144	8.000	4.000
15	225	3,375	3.873	2.466	65	4,225	274,625	8.062	4.021
16	256	4,096	4.000	2.520	66	4,356	287,496	8.124	4.041
17	289	4,913	4.123	2.571	67	4,489	300,763	8.185	4.062
18	324	5,832	4.243	2.621	68	4,624	314,432	8.246	4.082
19	361	6,859	4.359	2.668	69	4,761	328,509	8.307	4.102
20	400	8,000	4.472	2.714	70	4,900	343,000	8.367	4.121
21	441	9,261	4.583	2.759	71	5,041	357,911	8.426	4.141
22	484	10,648	4.690	2.802	72	5,184	373,248	8.485	4.160
23	529	12,167	4.796	2.844	73	5,329	389,017	8.544	4.179
24	576	13,824	4.899	2.884	74	5,476	405,224	8.602	4.198
25	625	15,625	5.000	2.924	75	5,625	421,875	8.660	4.217
26	676	17,576	5.099	2.962	76	5,776	438,976	8.718	4.236
27	729	19,683	5.196	3.000	77	5,929	456,533	8.775	4.254
28	784	21,952	5.292	3.037	78	6,084	474,552	8.832	4.273
29	841	24,389	5.385	3.072	79	6,241	493,039	8.888	4.291
30	900	27,000	5.477	3.107	80	6,400	512,000	8.944	4.309
31	961	29,791	5.568	3.141	81	6,561	531,441	9.000	4.327
32	1,024	32,798	5.657	3.175	82	6,724	551,368	9.055	4.344
33	1,089	35,937	5.745	3.208	83	6,889	571,787	9.110	4.362
34	1,156	39,304	5.831	3.240	84	7,056	592,704	9.165	4.380
35	1,225	42,875	5.916	3.271	85	7,225	614,125	9.220	4.397
36	1,296	46,656	6.000	3.302	86	7,396	636,056	9.274	4.414
37	1,369	50,653	6.083	3.332	87	7,569	658,503	9.327	4.481
38	1,444	54,872	6.164	3.362	88	7,744	681,472	9.381	4.448
39	1,521	59,319	6.245	3.391	89	7,921	704,969	9.434	4.465
40	1,600	64,000	6.325	3.420	90	8,100	729,000	9.487	4.481
41	1,681	68,921	6.403	3.448	91	8,281	753,571	9.539	4.498
42	1,764	74,088	6.481	3.476	92	8,464	778,688	9.592	4.514
43	1,849	79,507	6.557	3.503	93	8,649	804,357	9.644	4.531
44	1,936	85,184	6.633	3.530	94	8,836	830,584	9.695	4.547
45	2,025	91,125	6.708	3.557	95	9,025	857,375	9.747	4.563
46	2,116	97,336	6.782	3.583	96	9,216	884,736	9.798	4.579
47	2,209	103,823	6.856	3.609	97	9,409	912,673	9.849	4.595
48	2,304	110,592	6.928	3.634	98	9,604	941,192	9.900	4.610
49	2,401	117,649	7.000	3.659	99	9,801	970,299	9.950	4.626
50	2,500	125,000	7.071	3.684	100	10,000	1,000,000	10.000	4.642

SQUARE ROOTS OF NUMBERS (1.000 through 5.490)

Number	0	1	2	3	4	5	6	7	8	9
1.0	1.000	$1.00\bar{5}$	1.010	$1.01\bar{5}$	1.020	$1.02\bar{5}$	1.030	1.034	1.039	1.044
1.1	1.049	1.054	1.058	1.063	1.068	1.072	1.077	1.082	1.086	1.091
1.2	1.095	$1.10\underline{0}$	$1.10\bar{5}$	1.109	1.114	1.118	1.122	1.127	$1.13\underline{1}$	1.136
1.3	1.140	$1.14\bar{5}$	1.149	1.153	1.158	1.162	1.166	1.170	$1.17\bar{5}$	1.179
1.4	1.183	1.187	1.192	1.196	1.200	1.204	1.208	1.212	1.217	1.221
1.5	$1.22\bar{5}$	1.229	1.233	1.237	1.241	$1.24\bar{5}$	1.249	1.253	1.257	1.261
1.6	$1.26\bar{5}$	1.269	1.273	1.277	1.281	$1.28\bar{5}$	1.288	1.292	1.296	1.300
1.7	1.304	1.308	1.311	1.315	1.319	1.323	1.327	1.330	1.334	1.338
1.8	1.342	1.345	1.349	1.353	1.356	1.360	1.364	1.367	1.371	$1.37\bar{5}$
1.9	1.378	1.382	1.386	1.389	1.393	1.396	1.400	1.404	1.407	1.411
2.0	1.414	1.418	1.421	$1.42\bar{5}$	1.428	1.432	1.435	1.439	1.442	1.446
2.1	1.449	1.453	1.456	1.459	1.463	1.466	1.470	1.473	1.476	1.480
2.2	1.483	1.487	1.490	1.493	1.497	1.500	1.503	1.507	1.510	1.513
2.3	1.517	1.520	1.523	1.526	1.530	1.533	1.536	1.539	$1.54\underline{3}$	1.546
2.4	1.549	1.552	1.556	1.559	1.562	1.565	1.568	1.572	$1.57\bar{5}$	1.578
2.5	1.581	1.584	1.587	1.591	1.594	1.597	1.600	1.603	1.606	1.609
2.6	1.612	1.616	1.619	1.622	$1.62\bar{5}$	1.628	1.631	1.634	1.637	1.640
2.7	1.643	1.646	1.649	1.652	1.655	1.658	1.661	1.664	1.667	1.670
2.8	1.673	1.676	1.679	1.682	1.685	1.688	1.691	1.694	1.697	1.700
2.9	1.703	1.706	1.709	1.712	$1.71\bar{5}$	1.718	1.720	1.723	1.726	1.729
3.0	1.732	$1.73\bar{5}$	1.738	1.741	1.744	1.746	1.749	1.752	$1.75\bar{5}$	1.758
3.1	1.761	1.764	1.766	1.769	1.772	$1.77\bar{5}$	1.778	1.780	1.783	1.786
3.2	1.789	1.792	1.794	$1.79\underline{7}$	1.800	1.803	1.806	1.808	1.811	1.814
3.3	1.817	1.819	1.822	$1.82\bar{5}$	$1.82\underline{8}$	1.830	1.833	1.836	1.838	1.841
3.4	1.844	1.847	1.849	1.852	$1.85\bar{5}$	1.857	1.860	1.863	1.865	1.868
3.5	1.871	1.873	1.876	1.879	1.881	1.884	1.887	1.889	1.892	$1.89\bar{5}$
3.6	1.897	1.900	1.903	1.905	1.908	1.910	1.913	1.916	1.918	1.921
3.7	1.924	1.926	1.929	1.931	1.934	1.936	$1.93\underline{9}$	1.942	1.944	1.947
3.8	1.949	1.952	1.954	1.957	$1.96\underline{0}$	1.962	$1.96\bar{5}$	1.967	$1.97\underline{0}$	1.972
3.9	$1.97\bar{5}$	1.977	1.980	1.982	$1.98\bar{5}$	1.987	1.990	1.992	$1.99\bar{5}$	1.997
4.0	2.000	2.002	$2.00\bar{5}$	2.007	$2.01\underline{0}$	2.012	$2.01\bar{5}$	2.017	$2.02\underline{0}$	2.022
4.1	$2.02\bar{5}$	2.027	2.030	2.032	$2.03\bar{5}$	2.037	2.040	2.042	$2.04\bar{5}$	2.047
4.2	2.049	2.052	2.054	2.057	2.059	2.062	2.064	2.066	2.069	2.071
4.3	2.074	2.076	2.078	$2.08\underline{1}$	2.083	2.086	2.088	2.090	2.093	2.095
4.4	2.098	2.100	2.102	$2.10\bar{5}$	2.107	2.110	2.112	2.114	2.117	2.119
4.5	$2.12\underline{1}$	2.124	2.126	2.128	2.131	2.133	2.135	2.138	2.140	2.142
4.6	$2.14\bar{5}$	2.147	2.149	$2.15\underline{2}$	2.154	2.156	2.159	2.161	2.163	2.166
4.7	2.168	2.170	2.173	$2.17\bar{5}$	2.177	2.179	$2.18\underline{2}$	2.184	2.186	2.189
4.8	2.191	2.193	2.195	2.198	2.200	2.202	$2.20\bar{5}$	2.207	2.209	2.211
4.9	2.214	2.216	2.218	2.220	2.223	$2.22\bar{5}$	2.227	2.229	2.232	2.234
5.0	2.236	2.238	2.241	$2.24\underline{3}$	$2.24\bar{5}$	2.247	2.249	2.252	2.254	2.256
5.1	2.258	2.261	$2.26\underline{3}$	$2.26\bar{5}$	2.267	2.269	2.272	2.274	2.276	2.278
5.2	2.280	2.283	$2.28\bar{5}$	2.287	2.289	2.291	2.293	2.296	2.298	2.300
5.3	2.302	2.304	2.307	2.309	2.311	$2.31\underline{3}$	2.315	2.317	2.319	2.322
5.4	2.324	2.326	2.328	2.330	2.332	$2.33\bar{5}$	2.337	2.339	2.341	2.343
	0	1	2	3	4	5	6	7	8	9

SQUARE ROOTS OF NUMBERS (5.500 through 9.990)

Number	0	1	2	3	4	5	6	7	8	9
5.5	2.345	2.347	2.349	2.352	2.354	2.356	2.358	2.360	2.362	2.364
5.6	2.366	2.369	2.371	2.373	2.375	2.377	2.379	2.381	2.383	2.385
5.7	2.387	2.390	2.392	2.394	2.396	2.398	2.400	2.402	2.404	2.406
5.8	2.408	2.410	2.412	2.415	2.417	2.419	2.421	2.423	2.425	2.427
5.9	2.429	2.431	2.433	2.435	2.437	2.439	2.441	2.443	2.445	2.447
6.0	2.449	2.452	2.454	2.456	2.458	2.460	2.462	2.464	2.466	2.468
6.1	2.470	2.472	2.474	2.476	2.478	2.480	2.482	2.484	2.486	2.488
6.2	2.490	2.492	2.494	2.496	2.498	2.500	2.502	2.504	2.506	2.508
6.3	2.510	2.512	2.514	2.516	2.518	2.520	2.522	2.524	2.526	2.528
6.4	2.530	2.532	2.534	2.536	2.538	2.540	2.542	2.544	2.546	2.548
6.5	2.550	2.551	2.553	2.555	2.557	2.559	2.561	2.563	2.565	2.567
6.6	2.569	2.571	2.573	2.575	2.577	2.579	2.581	2.583	2.585	2.587
6.7	2.588	2.590	2.592	2.594	2.596	2.598	2.600	2.602	2.604	2.606
6.8	2.608	2.610	2.612	2.613	2.615	2.617	2.619	2.621	2.623	2.625
6.9	2.627	2.629	2.631	2.632	2.634	2.636	2.638	2.640	2.642	2.644
7.0	2.646	2.648	2.650	2.651	2.653	2.655	2.657	2.659	2.661	2.663
7.1	2.665	2.666	2.668	2.670	2.672	2.674	2.676	2.678	2.680	2.681
7.2	2.683	2.685	2.687	2.689	2.691	2.693	2.694	2.696	2.698	2.700
7.3	2.702	2.704	2.706	2.707	2.709	2.711	2.713	2.715	2.717	2.718
7.4	2.720	2.722	2.724	2.726	2.728	2.729	2.731	2.733	2.735	2.737
7.5	2.739	2.740	2.742	2.744	2.746	2.748	2.750	2.751	2.753	2.755
7.6	2.757	2.759	2.760	2.762	2.764	2.766	2.768	2.769	2.771	2.773
7.7	2.775	2.777	2.778	2.780	2.782	2.784	2.786	2.787	2.789	2.791
7.8	2.793	2.795	2.796	2.798	2.800	2.802	2.804	2.805	2.807	2.809
7.9	2.811	2.812	2.814	2.816	2.818	2.820	2.821	2.823	2.825	2.827
8.0	2.828	2.830	2.832	2.834	2.835	2.837	2.839	2.841	2.843	2.844
8.1	2.846	2.848	2.850	2.851	2.853	2.855	2.857	2.858	2.860	2.862
8.2	2.864	2.865	2.867	2.869	2.871	2.872	2.874	2.876	2.877	2.879
8.3	2.881	2.883	2.884	2.886	2.888	2.890	2.891	2.893	2.895	2.897
8.4	2.898	2.900	2.902	2.903	2.905	2.907	2.909	2.910	2.912	2.914
8.5	2.915	2.917	2.919	2.921	2.922	2.924	2.926	2.927	2.929	2.931
8.6	2.933	2.934	2.936	2.938	2.939	2.941	2.943	2.944	2.946	2.948
8.7	2.950	2.951	2.953	2.955	2.956	2.958	2.960	2.961	2.963	2.965
8.8	2.966	2.968	2.970	2.972	2.973	2.975	2.977	2.978	2.980	2.982
8.9	2.983	2.985	2.987	2.988	2.990	2.992	2.993	2.995	2.997	2.998
9.0	3.000	3.002	3.003	3.005	3.007	3.008	3.010	3.012	3.013	2.015
9.1	3.017	3.018	3.020	3.022	3.023	3.025	3.027	3.028	2.030	3.032
9.2	3.033	3.035	3.036	3.038	3.040	3.041	3.043	3.045	3.046	3.048
9.3	3.050	3.051	3.053	3.055	3.056	3.058	3.059	3.061	3.063	3.064
9.4	3.066	3.068	3.069	3.071	3.072	3.074	3.076	3.077	3.079	3.081
9.5	3.082	3.084	3.085	3.087	3.089	3.090	3.092	3.094	3.095	3.097
9.6	3.098	3.100	3.102	3.103	3.105	3.106	3.108	3.110	3.111	3.113
9.7	3.114	3.116	3.118	3.119	3.121	3.122	3.124	3.126	3.127	3.129
9.8	3.130	3.132	3.134	3.135	3.137	3.138	3.140	3.142	3.143	3.145
9.9	3.146	3.148	3.150	3.151	3.153	3.154	3.156	3.158	3.159	3.161
	0	1	2	3	4	5	6	7	8	9

SQUARE ROOTS OF NUMBERS (10.00 through 54.90)

Number	0	1	2	3	4	5	6	7	8	9
10.	3.162	3.178	3.194	3.209	$3.22\bar{5}$	3.240	3.256	3.271	3.286	3.302
11.	3.317	3.332	3.347	3.362	3.376	3.391	3.406	3.421	3.435	$3.45\bar{0}$
12.	3.464	3.479	3.493	3.507	3.521	3.536	$3.55\bar{0}$	3.564	3.578	3.592
13.	3.606	3.619	3.633	3.647	3.661	3.674	3.688	3.701	$3.71\bar{5}$	3.728
14.	3.742	$3.75\bar{5}$	3.768	3.782	$3.79\bar{5}$	3.808	3.821	3.834	3.847	3.860
15.	3.873	3.886	3.899	3.912	3.924	3.937	$3.95\bar{0}$	3.962	$3.97\bar{5}$	3.987
16.	4.000	4.012	$4.02\bar{5}$	4.037	$4.05\bar{0}$	4.062	4.074	4.087	4.099	4.111
17.	4.123	4.135	4.147	4.159	4.171	4.183	4.195	4.207	4.219	4.231
18.	4.243	4.254	4.266	4.278	4.290	4.301	4.313	4.324	4.336	4.347
19.	4.359	4.370	4.382	4.393	$4.40\bar{5}$	4.416	4.427	4.438	4.450	4.461
20.	4.472	4.483	4.494	4.506	4.517	4.528	4.539	$4.55\bar{0}$	4.561	4.572
21.	4.583	4.593	4.604	4.615	4.626	4.637	4.648	4.658	4.669	4.680
22.	4.690	4.701	4.712	4.722	4.733	4.743	4.754	4.764	$4.77\bar{5}$	4.785
23.	4.796	4.806	4.817	4.827	4.837	4.848	4.858	4.868	4.879	4.889
24.	4.899	4.909	4.919	4.930	4.940	$4.95\bar{0}$	4.960	4.970	4.980	4.990
25.	5.000	5.010	5.020	5.030	5.040	$5.05\bar{0}$	5.060	5.070	5.079	5.089
26.	5.099	5.109	5.119	5.128	5.138	5.148	5.158	5.167	5.177	5.187
27.	5.196	5.206	5.215	$5.22\bar{5}$	$5.23\bar{5}$	5.244	5.254	5.263	2.273	5.282
28.	5.292	5.301	5.310	5.320	5.329	5.339	5.348	5.357	5.367	5.376
29.	5.385	5.394	5.404	5.413	5.422	5.431	5.441	$5.45\bar{0}$	5.459	5.468
30.	5.477	5.486	5.495	$5.50\bar{5}$	5.514	5.523	5.532	5.541	$5.55\bar{0}$	5.559
31.	5.568	5.577	5.586	$5.59\bar{5}$	5.604	5.612	5.621	5.630	5.639	5.648
32.	5.657	5.666	$5.67\bar{5}$	5.683	5.692	5.701	5.710	5.718	5.727	5.736
33.	$5.74\bar{5}$	5.753	5.762	5.771	5.779	5.788	5.797	5.805	5.814	5.822
34.	5.831	5.840	5.848	5.857	5.865	5.874	5.882	5.891	5.899	5.908
35.	5.916	$5.92\bar{5}$	5.933	5.941	$5.95\bar{0}$	5.958	5.967	$5.97\bar{5}$	5.983	5.992
36.	6.000	6.008	6.017	$6.02\bar{5}$	6.033	6.042	$6.05\bar{0}$	6.058	6.066	$6.07\bar{5}$
37.	6.083	6.091	6.099	6.107	6.116	6.124	6.132	6.140	6.148	6.156
38.	6.164	6.173	6.181	6.189	6.197	$6.20\bar{5}$	6.213	6.221	6.229	6.237
39.	$6.24\bar{5}$	6.253	6.261	6.269	6.277	$6.28\bar{5}$	6.293	6.301	6.309	6.317
40.	$6.32\bar{5}$	6.332	6.340	6.348	6.356	6.364	6.372	6.380	6.387	6.395
41.	6.403	6.411	6.419	6.427	6.434	6.442	$6.45\bar{0}$	6.458	6.465	6.473
42.	6.481	6.488	6.496	6.504	6.512	6.519	6.527	$6.53\bar{5}$	6.542	$6.55\bar{0}$
43.	6.557	6.565	6.573	6.580	6.588	6.595	6.603	6.611	6.618	6.626
44.	6.633	6.641	6.648	6.656	6.663	6.671	6.678	6.686	6.693	6.701
45.	6.708	6.716	6.723	6.731	6.738	6.745	6.753	6.760	6.768	$6.77\bar{5}$
46.	6.782	6.790	6.797	6.804	6.812	6.819	6.826	6.834	6.841	6.848
47.	6.856	6.863	6.870	6.877	$6.88\bar{5}$	6.892	6.899	6.907	6.914	6.921
48.	6.928	6.935	6.943	$6.95\bar{0}$	6.957	6.964	6.971	6.979	6.986	6.993
49.	7.000	7.007	7.014	7.021	7.029	7.036	7.043	$7.05\bar{0}$	7.057	7.064
50.	7.071	7.078	7.085	7.092	7.099	7.106	7.113	7.120	7.127	7.134
51.	7.141	7.148	7.155	7.162	7.169	7.176	7.183	7.190	7.197	7.204
52.	7.211	7.218	$7.22\bar{5}$	7.232	7.239	7.246	7.253	7.259	7.266	7.273
53.	7.280	7.287	7.294	7.301	7.308	7.314	7.321	7.328	$7.33\bar{5}$	7.342
54.	7.348	7.355	7.362	7.369	7.376	7.382	7.389	7.396	7.403	7.409
	0	1	2	3	4	5	6	7	8	9

SQUARE ROOTS OF NUMBERS (55.00 through 99.90)

Numbers	0	1	2	3	4	5	6	7	8	9
55.	7.416	7.423	7.430	7.436	7.443	7.450	7.457	7.463	7.470	7.477
56.	7.483	7.490	7.497	7.503	7.510	7.517	7.523	7.530	7.537	7.543
57.	7.550	7.556	7.563	7.570	7.576	7.583	7.589	7.596	7.603	7.609
58.	7.616	7.622	7.629	7.635	7.642	7.649	7.655	7.662	7.668	7.675
59.	7.681	7.688	7.694	7.701	7.707	7.714	7.720	7.727	7.733	7.740
60.	7.746	7.752	7.759	7.765	7.772	7.778	7.785	7.791	7.797	7.804
61.	7.810	7.817	7.823	7.829	7.836	7.842	7.849	7.855	7.861	7.868
62.	7.874	7.880	7.887	7.893	7.899	7.906	7.912	7.918	7.925	7.931
63.	7.937	7.944	7.950	7.956	7.962	7.969	7.975	7.981	7.987	7.994
64.	8.000	8.006	8.012	8.019	8.025	8.031	8.037	8.044	8.050	8.056
65.	8.062	8.068	8.075	8.081	8.087	8.093	8.099	8.106	8.112	8.118
66.	8.124	8.130	8.136	8.142	8.149	8.155	8.161	8.167	8.173	8.179
67.	8.185	8.191	8.198	8.204	8.210	8.216	8.222	8.228	8.234	8.240
68.	8.246	8.252	8.258	8.264	8.270	8.276	8.283	8.289	8.295	8.301
69.	8.307	8.313	8.319	8.325	8.331	8.337	8.343	8.349	8.355	8.361
70.	8.367	8.373	8.379	8.385	8.390	8.396	8.402	8.408	8.414	8.420
71.	8.426	8.432	8.438	8.444	8.450	8.456	8.462	8.468	8.473	8.479
72.	8.485	8.491	8.497	8.503	8.509	8.515	8.521	8.526	8.532	8.538
73.	8.544	8.550	8.556	8.562	8.567	8.573	8.579	8.585	8.591	8.597
74.	8.602	8.608	8.614	8.620	8.626	8.631	8.637	8.643	8.649	8.654
75.	8.660	8.666	8.672	8.678	8.683	8.689	8.695	8.701	8.706	8.712
76.	8.718	8.724	8.729	8.735	8.741	8.746	8.752	8.758	8.764	8.769
77.	8.775	8.781	8.786	8.792	8.798	8.803	8.809	8.815	8.820	8.826
78.	8.832	8.837	8.843	8.849	8.854	8.860	8.866	8.871	8.877	8.883
79.	8.888	8.894	8.899	8.905	8.911	8.916	8.922	8.927	8.933	8.939
80.	8.944	8.950	8.955	8.961	8.967	8.972	8.978	8.983	8.989	8.994
81.	9.000	9.006	9.011	9.017	9.022	9.028	9.033	9.039	9.044	9.050
82.	9.055	9.061	9.066	9.072	9.077	9.083	9.088	9.094	9.099	9.105
83.	9.110	9.116	9.121	9.127	9.132	9.138	9.143	9.149	9.154	9.160
84.	9.165	9.171	9.176	9.182	9.187	9.192	9.198	9.203	9.209	9.214
85.	9.220	9.225	9.230	9.236	9.241	9.247	9.252	9.257	9.263	9.268
86.	9.274	9.279	9.284	9.290	9.295	9.301	9.306	9.311	9.317	9.322
87.	9.327	9.333	9.338	9.343	9.349	9.354	9.359	9.365	9.370	9.375
88.	9.381	9.386	9.391	9.397	9.402	9.407	9.413	9.418	9.423	9.429
89.	9.434	9.439	9.445	9.450	9.455	9.460	9.466	9.471	9.476	9.482
90.	9.487	9.492	9.497	9.503	9.508	9.513	9.518	9.524	9.529	9.534
91.	9.539	9.545	9.550	9.555	9.560	9.566	9.571	9.576	9.581	9.586
92.	9.592	9.597	9.602	9.607	9.612	9.618	9.623	9.628	9.633	9.638
93.	9.644	9.649	9.654	9.659	9.664	9.670	9.675	9.680	9.685	9.690
94.	9.695	9.701	9.706	9.711	9.716	9.721	9.726	9.731	9.737	9.742
95.	9.747	9.752	9.757	9.762	9.767	9.772	9.778	9.783	9.788	9.793
96.	9.798	9.803	9.808	9.813	9.818	9.823	9.829	9.834	9.839	9.844
97.	9.849	9.854	9.859	9.864	9.869	9.874	9.879	9.884	9.889	9.894
98.	9.899	9.905	9.910	9.915	9.920	9.925	9.930	9.935	9.940	9.945
99.	9.950	9.955	9.960	9.965	9.970	9.975	9.980	9.985	9.990	9.995
	0	1	2	3	4	5	6	7	8	9

CIRCUMFERENCES AND AREAS (0.2 to 9.8; 10 to 99)

Diameter	Circum.	Area	Diameter	Circum.	Area
0.2	0.628	0.0314	31	97.39	754.8
0.4	1.26	0.1256	32	100.5	804.2
0.6	1.88	0.2827	33	103.7	855.3
0.8	2.51	0.5026	34	106.8	907.9
1	3.14	0.7854	35	110	962.1
1.2	3.77	1.131	36	113.1	1,017.9
1.4	4.39	1.539	37	116.2	1,075.2
1.6	5.02	2.011	38	119.4	1,134.1
1.8	5.65	2.545	39	122.5	1,194.6
2	6.28	3.142	40	125.7	1,256.6
2.2	6.91	3.801	41	128.8	1,320.3
2.4	7.53	4.524	42	131.9	1,385.4
2.6	8.16	5.309	43	135.1	1,452.2
2.8	8.79	6.158	44	138.2	1,520.5
3	9.42	7.069	45	141.4	1,590.4
3.2	10.05	7.548	46	144.5	1,661.9
3.4	10.68	8.553	47	147.7	1,734.9
3.6	11.3	10.18	48	150.8	1,809.6
3.8	11.93	11.34	49	153.9	1,885.7
4	12.57	12.57	50	157.1	1,963.5
4.2	13.19	13.85	51	160.2	2,042.8
4.4	13.82	15.21	52	163.4	2,123.7
4.6	14.45	16.62	53	166.5	2,206.2
4.8	15.08	18.1	54	169.6	2,290.2
5	15.7	19.63	55	172.8	2,375.8
5.2	16.33	21.24	56	175.9	2,463
5.4	16.96	22.9	57	179.1	2,551.8
5.6	17.59	24.63	58	182,2	2,642.1
5.8	18.22	26.42	59	185.4	2,734
6	18.84	28.27	60	188.5	2,827.4
6.2	19.47	30.19	61	191.6	2,922.5
6.4	20.1	32.17	62	194.8	3,019.1
6.6	20.73	34.21	63	197.9	3,117.3
6.8	21.36	36.32	64	201.1	3,217
7	21.99	38.48	65	204.2	3,318.3
7.2	22.61	40.72	66	207.3	3,421.2
7.4	23.24	43.01	67	210.5	3,525.7
7.6	23.87	45.36	68	213.6	3,631.7
7.8	24.5	47.78	69	216.8	3,739.3
8	25.13	50.27	70	219.9	3,848.5
8.2	25.76	52.81	71	223.1	3,959.2
8.4	26.38	55.42	72	226.2	4,071.5
8.6	27.01	58.09	73	229.3	4,185.4
8.8	27.64	60.82	74	232.5	4,300.8
9	28.27	63.62	75	235.6	4,417.9
9.2	28.9	66.48	76	238.8	4,536.5
9.4	29.53	69.4	77	241.9	4,656.6
9.6	30.15	72.38	78	245	4,778.4
9.8	30.78	75.43	79	248.2	4,901.7
10	31.41	78.54	80	251.3	5,026.6
11	34.55	95.03	81	254.5	5,153
12	37.69	113	82	257.6	5,281
13	40.84	132.7	83	260.8	5,410.6
14	43.98	153.9	84	263.9	5,541.8
15	47.12	176.7	85	267.0	5,674.5
16	50.26	201	86	270.2	5,808.8
17	53.4	226.9	87	273.3	5,944.7
18	56.54	254.4	88	276.5	6,082.1
19	59.69	283.5	89	279.6	6,221.2
20	62.83	314.1	90	282.7	6,361.7
21	65.97	346.3	91	285.9	6,503.9
22	69.11	380.1	92	289.0	6,647.6
23	72.25	415.4	93	292.2	6,792.9
24	75.39	452.3	94	295.2	6,939.8
25	78.54	490.8	95	298.5	7,088.2
26	81.68	530.9	96	301.6	7,238.2
27	84.82	572.5	97	304.7	7,389.8
28	87.96	615.7	98	307.9	7,543.0
29	91.1	660.5	99	311.9	7,697.7
30	94.24	706.8			

4/87 (6C1035B)